Advanced Structured Materials

Volume 126

Common engineering materials reach in many applications their limits and new developments are required to fulfil increasing demands on engineering materials. The performance of materials can be increased by combining different materials to achieve better properties than a single constituent or by shaping the material or constituents in a specific structure. The interaction between material and structure may arise on different length scales, such as micro-, meso- or macroscale, and offers possible applications in quite diverse fields.

This book series addresses the fundamental relationship between materials and their structure on the overall properties (e.g. mechanical, thermal, chemical or magnetic etc) and applications.

The topics of *Advanced Structured Materials* include but are not limited to

- classical fibre-reinforced composites (e.g. glass, carbon or Aramid reinforced plastics)
- metal matrix composites (MMCs)
- micro porous composites
- micro channel materials
- multilayered materials
- cellular materials (e.g., metallic or polymer foams, sponges, hollow sphere structures)
- porous materials
- truss structures
- nanocomposite materials
- biomaterials
- nanoporous metals
- concrete
- coated materials
- smart materials

Advanced Structured Materials is indexed in Google Scholar and Scopus.

More information about this series at http://www.springer.com/series/8611

Shakeel Ahmed · Wazed Ali
Editors

Green Nanomaterials

Processing, Properties, and Applications

Springer

Editors
Shakeel Ahmed
Department of Chemistry
Government Degree College Mendhar
Jammu, India

Wazed Ali
Indian Institute of Technology Delhi
New Delhi, India

ISSN 1869-8433 ISSN 1869-8441 (electronic)
Advanced Structured Materials
ISBN 978-981-15-3559-8 ISBN 978-981-15-3560-4 (eBook)
https://doi.org/10.1007/978-981-15-3560-4

This Springer imprint is published by the registered company Springer Nature Singapore Pte Ltd.
The registered company address is: 152 Beach Road, #21-01/04 Gateway East, Singapore 189721, Singapore

Preface

Use of nanomaterials as a multifunctional agent is a revolutionary paradigm shift in all areas of advancements including transdisciplinary segments. In this direction, the research community is engaged in exploring various innovative routes to synthesize various nanomaterials for a range of value-added products. Most recently, it has been proven that either the ways of synthesis or the nanomaterials themselves are not meeting the expectation of green movement in terms of exclusion of toxicity parameters in new innovation to achieve holistic development to protect sustainability in our ecosystem. Nevertheless, significant efforts have been put together to discover and open up the wonder properties of nanomaterials where researchers have very wisely adopted eco-friendly greener approaches to provide a green solution in totality. The current edited book is the most recent glimpse of the same approach, whereas the context of various chapters is designed to record and analyze the recent green inventions in the world of green nanomaterials. The chapters are composed of eminent researchers and academicians in their own expertise. We are very sure that this book entitled 'Green Nanomaterials' will serve as a reference handy for the readers and followers aspired in the exploration of the potential of nanomaterials in sustainable ways.

We are very much thankful to all the contributors for their positive consents and efforts to compose this book with the most recent findings in this area. Sincere thanks go to '**Springer Nature**' for publishing this book in one of the most demanded fields at the right time.

Jammu, India — Shakeel Ahmed
New Delhi, India — Wazed Ali

Contents

About the Editors

Dr. Shakeel Ahmed is Assistant Professor at Higher Education Department, Govt. of Jammu and Kashmir, Jammu, India. He obtained his PhD in the area of biopolymers and bio-nanocomposites. He has published several research articles in the area of green nanomaterials and biopolymers for various applications including biomedical, packaging, sensors, and water treatment. He has also contributed several book chapters, and written review articles on green synthesis of metal/metal oxides and nanomaterials. He is member of Royal Society of Chemistry (UK) and American Chemical Society. He has published several books in the area of green materials with publishers of international repute.

Dr. Wazed Ali is Assistant Professor at the Department of Textile Technology, Indian Institute of Technology (IIT) Delhi. After obtaining his Master's and PhD from IIT Delhi in 2006 and 2011, respectively, he joined GE India Technology Centre Pvt. Ltd. (JFWTC), Bangalore as a research scientist. He was a recipient of the Bolton Fellowship from the University of Bolton, UK for three year's PhD program at IIT Delhi. He was awarded the prestigious Commonwealth Split-Site Doctoral Fellowship 2008 from the Commonwealth Fellowship Commission, London to pursue one-year research work in the UK during his PhD. He has published several research articles in international journals and has also filed for a US patent. He is a life member of the Indian Natural Fibre Society.

Introduction to Green Nanomaterials

Pintu Pandit and T. Nadathur Gayatri

Abstract This chapter presents an introduction to nanomaterials, which can be synthesized by green chemistry, or nano-sized functionally advanced materials which have high-performance applications in energy generation and storage, carbon dioxide fixation, electronic devices and are sustainable in terms of production and application with respect to the environment. Methods in brief of preparation of nanoparticles and nanofibres, advantages of green synthesis, and limitations of nanomaterials are discussed. This chapter also provides information related to recent research work on green nanomaterials and the available methods for their synthesis. It also gives a comprehensive overview of the recent status and suggests future directions for employing green nanomaterials for possible various application mainly in the biotechnology, agriculture and biomedical areas. The sustainability of major natural resources utilized in green nanomaterials' synthesis is considered.

Keywords Green nanomaterial · Nanotechnology · Sustainability · Raw materials

1 Introduction

1.1 Nanotechnology and Nanomaterials

The word 'nano' refers to the dimension of length of the order of 10^{-9} m (Tolle 2007). Nanotechnology considers the knowledge area of engineering and understanding of nanoscaled materials and devices, wherein materials exhibit unusual phenomena leading to special applications. Comparison of the dimensions of objects occurring in the living world can be done by considering the red blood cell—7000 nm wide

P. Pandit (✉)
Department of Textile Design, National Institute of Fashion Technology, Ministry of Textiles, Government of India, NIFT Campus, Mithapur Farms, Patna 800001, India
e-mail: pintupanditict@gmail.com; pintu.pandit@nift.ac.in

T. N. Gayatri
Department of Fibres and Textile Processing Technology, Institute of Chemical Technology, Matunga, Mumbai 400019, India

S. Ahmed and W. Ali (eds.), *Green Nanomaterials*,
Advanced Structured Materials 126,
https://doi.org/10.1007/978-981-15-3560-4_1

and 2000 nm long, rhinovirus which causes the common cold, being 25 nm wide, DNA molecule being 2 nm thick and silicon atom with a diameter of 0.2 nm.

1.2 Properties of Nanomaterials

The chemical reactivity of a material is highly dependent on the ratio of its surface area to volume. The surface area of a nanoparticle is exponentially larger than its unit volume. When a material block is dismantled and cut down to nano-size, the combined surface area of the particles is vastly greater than the surface area of the starting block. This finds immediate application in the surface area sensitive processes like catalysis, adsorptive removal of pollutants and sustained drug release and wound healing in medicine where chemical reactivity and coverage are linked.

Products can be coated with thin films of nanomaterials to exhibit specific properties, like hydrophobicity for water repellency or resistance to microbial or fungal growth while leaving the product appearance unchanged.

Newtonian physics is not applicable to the behaviour of nanomaterial as such rules apply to bulk matter while at nano-level physical properties like electric conductivity, colour, strength, weight and chemical properties are drastically different. One illustration is bulk silver which is non-toxic, but silver nanoparticles can kill viruses on contact. Normally, brittle ceramics can become more elastic when the size of their component grains is reduced to minuscule nanometre diameter. A gold particle of 1 nm diameter appears red in colour. Both graphite and diamond are constitutionally of only the carbon element, but the sheet-like arrangement of atoms in graphite makes it flaky while the tetrahedral arrangement of atoms internally makes diamond hard.

1.3 Classification of Nanomaterials

Nanomaterials are classified on the basis of the order of dimensions to which their electron is restricted, ranging from one, two or three dimensions (Suresh 2013).

- **Zero-dimensional (0-D) nanomaterial**

Zero-dimensional shapes are the most basic and symmetric, including spheres, cubes and polygon, with nano-dimensions along all the axes of x, y and z. Metallic nanoparticles of gold and silver and semiconductor like quantum dots are examples of this class, being spherical in shape and with diameters ranging from 1–50 nm.

- **One-dimensional (Quasi 1D) nanomaterial**

In these nanostructures, one dimension of the nanostructure will lie beyond the nanometre range, such as observed in nanowires, nanorods and nanotubes from metals and metal oxides. These materials are several micrometres long but are just a few nanometres across the diameter (Xia et al. 2003).

- **Two dimensional (2-D) nanomaterial**

In nanomaterials, like nano films and thin-film multilayers, nanosheets or nanowalls two dimensions lie beyond the nanometre range. The surface area of nano films could be in the range of several square micrometres, but the thickness remains at the nanoscale. There exist fewer types and of inorganic nano-objects in the 2D class.

- **Three-dimensional (3-D) nanomaterial**

Objects having a cumulative size in the non-nanometre range of the order of micrometres or millimetres but exhibiting nanometric features like nano-sized confinement spaces or those formed by the periodic arrangement and assembly of nano-sized units can be considered '3D nanosystems.' With all dimensions beyond the nano range, they still exhibit unusual molecular and bulk properties, reflecting their component units which are in the nanoscale (1–100 nm). Three-dimensional nanocrystals are formed from the organization of 0D spheres, and 1D rods or 2D plates give rise to unusual superstructures, such as in box-shaped graphene or mesostructured platinum films.

1.4 Nanoparticles

1.4.1 Carbon-Based Nanoparticles

Carbon-based polymers such as fullerenes, carbon dots, nano-diamonds and nanofoams are a class of nanomaterial, which are much researched. Carbon nanotubes are just layers of graphite rolled into a tube form, which could be single- or multi-walled, depending on the number of layers. Fullerenes consisting of C_{60} (Pal et al. 2011), single-walled nanotubes and multi-walled nanotubes have been suggested for use in composites as structural reinforcements to conventional materials and even as drug carriers (Prasek et al. 2011). Graphene is also a carbon-based nanomaterial made of a one carbon atom thick, single layer sheet structure similar to natural graphite (Naahidi et al. 2013).

1.4.2 Dendrimers

Dendrimers are highly branched polymers whose shape, size, branching length, generation number and surface functionality can be controlled at the synthesis stage.

Fig. 1 TEM image of **a** Carbon nanotube, **b** PAMAM dendrimer and **c** Silver nanoparticles

From a nano-sized centre, polymeric branches of defined length and structure grow outward, generating porous enclosures, in the whole molecule, which allows for the trapping or encapsulating drug molecules, fragrance molecules or even biological cells. TEM image of PAMAM dendrimer is given in Fig. 1b (Anandalakshmi and Venugobal 2017).

1.4.3 Metal Nanoparticles

Nanogold, nanosilver and metal oxides, like titanium dioxide and closely packed semiconductor like quantum dots, derived from metal atoms, are metal-based nanoparticles. Metallic nanoparticles are often formed as solid colloidal solutions. Further, they are amenable to surface modification to achieve target functionality as they possess the large surface area to volume ratio. They can be synthesized in different forms like a rod, dot or cubes and often exhibit special optical properties like fluorescence or surface enhanced resonance, allowing for easy detection at low concentrations, and hence, their imaging applications. TEM image of the silver nanoparticle is given in Fig. 1c

1.5 Approach for Synthesis of Nanoparticles

Synthetic approaches to nanoparticles could be configured to be either top-down or bottom-up methods as shown in Fig. 2 (Ahmed 2015). When thin films or bulk materials are scaled down by various lithographic and mechanical techniques like grinding, milling and thermal/laser ablation, it is a top-down method. Bottom-up methods use basic building blocks (atoms, molecules, nanoparticles, etc.) to self-organize from nuclei to grow into superstructures, of greater complexity. Most chemical or biological synthetic procedures which make complex 3D structures at low cost and in bulk fall into the bottom-up class of methods (Colson et al. 2013).

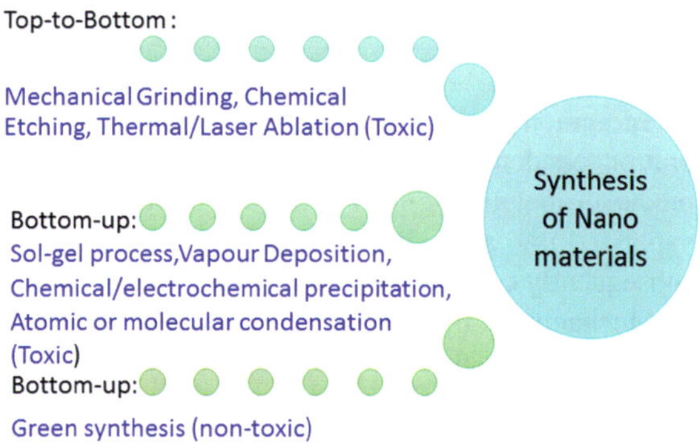

Fig. 2 Approach for nanomaterial synthesis

1.6 Global Market Scenarios of Nanoparticles

The analysis of the global market for nanoparticles showed that its value is projected to rise to USD 91.1 million by 2020, at CAGR of 5.4% from 2015 to 2020. Increasing investments in nanotechnology research and development with applications in pharmaceuticals, textiles and agriculture in the high growth rate developing nations are expanding the market (Rai et al. 2009; Chaudhry et al. 2010). Production of nanoscale silver, (AgNPs), has allowed their potential use in electronic and transparent conductor applications, in antimicrobial consumer goods and medical products causing the nanosilver market to rise in value from US$290 million in 2011 to around US$1.2 billion by 2016 (Murphy et al. 2015). The use of AgNPs to make bactericidal cotton fibres and hence antimicrobial fabric has been commercialized, along with nanosilver-based wall paint, for mould-resistant constructions, which is another high volume consumer product. There is much interest in nanofinished textiles for medical use, which need to be antimicrobial, along with odour control inner garments and bedding fabrics, where AgNPs can be easily used.

The composition of nanoparticles mostly utilized in the medical industry in devices and products is of metals such as gold and silver, metal oxides of zinc and titanium and salts such as calcium phosphate. These particles make up a market share of over 30%.

2 Synthesis of Metal Nanoparticles

Metal nanoparticles can be prepared by three different methods—physical, chemical and green process—which are described below.

2.1 Physical Synthesis

One common physical process, through which nanoparticles are synthesized, is evaporation–condensation that is performed in a tube furnace at atmospheric pressure. The precursor metallic or metal oxide material, placed in a boat at the furnace centre, is vaporized into an inert carrier gas medium like argon, nitrogen or even air and is subsequently condensed to nanoparticles (El-Nour et al. 2010).

Mechanical milling performed using ball milling machinery initiates a solid-state displacement reaction. Milling/grinding reduces the particle size, and blends of particles into new phases mechanochemical processes modify the chemical composition of precursors, by pressure-induced reactions on stressed solid phase (Ghorbani 2014).

Pulsed laser ablation (PLA) has the advantage of giving nanoparticle product with a uniform size distribution, i.e. low dispersity. Initially, a laser pulse vaporizes a target to form a plasma plume, which then expands adiabatically in the closed volume; as the expansion ends, condensation to nanoparticles begins (Hassan and Wagner 2017). Process parameters like energy, wavelength and duration of pulse of laser along with the substrate phase (liquid or gas) and duration of ablation can modify the product properties (Amendola and Meneghetti 2009).

2.2 Chemical Synthesis

Chemical approach involves reduction of a solution of metal ions, at specific temperature, with a reagent of defined concentration and with an ultrasonic vibration or stirring, or under microwave irradiation, to finally form microscopic metal clusters or aggregates (El-Nour et al. 2010). The wet chemical process of nanoparticle synthesis is a bottom-up strategy for synthesizing nanoparticles from the atoms of elements of the starting material. Sol–gel process is another physico-chemical process in which a network is formed from a colloidal suspension (sol) of metal oxide or other metal derivatives that then condenses through gelation of the sol solution to form a gel in liquid phase (Zhang et al. 2014). Solvothermal synthesis starts from precursor molecules in a solvent medium, heat treated to generate nanoparticles bottom-up. The reaction vessel is closed to promote decomposition or further chemical reaction of the precursors, under conditions of high temperature and/or high pressure. When the solvent is water, the process is ‘hydrothermal synthesis.’

2.3 Green Synthesis

Reducing agents from natural sources like polysaccharides or plants extract or biological micro-organism like bacteria and fungus can be used to synthesize nanoparticles.

2.3.1 Biological Synthesis

Biosynthesis metallic nanoparticles from microorganisms such as bacteria, actinomycetes, fungi and algae which is green and eco-friendly process. The synthesis of nanoparticles may be internal to the cell or an external reaction, depending on the location of the nanoparticle product (Hasan 2015).

2.3.2 Plant Extract Mediated Synthesis

The mode of usage of plant extracts to produce nanoparticles is simple, in that, the extract is mixed with a solution of the metal salt at different temperature and kept under constant stirring, sometimes under ultrasonic or microwave irradiation for a specific time duration. The single-step biogenic reduction of metal ion to base metal is fast, may be carried out at room temperature and atmospheric pressure, such that scale-up is easily possible. The reducing agents identified are plant secondary metabolites, chemically classified as alkaloids, tannins, terpenoids and co-enzymes (Makarov 2014). Extracts of varied plant species have been employed in synthesizing metallic nanoparticles; alternately, live plants have also been used (Mittal et al. 2013). Some of them are presented in Table 1.

2.4 Advantage of Green Synthesis Over Conventional Process

Green synthesis of metal nanoparticles is a cheaper, simpler and environmental benign alternative to conventional chemical and physical methods, with good yields that do not require complex reagents, or high-end apparatus (Malik 2014). Some of the herbs used in green synthesis of metallic nanoparticles are considered briefly.

3 Herbs Used for Green Synthesis of Metallic Nanoparticles

Exploratory research indicates that the following herbs (leaf and fruit) are rich in biomolecules which can be utilized for synthesis of nanoparticle.

3.1 Jackfruit Leaf

Artocarpus heterophyllus, commonly known as jackfruit, common in India and Southeast Asia, consists of dark green, alternate, fairly large and oval-shaped leaves which are deeply lobed on young shoots. The leaf is shown in Fig. 3.

Table 1 Plant extract used in the synthesis of nanoparticles

Plant species	Nanoparticle size	Shape
Ag nanoparticle		
Acalypha indica	20–30 nm	Spherical
Allium sativum (garlic clove)	4–22 nm	Spherical
Aloe vera	50–350 nm	Triangular
Apiin extracted (henna leaves)	7.5–65 nm	Triangular
Boswellia ovalifoliolata	30–40 nm	
Calotropis procera	150–1000 nm	
Camelia sinensis	30–40 nm	
Carica papaya	25–50 nm	
Catharanthus roseus	48–67 nm	
Citrus sinensis peel	10–35 nm	Spherical
Datura metel	16–40 nm	Quasilinear
Emblica officinalis	10–20 nm	
Euphorbiaceae latex	10–20 nm	
Nelumbo nucifera (lotus)	25–80 nm	Spherical, triangular
Mentha piperita (peppermint)	5–150 nm	Spherical
Moringa oleifera	57 nm	
Tanacetum vulgare	16 nm	
Syzygium cumini	29–92	Spherical
Au nanoparticle		
Aloe vera	50–350 nm	Spherical
Apiin extracted (henna leaves)	7.5–65 nm	Spherical
Cinnamomum camphora	3.2–20 nm	Cubic
Camelia sinensis	30–40 nm	
Cymbopogon sp. (lemon grass)	200–500 nm	Spherical
Emblica officinalis	15–25 nm	
Mentha piperita (peppermint)	5–150 nm	Spherical
Psidium guajava	25–30 nm	Spherical

(continued)

Table 1 (continued)

Plant species	Nanoparticle size	Shape
Tanacetum vulgare	11 nm	
Pyrus sp. (pear fruit extract)	200–500 nm	Triangular
Parthenium leaf	50 nm	Face-centred cubic
Ag/Au bimetallic nanoparticle		
Anacardium occidentale	~6 nm at 27 °C; 17 nm at 100 °C	
Azadirachta indica (neem)	50–100 nm	
Swietenia mahogani	50 nm	
Pb nanoparticle		
Jatropha curcas L. latex	10–12.5 nm	
Cu nanoparticle		
Euphorbiaceae latex	10.5 nm	
Indium oxide nanoparticle		
Aloe vera	5–50 nm	Spherical
Pd nanoparticle		
Cinnamomum camphora	3.2–20 nm	Hexagonal
Pt nanoparticle		
Diopyros kaki	15–19 nm	
Hybrid Ag nanoparticle		
Eucalyptus	50–150 nm	

Fig. 3 Panasa (*Artocarpus heterophyllus*) leaves

Fig. 4 Narikela (*Cocos nucifera*) leaves

All parts of the leaf contain sticky white latex (Crane et al. 2005). Phytochemical screening has revealed that the hot water extract of leaf contains flavonoids, leucoanthocyanins, anthocyanins and tannins as components (Chandrika et al. 2006).

3.2 Coconut Leaf

Cocos nucifera plant, commonly known as coconut, has pinnate leaves of 4–6 m (13–20 ft) long with pinnae 60–90 cm long as shown in Fig. 4. Its potential for antioxidant, antibiotics and free radical scavenging activity is much investigated. Total phenol content (TPC) of coconut leaf extract is 0.59–2.22 mg/g. Phenolic compounds function as a green reducing agent.

3.3 Night Flowering Jasmine Leaf

Nyctanthes arbortristis commonly known as a night flowering jasmine tree and the leaf is shown in Fig. 5, growing in India and other tropical and subtropical region is a medicinal plants with a wide spectrum of biological activity. An alkaloid named nactanthine is also found in leaves. Leaves also contain mannitol, astringent, resinous substances, ascorbic acid, sugar, tannic acid, methyl salicylate and trace of volatile oils.

Fig. 5 Parijata (*Nyctanthes arbortristis*) leaves

3.4 Indian Gooseberry Fruit

Phyllanthus emblica, commonly known as Indian gooseberry (amla), originated in India, and the fruit is shown in Fig. 6. The fruit extract of amla and its individual constituents has shown to impart antioxidant effect. Amla is a rich source of Vitamin C, a water-soluble antioxidant that acts as a scavenger of free radicals. Amla also contains a variety of active constituents like polyphenols, flavones, tannins, anthocyanins, flavonols, ellagic acid and derivatives (Singh et al. 2012).

4 Silver Nanoparticles

Silver nanoparticles, of the class of noble metal nanoparticles, have been of interest since they exhibit a range of bactericidal and fungicidal activities making them

Fig. 6 Amalaki (*Phyllanthus emblica*) fruits

popular for common use in a wide variety of consumer products, including plastics, soaps, pastes, food and textiles (Rauwel 2015). Conventionally, silver nanoparticles are synthesized using synthetic chemicals as reducing agents which later, during the use of the product, cause various biological risks due to their general toxicity, even at residual levels. Feasible, alternate and superior reagents have been found in biological molecules derived from plant sources in the form of extracts and used in green synthesis. An example is the synthesis of silver nanoparticles using aqueous leaf extract of *Azadirachta indica*, as reducing agent and capping agent. Well dispersed, spherically shaped silver nanoparticles with a diameter of 34 nm were formed from silver ions, at the end of 15 min of reduction, without using toxic chemicals. The silver nanoparticles were active against both Gram-positive (*Staphylococcus aureus*) and Gram-negative (*Escherichia coli*) bacteria (Ahmed et al. 2016).

Khalil et al. (2014) prepared quasi-spherical silver nanoparticles of 20–25 nm size by means of using hot water olive leaf extract as reductant and stabilizer on silver nitrate solution. The silver nanoparticles were mostly spherical with an average size of 20–25 nm. Silver nanoparticles showed better antibacterial efficacy, against multi-drug-resistant *S. aureus*, *Pseudomonas aeruginosa* and *E. coli* compared to OLE extract alone (Fig. 7).

The synthesis of silver nanoparticles using silver nitrate as precursor and leaf extracts of three plants, *Musa balbisiana* (banana), *A. indica* (neem) and *Ocimum tenuiflorum* (black tulsi), as reducing and stabilizing agent has been achieved. Silver nanoparticles showed much stronger antimicrobial activity against *E. coli* and *Bacillus sp.* relative to both silver nitrate and raw plant extracts. Moong bean (*Vigna radiata*) and chickpea (*Cicer arietinum*) seeds treated with silver nanoparticles showed faster and more successful germination pointing to the low toxicity of silver nanoparticles (Banerjee et al. 2014).

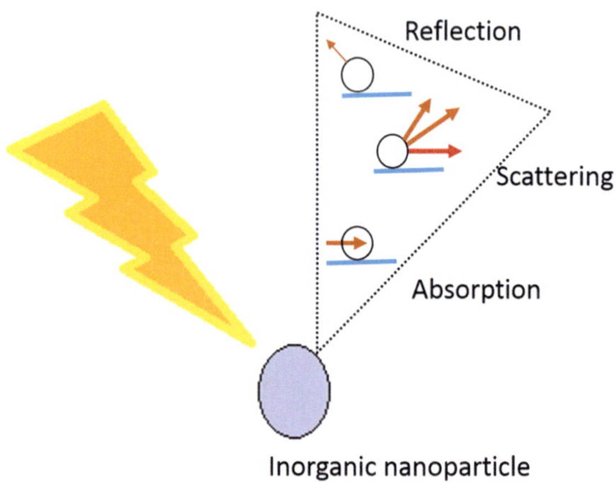

Fig. 7 Action mode of inorganic UV filters

5 Copper Nanoparticles

Copper nanoparticles have garnered attention for their special optical and electrical properties, along with good thermal conductivity and corrosion resistance. Copper is cheaper compared to noble metals like gold and silver and also has effective antimicrobial action. Well dispersed, spherical and monodisperse copper nanoparticles have found application in lubricants and also as nano-fluids, etc. (Khodashenas and Ghorbani 2014).

Copper nanoparticles are susceptible to aggregation and oxidation, which could be by adding a suitable stabilizer for capping of copper nanoparticles. Copper nanoparticles are widely used in wound dressings for their biocidal properties, as well as in the manufacturing industry as gas sensors, in catalytic process, high-temperature superconductors and solar cells.

Environmentally safe copper nanoparticles/perlite composites are prepared without the use of any toxic reductants or capping agents. Leaves of *Euphorbia esula* L. functioned as both reductant and stabilizer for the formation of copper nanoparticles with a size lower than 32 nm. The Cu NPs/perlite composites showed good response in the catalytic reduction of 4-nitrophenol and were reusable without loss of the catalytic function.

Nanostructured copper particles are prepared using *Garcinia mangostana* leaf extract (Prabhu et al. 2017) as a reducing agent with precursor metal ions from copper nitrate. Spherical copper nanoparticles agglomerated to a particle size of 20–25 nm were obtained. These Cu nanoparticles had good antibacterial action against *E. coli* and *S. aureus*.

In another study, authors presented a novel biological approach for the formation of copper nanoparticles using precursor ions from copper sulphate with clove as the reducing agent. The particles size of obtained copper nanoparticles was in the range of 5–40 nm (Subhankari and Nayak 2013).

Khan et al. (2016) developed zero-valent copper (Cu) nanoparticles, protected with starch, by chemical reduction of copper salt solution using reducing agent, ascorbic acid at low temperature. Copper nanocrystals were found to have average crystallite size of 28.73 nm.

Fabrication of copper nanoparticle using *catharanthus roseus* by green synthesis and act as dual purpose ‘reducing and capping’ agent. The size of nanoparticle ranges between 40 and 47 nm and appears like coral reef. Copper nanoparticles also exhibited excellent antibacterial activity (Birla et al. 2009).

6 Zinc-Based Nanoparticles

Zinc nanoparticles have an unusual combination of physical and chemical properties, such as high chemical stability, high electrochemical coupling coefficient, wide range of radiation absorption and strong photostability, leading to its multifunctional usage. Zinc oxide behaves as a semiconductor, whose covalence is on the boundary separating ionic and covalent semiconductors. It possesses a broad energy band gap (3.37 eV), large free exciton binding energy (60 meV) and high thermal and mechanical stability at room temperature, which allows to its potential use in optoelectronics, as a piezoelectric ceramic switch and in laser technology. The piezo- and pyroelectric properties of ZnO allow for its use as a sensor, transducer, energy generator and photocatalyst for hydrogen production. The low toxicity, biocompatibility and biodegradability of zinc oxide favour its application in biomedicine and in agriculture (Kołodziejczak-Radzimska and Jesionowski 2014). Zinc oxide nanoparticles show exciting photoluminescence properties and resistance to oxidation. The ZnO particles also were found to emit light in the UV to blue range by suggesting that it can be used for UV protection. In a typical process, zinc chloride was dissolved in a phenylether and complexed with oleylamine to form Zn^{2+}, after which it nucleates into the final structure. Hexagonal zinc nanoparticles having a diameter of 250–350 nm were formed.

Pavani et al. (2011) studied the synthesis of zinc nanoparticles by extracellular synthesis method using *Aspergillus species*. The synthesized zinc nanoparticles showed absorption maxima at 230 nm. Zinc nanoparticles have a size in the range of 50–120 nm. Sangeetha et al. (2011) prepared stable and spherical nanostructured zinc oxide by green chemical method using zinc nitrate and aloe vera leaf extract. The yield from precursor was greater than 95% having polydispersed zinc oxide nanoparticles with an average particle size between 25 and 40 nm. The particle size could be modified by changing the concentrations of leaf extract.

7 Green Synthesis of Nanofibres

Nanofibers prepared by using ecological solvents or naturally occurring biodegradable polymers have found a niche use in medical devices for wound healing and controlled drug release, especially because of their high surface area and membrane-like structure. Electrospun nanofibers mats have been green synthesized using hydroxypropyl cellulose (HPC) separately or by addition of fibre forming polymer like poly(vinyl alcohol) (PVA) or polyvinylpyrrolidone (PVP) to enhance the mechanical properties of the nanofibers mats. Favourable properties like thermal stability and visual appearance and mechanical properties of PVA or PVP were dramatically enhanced with the addition of HPC. Drug loading on such nanofibres allowed for sustained release of drug, when tested in vitro (El-Newehy et al. 2018).

A green nanofiber prepared by using H_2O_2-assisted water-soluble chitosan/polyvinyl alcohol (WSCHT/PVA) dissolved in water, an eco-friendly solvent, was done. This is necessary to replace organic solvents used to solubilize water-insoluble polymers, with aqueous spinnable solutions. It is essential for end applications of nanofibres in environmental solutions or in biological systems that no toxic chemicals be used in their processing. The method developed produces uniform nanofibers, with PVA, polyvinyl alcohol as a supporting polymer and chitosan made water soluble by a heterogeneous reaction (Pervez and Stylios 2018).

Carbon nanofibers (CNF) have been commonly prepared from synthetic carbon polymer precursors, and organic solvents like methylene dichloride and dimethylsulphoxide used to dissolve these are toxic and hazardous. But their preparation from naturally available precursors with benign solvents is less common. Porous carbon nanofibers (CNF) having high surface area were electrospun from a solution of polymerized sucrose and poly (vinyl alcohol) (PVA) in water. Thermal treatment for stabilization, at 350 °C, followed by carbonization of the electrospun polymer at 950 °C in reducing environment formed semigraphitic CNF. In situ doping of Ag nanoparticles in CNF increased electrical conductivity tenfold with just 0.1 wt% Ag doping (Chakravarty et al. 2017).

8 Application of Nanoparticles in Textile Finishing

Conventional textile finishes that impart advantages such as water repellency and stain repellency, to the fabric are rarely durable to frequent laundering. Nanoparticles with their characteristic large surface area and high surface energy bind to fabrics, with greater strength conferring better durability of the desired textile property (Kathirvelu et al. 2009). Nanoparticles provide functional application like chromic behaviour, antibacterial properties, self-cleaning and UV protective, etc., but most significant to textile use are antibacterial, UV protective and colouration properties.

8.1 Antibacterial Finishing

Odour control, stain control and restricting spread of infection in densely populated areas (clean human environment being a civilizational need), all of these require antibacterial coatings on medical devices, hygienic products, water purification systems, medical surgery equipment, textiles, food packaging and storage material. Hygiene finishes protect the textile user from pathogenic or odour causing microorganisms in common textile garments like socks, sportswear and working clothes as well as mattresses, floor coverings and shoe linings. Protection of the textile itself from damage caused by fungi or rot producing microorganisms is another goal. Quaternary ammonium compounds, like Triclosan (2, 4, 4-hydrophenyl trichloro (II) ether), having action against Gram-negative and Gram-positive bacteria are often used as additives in laundering formula (Fouda 2012). Natural antibacterial compounds exist, and like chitosan [poly-(1-4)-d-glucosamine], a cationic polysaccharide, is obtained by alkaline deacetylation of chitin, the major component in crustacean shells (Ye et al. 2006). Natural antimicrobial compounds like sericin, neem extract, natural dyes, aloe vera, tea extract, coconut shell extract, *Sterculia foetida*

Table 2 Grades and classification of UPF

UPF	Transmittance (%)	Protection level	Grade
>40	<2.5	Excellent protection	III
25–40	3.3–2.5	Very good protection	II
15–25	5.0–2.5	Good	I

fruit shell extract, *Delonix regia* stem shell, eucalyptus oil, tulsi leaves and cloves oil, etc., have found use as antibacterial finishing agents (Teli et al. 2018; Joshi 2009; Teli and Pandit 2017a, b, c, d, e 2018).

A new generation of biocides based on nanoparticle products from nano-sized metals and metal oxides, of silver (Ag), titanium dioxide (TiO_2), zinc oxide (ZnO) and cooper II oxide (CuO) has been developed. Silver is the most biocidal against different pathogenic bacteria. Immobilization of silver nanoparticles on synthetic and natural fibres for odour control and biocidal activity has been achieved (Rivero et al. 2015). The antibacterial effect is obtained at extremely low loads of the silver component. The AATCC Technical Manual12 contains numerous test methods to evaluate antimicrobial finishes on textiles.

8.2 UV Protective Finishing

The frequency of occurrence of skin cancer has risen worldwide due to excessive exposure to ultraviolet rays of sunlight. The longer wavelength (320–400 nm) ultraviolet rays (UV-A) cause pigmentation of the skin with more melanin, while higher energy ultraviolet rays (UV-B) penetrate the skin surface, causing acute chronic skin eruption and damages such as skin reddening (erythema) or sunburn (Das 2010). Use of sunscreens and wearing clothing that covers as much of the skin surface maximizes UV protection. Textile clothing treated with nanoparticles of zinc, zinc oxide and TiO_2 acts as UV protection barriers on skin. Nanoparticles of TiO_2, ZnO and metallic zinc in size of 20–50 nm are photostable and offer a broad spectrum protection, but skin sensitivity and allergic reactions have to be considered, for the optimum load of UV protective finishing on fabric (Manaia et al. 2013).

In accordance with AATCC TM 183 or AS/NZS 4399 standards, the performance criteria recommended as UV protective in terms of grade of Ultraviolet Protection Factor (UPF) are given in Table 2 (Saravanan 2007).

8.3 Colouration Properties

Colouration of textile is achieved by dyeing with either natural dyes or synthetic dyes, but colour changes due to UV light damage, repeated washing and surface abrasion during wear are factors that deteriorate the colourfastness and overall quality of the fabrics and textile. Colourfastness is essential for high-quality fabrics and textiles for high value products (Johnston 2008). Coloured metallic nanoparticles (NPs), whose colour depends on the size of the particle, can be synthesized. The ultraviolet (UV) light stability of metallic nanoparticles makes them suitable to colour textiles that may undergo exposure to extreme levels of UV light (Teli and Pandit 2017d, e, 2018). Colouration properties are also imparted with the help of natural resources (Pandey 2017; Pandit and Nadathu 2018; Pandit et al. 2018b).

Gold and silver nanoparticles exhibit localized surface plasmon resonance (LSPR) phenomena that give bright colours and also act as catalysts (Tang et al. 2011). The colouration of fibres has been done through the LSPR effect of silver and gold nanoparticles that also provides UV protection and antibacterial action (Johnston 2008; Tang et al. 2015). Gold and silver nanoparticles with different size provide stable colourfast colourants which can be applied in high-quality fabrics and textiles (Tang et al. 2012).

9 Methods of Application of Nanoparticles on Textile

In recent past years, the demand for different multifunctional textiles with properties such as self-cleaning and antibacterial was high. In order to obtain textile materials with functional performances, many procedures are used for impregnation of nanoparticles on the fibrous material such as conventional technique like pad dry cure, coating by the help of binders and non-conventional process like grafting through plasma treatment (Pandit and Nadathu 2018; Pandit et al. 2018a).

Pad dry cure and coating are polluting multi-steps wet chemical processes. If durability is a concern effects of pad dry cure performs and the coating diminishes results better by imparting stiffness.

Plasma is a mixture of varying concentration of ions, electrons, neutrons, photons, free radicals, metastable excited species, molecular and polymer fragments. Plasma can be created under low pressure or atmospheric pressure (Teli 2015; Teli et al. 2015). It is a good tool to modify the structure and topography of the surface of different bulk materials. Plasma methods involving radio frequency plasma (RF-plasma) and microwave plasma (MW-plasma) have been used in pre-treatment and cleaning of surfaces to better deposition process and the subsequent loading of nanoparticles on the textile surface (Kiumarsi et al. 2017). This technique can reduce the usage of water and solvents and minimize chemical usage to achieve the same level of effective coating as conventional wet methods, but control and quenching of free radicals, radiation and heat generated are essential from the environmental perspective. Grafting is a surface modification technique, which deposits nanoparticles by creating polar groups on textile substrate. But this technique consumes costly and hazardous chemical and solvent.

This has led to greener methods of processing of textiles with the development of new technologies for deposited of nanoparticles on textiles such as layer-by-layer technology or polyelectrolyte multilayer (PEM).

Layer-by-layer (LBL) technique, based on the principle of self-assembled nano layers, can produce a thin layer film, on numerous types of surfaces. LBL process allows control of film thickness and also the choice of molecule as precursors can direct the formation of the layered assembly. Electrostatic self-assembly (ESA) is favoured for its simplicity and efficiency. In LBL method to coat textile fabric, suitable polyelectrolytes with opposite charge were deposited alternately, interrupted with wash steps in between to remove excess unbound polyelectrolyte, on the fabric surface. The thickness is determined by the number of cycles of adsorption. LBL technique enabled deposition of a thin nanocomposite on the fabric surface and different properties like colour, hydrophobicity, fire retardancy and UV photoprotection have been imparted to fabric (Srikulkit et al. 2017). Studies on durability indicate stability even after repeated washes. This technique is an economical, eco-friendly and zero energy-consuming process. Therefore, it can be implemented at the grass level (Dubas et al. 2006).

10 Future of Nanomaterials—Limits and Opportunities

Engineered nanomaterials are a crucial component of sustainable solutions to challenges in the field of non-fossil fuel-based energy generation, storage and saving, maintaining a clean environment, recycling wastewater, rapid and cheap mass transportation, applications in health care like lab-on-chip diagnosis, fulfilling infrastructure and food security needs for a huge population. The application of nanomaterials in an environmentally safe manner requires a sufficiently deep understanding of the complexities of nano-biological interactions that are mediated by chemical functionalities on cell surfaces and the functional reactivity, which is also shape and size dependent, of the nanomaterial. They often involve dynamic interactions at solid–liquid interfaces that have implications in different areas of application like drug release, filtration and adsorption, material resistance to corrosion and damage and mechanical strength of composite materials. Nanomaterials also are specialized complex structures that require specific handling and usage conditions such that functionality is retained. Better knowledge of the toxicity of nanomaterials in biological systems would help in the establishment of guidelines for their use in applications. Green technologies for synthesis of nanomaterials also require successful scale-up and adaptability to current manufacturing modes. Research in advanced materials is progressing towards competent solutions to resolve these obstacles (Hamers 2017).

References

Ahmed EM (2015) Hydrogel: Preparation, characterization, and applications: a review. J Adv Res 6(2):105–121 (Elsevier)

Ahmed S et al (2016) A review on plants extract mediated synthesis of silver nanoparticles for antimicrobial applications: a green expertise. J Adv Res 7(1):17–28 (Elsevier)

Amendola V, Meneghetti M (2009) Laser ablation synthesis in solution and size manipulation of noble metal nanoparticles. Phys Chem Chem Phys 11(20):3805–3821 (Royal Society of Chemistry)

Anandalakshmi K, Venugobal J (2017) Green synthesis and characterization of silver nanoparticles using *Vitex negundo* (Karu Nochchi) leaf extract and its antibacterial activity. Med Chem 7:218–225

Banerjee P et al (2014) Leaf extract mediated green synthesis of silver nanoparticles from widely available Indian plants: synthesis, characterization, antimicrobial property and toxicity analysis. Bioresour Bioprocess 1(1):3 (Springer)

Birla SS et al (2009) Fabrication of silver nanoparticles by Phoma glomerata and its combined effect against *Escherichia coli, Pseudomonas aeruginosa* and *Staphylococcus aureus*. Lett Appl Microbiol 48(2):173–179 (Wiley Online Library)

Chakravarty A et al (2017) Green synthesis of electrospun porous carbon nanofibers from sucrose and doping of Ag nanoparticle with improved electrical and electrochemical properties. ChemistrySelect 2(7):2265–2276 (Wiley Online Library)

Chandrika UG et al (2006) Hypoglycaemic action of the flavonoid fraction of. Afr J Tradit Complement Altern Med 3(2):42–50

Chaudhry Q, Castle L, Watkins DR (2010) Nanotechnologies in food. Royal Society of Chemistry London

Colson P, Henrist C, Cloots R (2013) Nanosphere lithography: a powerful method for the controlled manufacturing of nanomaterials. J Nanomater 2013:21 (Hindawi Publishing Corp)

Crane JH, Balerdi C, Maguire I (2005) Jackfruit growing in the Florida home landscape. Fact Sheet HS-882, pp 1–10

Das BR (2010) UV radiation protective clothing. Open Text J 3:14–21 (Bentham Science Publishers B. V., P. O. Box 1673 Hilversum 1200 BR The Netherlands)

Dubas ST, Kumlangdudsana P, Potiyaraj P (2006) Layer-by-layer deposition of antimicrobial silver nanoparticles on textile fibers. Colloids Surf A Physicochem Eng Aspects 289(1–3):105–109 (Elsevier)

El-Newehy MH et al (2018) Green electrospining of hydroxypropyl cellulose nanofibres for drug delivery applications. J Nanosci Nanotechnol 18(2):805–814 (American Scientific Publishers)

El-Nour KMMA et al (2010) Synthesis and applications of silver nanoparticles. Arab J Chem 3(3):135–140 (Elsevier)

Fouda MMG (2012) Antibacterial modification of textiles using nanotechnology. In: A search for antibacterial agents. InTech

Ghorbani HR (2014) A review of methods for synthesis of Al nanoparticles. Orient J Chem 30(4):1941–1949

Hamers RJ (2017) Nanomaterials and global sustainability. Acc Chem Res 50(3):633–637 (ACS Publications)

Hasan S (2015) A review on nanoparticles: their synthesis and types. Res J Recent Sci. ISSN 2277, p 2502

Hassan MM, Wagner MH (2017) Surface modification of natural fibers for reinforced polymer composites. In: Progress in adhesion and adhesives. John Wiley & Sons Inc, pp 1–44

Johnston JH et al (2008) Gold nanoparticles as colourants in high fashion fabrics and textiles 1:712–715

Joshi M et al (2009) Ecofriendly antimicrobial finishing of textiles using bioactive agents based on natural products. CSIR

Kathirvelu S, D'souza L, Dhurai B (2009) UV protection finishing of textiles using ZnO nanoparticles. CSIR
Khalil MMH et al (2014) Green synthesis of silver nanoparticles using olive leaf extract and its antibacterial activity. Arab J Chem 7(6):1131–1139 (Elsevier)
Khan A et al (2016) A chemical reduction approach to the synthesis of copper nanoparticles. Int Nano Letters 6(1):21–26 (Springer)
Khodashenas B, Ghorbani HR (2014) Synthesis of copper nanoparticles: an overview of the various methods. Korean J Chem Eng 31(7):1105–1109 (Springer)
Kiumarsi A et al (2017) 'Extraction of dyes from Delphinium Zalil flowers and dyeing silk yarns. J Text Inst 108(1):66–70 (Taylor & Francis)
Kołodziejczak-Radzimska A, Jesionowski T (2014) Zinc oxide—from synthesis to application: a review. Materials 7(4):2833–2881 (Multidisciplinary Digital Publishing Institute)
Makarov VV et al (2014) "Green" nanotechnologies: synthesis of metal nanoparticles using plants. Acta Naturae (англоязычная версия). Общество с ограниченной ответственностью Парк-медиа 6(1(20))
Malik P et al (2014) Green chemistry based benign routes for nanoparticle synthesis. J Nanopart (Hindawi)
Manaia EB et al (2013) Inorganic UV filters. Braz J Pharm Sci 49(2):201–209 (SciELO Brasil)
Mittal AK, Chisti Y, Banerjee UC (2013) Synthesis of metallic nanoparticles using plant extracts. Biotechnol Adv 31(2):346–356 (Elsevier)
Murphy M et al (2015) Current development of silver nanoparticle preparation, investigation, and application in the field of medicine. J Nanomater 2015:5 (Hindawi Publishing Corp)
Naahidi S et al (2013) Biocompatibility of engineered nanoparticles for drug delivery. J Controlled Release 166(2):182–194 (Elsevier)
Pal SL et al (2011) Nanoparticle: an overview of preparation and characterization. J Appl Pharm Sci 1(6):228–234
Pandey R et al (2017) Colouration of textiles using roasted peanut skin-an agro processing residue. J Cleaner Prod (Elsevier)
Pandit P et al (2018a) Applications of textile materials using emerging sources and technology: a new perspective. Green Sustain Adv Mater Appl 2:49–83 (John Wiley & Sons, Inc. Hoboken, NJ, USA)
Pandit P, Gayatri TN, Maiti S (2018b) Green and sustainable textile materials using natural resources. Green Sustain Adv Mater Process Charact 1:213–261 (John Wiley & Sons, Inc. Hoboken, NJ, USA)
Pandit P, Nadathu GT (2018) Characterization of green and sustainable advanced materials. Green Sustain Adv Mater Process Charact 1:35–66 (John Wiley & Sons, Inc. Hoboken, NJ, USA)
Pavani KV, Balakrishna K, Cheemarla NK (2011) Biosynthesis of zinc nanoparticles by Aspergillus species. Int J Nanotechnol Appl 5(1):27–36
Pervez MN, Stylios GK (2018) An experimental approach to the synthesis and optimisation of a "Green" Nanofibre. Nanomaterials 8(6) (Multidisciplinary Digital Publishing Institute (MDPI))
Prabhu YT et al (2017) A facile biosynthesis of copper nanoparticles: a micro-structural and antibacterial activity investigation. J Saudi Chem Soc 21(2):180–185 (Elsevier)
Prasek J et al (2011) Methods for carbon nanotubes synthesis. J Mater Chem 21(40):15872–15884 (Royal Society of Chemistry)
Rai M, Yadav A, Gade A (2009) Silver nanoparticles as a new generation of antimicrobials. Biotechnol Adv 27(1):76–83 (Elsevier)
Rauwel P et al (2015) Silver nanoparticles: synthesis, properties, and applications. In: Advances in materials science and engineering. Hindawi
Rivero PJ et al (2015) Nanomaterials for functional textiles and fibers. Nanoscale Res Lett 10(1):501 (Springer)
Sangeetha G, Rajeshwari S, Venckatesh R (2011) Green synthesis of zinc oxide nanoparticles by aloe barbadensis miller leaf extract: structure and optical properties. Mater Res Bull 46(12):2560–2566 (Elsevier)

Saravanan D (2007) UV protection textile materials. AUTEX Res J 7(1):53–62
Singh M et al (2012) Natural minerals and cancer. J Appl Pharm Sci 2(04):158–165
Srikulkit K, Iamsamai C, Dubas ST (2017) Development of flame retardant polyphosphoric acid coating based on the polyelectrolyte multilayers technique. J Metals Mater Miner 16(2)
Subhankari I, Nayak PL (2013) Synthesis of copper nanoparticles using *Syzygium aromaticum* (Cloves) aqueous extract by using green chemistry. World J Nano Sci Technol 2(1):14–17
Suresh S (2013) Semiconductor nanomaterials, methods and applications: a review. Nanosci Nanotechnol 3(3):62–74 (Scientific & Academic Publishing)
Tang B et al (2011) Using hydroxy carboxylate to synthesize gold nanoparticles in heating and photochemical reactions and their application in textile colouration. Chem Eng J 172(1):601–607 (Elsevier)
Tang B et al (2012) Coloration of cotton fibers with anisotropic silver nanoparticles. Ind Eng Chem Res 51(39):12807–12813 (ACS Publications)
Tang B et al (2015) Functional application of noble metal nanoparticles in situ synthesized on ramie fibers. Nanoscale Res Lett 10(1):366 (Springer)
Teli MD et al (2015) Hydrophobic silk fabric using atmospheric pressure plasma. Int J Bioresour Sci 2(1):15 (Renu Publishers)
Teli MD et al (2015) Low-temperature dyeing of silk fabric using atmospheric pressure helium/nitrogen plasma. Fibers Polym 16(11): 2375–2383 https://doi.org/10.1007/s12221-015-5166-4.
Teli MD, Pandit P (2017a) A novel natural source *Sterculia foetida* fruit shell waste as colorant and ultraviolet protection for linen. J Nat Fibers, 15(3): 337—343 (Taylor & Francis)
Teli MD, Pandit P (2017b) Development of thermally stable and hygienic colored cotton fabric made by treatment with natural coconut shell extract. J Ind Text 48(1):87–118. (SAGE Publications Sage UK: London, England, p. 1528083717725113)
Teli MD, Pandit P (2017c) Multifunctionalised silk using Delonix regia stem shell waste. Fibers Polym 18(9):1679–1690 (Springer)
Teli MD, Pandit P (2017d) Multifunctionalised silk using Delonix regia stem shell waste. Fibers Polym 18(9):1679–1690. https://doi.org/10.1007/s12221-017-1228-0
Teli MD, Pandit P (2017e) Novel method of ecofriendly single bath dyeing and functional finishing of wool protein with coconut shell extract biomolecules. ACS Sustain Chem Eng 5(9):8323–8333 (ACS Publications)
Teli MD, Pandit P (2018) Application of *Sterculia foetida* fruit shell waste biomolecules on silk for aesthetic and wellness properties. Fibers Polym 19(1):41–54. https://doi.org/10.1007/s12221-018-7315-4
Teli MD, Pandit P, Basak S (2018) Coconut shell extract imparting multifunction properties to ligno-cellulosic material, J Ind Text 47(6):1261–1290. https://doi.org/10.1177/1528083716686937
Tolle R et al (2007) 'Lloyd's of London report on nanotechnology: recent development, risks and opportunities. London, England, United Kingdom, Lloyd's of London Emerging Risk Team
Xia Y et al (2003) One-dimensional nanostructures: synthesis, characterization, and applications. Adv Mater 15(5):353–389 (Wiley Online Library)
Ye W et al (2006) Durable antibacterial finish on cotton fabric by using chitosan-based polymeric core-shell particles. J Appl Polym Sci 102(2):1787–1793 (Wiley Online Library)
Zhang Z, Decker EA, McClements DJ (2014) Encapsulation, protection, and release of polyunsaturated lipids using biopolymer-based hydrogel particles. Food Res Int 64:520–526 (Elsevier)

Green Nanomaterials: A Sustainable Perspective

Sukanchan Palit and Chaudhery Mustansar Hussain

Abstract Nanotechnology is the most innovative and versatile scientific discipline of the twenty-first century. Applications of nanotechnology crossed scientific frontiers from medicine to electronics, to modern manufacturing. Likewise, it has the potential and vision to significantly modify modern lifestyles, society and whole global economics. However, most of the materials and methods presently employed in this technology rely on nonrenewable resources as well as produce hazardous wastes. Techniques of green chemistry and green technology can play a vital and decisive role in deciding the future of nanoscience and nanotechnology applications. Natural materials to nanomaterials and design of naturally benign synthetic processes have been widely investigated, whereas most of the green nanomaterials are now routinely used from the laboratory to commercial scale, but they are also facing substantial challenges. This chapter reviews the most recent developments and challenges during applications of green nanomaterials. Green chemistry, green nanotechnology and green engineering are the ultimate needs of civilization today. In this chapter, the authors elucidate on the vision of the science of sustainability and its interface with green nanotechnology and green nanomaterials applications. This well-researched treatise will surely open up new knowledge dimensions in the domains of green nanomaterials, nanoscience, nanotechnology and green chemistry in decades to come.

Keywords Green · Nanotechnology · Nanomaterial · Sustainability · Engineering · Chemistry · Vision

S. Palit
Department of Chemical Engineering, University of Petroleum and Energy Studies, Energy Acres, Post-Office-Bidholi via Premnagar, Dehradun, Uttarakhand 248007, India
e-mail: sukanchan68@gmail.com; sukanchan92@gmail.com; sukanchanp@rediffmail.com

C. M. Hussain (✉)
Department of Chemistry and Environmental Sciences, New Jersey Institute of Technology, University Heights, Newark, NJ 07102, USA
e-mail: chaudhery.m.hussain@njit.edu

S. Ahmed and W. Ali (eds.), *Green Nanomaterials*,
Advanced Structured Materials 126,
https://doi.org/10.1007/978-981-15-3560-4_2

1 Introduction

Green nanotechnology and the vast world of green chemistry are in the roads toward new scientific ingenuity and scientific vision. Global warming, global climate change and depletion of fossil fuel resources are immensely challenging the vast scientific vision of environmental engineering and petroleum engineering. Today, technology has few answers to the scientific provenance and scientific subtleties of heavy metal poisoning of drinking water in developing and industrialized nations around the world. Thus the need of green nanomaterials, green nanotechnology and green chemistry. Nanotechnology, green chemistry and green engineering can solve vast scientific issues globally at this critical juncture of history and time. This challenge is taken by the author as science and technology move forward. Nanotechnology and green engineering are two opposite sides of the coin today. Environmental remediation is a major issue in the scientific firmament today. Civilization's immense scientific grit, determination and understanding the needs of human development and the vision of nanotechnology will all lead an effective way in the true emancipation of science and technology globally. In this treatise, the author delineates scientific ingenuity in the field of nanotechnology, green nanomaterials and green chemistry. The author also focuses on the deep understanding in the field of water treatment, wastewater treatment and other areas of industrial pollution control.

2 The Vision of This Study

Today, science and technology are moving at a drastic way surpassing one idealistic approach over another. This well-researched treatise reviews recent advancements in the field of nanotechnology, green nanomaterials and green chemistry with the sole vision toward greater emancipation of science. The vast domain of science and engineering today stands in the middle of ingenuity, vision and scientific divination. Nanotechnology and green engineering will veritably solve intricate scientific problems such as global warming, environmental engineering issues and barriers in petroleum engineering science. The main thrust of this chapter is to open new avenues of scientific advances in the field of green nanotechnology. The other areas of research are of groundwater contamination by heavy metals and the latest research quest in the field of sustainability. Today, global warming is a veritably challenging vast scientific firmament of human civilization and scientific progress leading civilization to a state of immense devastation. There is an endless chain of environmental catastrophes such as arsenic-contaminated drinking water in South Asia, and other developing and industrialized nations around the world. Materials science applications are the marvels of science in present-day human civilization today. Engineered nanomaterials are the visionary novel and efficient materials of today. Green nanomaterials and green nanotechnology will surely unravel new future thoughts and new visionary avenues of conscience and ingenuity in the field of science and technology globally. This chapter also deals proactively on the scientific ingenuity and vision in the field of water treatment and the larger path-breaking areas of environmental protection. This chapter will surely be a visionary eye-opener toward the scientific importance and the technological far-sightedness in green nanomaterial applications in water purification science (Palit and Hussain 2018a, b, c, d, e, f, g, h, i, j).

3 The Need and the Rationale of This Study

Today, scientific progress is developing in the route of new scientific determination. Today, engineering and technology have few answers toward the monstrous and ever-growing issues of poisoning of groundwater and drinking water in many developing and industrialized nations. Green nanomaterials and green nanotechnology have vast and intricate answers to the growing concerns of water purification and industrial wastewater treatment. Thus, the immediate need of the concerted efforts toward global research forays in water technology and green nanotechnology. Thus, also the immediate need of scientific vision and a sustainable initiative in green nanomaterials applications. Nanotechnology, space technology and nuclear technology are overcoming engineering and scientific frontiers. Environmental and energy sustainability are immensely needed as civilization moves forward. The need of this research is that sustainable development should go along with human scientific progress. Renewable energy and energy engineering are in the avenues of new engineering ingenuity and scientific affirmation. Green nanomaterials and its applications are today changing the face of scientific research pursuit as civilization confronts the ravages of nature such as climate change, environmental catastrophes and loss of fossil fuel resources. In this chapter, the author deeply pronounces the need of nanotechnology, environmental engineering science and renewable energy technology in the true unraveling of science and technology research questions. This chapter is an eye-opener to the new generation of scientists, engineers and the civil society. The author also targets the success and ingenuity in sustainability and green nanomaterials applications to human society (Palit and Hussain 2018a, b, c, d, e, f, g, h, i, j). Green nanomaterials and its applications are the marvels of science and engineering as it pertains to green chemistry and green technology.

4 Sources of Green Nanomaterial

The application of green nanomaterials and green nanotechnology is of immense importance as science and engineering move forward toward a new age. Human civilization's vast girth, the ingenuity and the futuristic vision of material science will all be the forerunners toward the true emancipation of science globally. Green nanomaterial is today an area of immense engineering research and vast emancipation. The challenges of the domain of green nanotechnology, the domain of green nanomaterials and the future of the science of nanotechnology are all the pallbearers toward a path-breaking era in the field of science and engineering. Sources of green nanomaterials are vast and versatile. Nature provides us with numerous chemical substances that serve as reducing agents including plant extracts, biopolymers, vitamins, proteins, peptides (e.g., glutathione) and sugars (e.g., glucose, fructose). Plant derivatives are the most studied so far. One area of particular scientific success is the manufacture of metal nanoparticles, useful in electronics and medicinal applications, using plant extracts as reducing agents. Biomedical applications such as drug and gene delivery have recently become a promising area of visionary research endeavor. Biopolymers are another family of natural sources used for metal nanoparticle synthesis. These polymeric carbohydrate molecules have already been used in a wide range of applications and thus are available for the large-scale production of nanoparticles (Palit and Hussain 2018a, b, c, d, e, f, g, h, i, j).

5 Nanocellulose—A Green Nanomaterial

Nanocellulose is defined as nano-structured cellulose. This encompasses cellulose nanofibers, microfibrillated cellulose, nanocrystalline cellulose or bacterial nanocellulose which refer to nano-structured cellulose produced by bacteria. Enzymatic treatment during nanocellulose processing is a vastly promising area of science and technology today. The property of nanocellulose makes it a promising and important material for many applications and the potential to form a rich and affluent industry (Palit and Hussain 2018a, b, c, d, e, f, g, h, i, j).

6 Green Processing Systems

Green processing systems are today in the middle of vast revamping. Green engineering, green chemistry and green nanotechnology are the immediate needs of the hour as science and technology surge forward. The world today is in the middle of immense scientific disasters as global warming and loss of fossil fuel resources challenges the vast scientific fabric of immense girth and determination. Water purification, drinking water and groundwater treatment and integrated water resource management are the immediate need of the scientific vision and ingenuity of human progress. Nanotechnology and applied green chemistry are the new arenas of scientific research pursuit. Green nanotechnology envisions the field of green chemistry and green engineering. Green nanoscience and nanotechnology refer to the field of nanotechnology which envisions environmental sustainability and maintains green environment. The chemical industry plays a pivotal role in sustaining the world's economy and envisions future scientific advancements in novel materials, less toxic products, renewable energy sources, environmental protection, industrial processes with energy efficiency and renewable green materials. Green chemistry can be applied to design environmentally benign products, to produce life-saving medicines, environmentally friendly agrochemicals, new enzymes, new renewable energy processes, energy efficiency in chemical engineering and innovative smart materials. Rapid and visionary advances of human mankind in the twentieth and twenty-first century forced scientists and engineers to realize that further development of human mankind and fulfillment of socioeconomic needs will only be possible if the natural resources are properly managed. Green processing systems are changing the scenario of human scientific progress today. Environmentally benign processes are the necessities and utmost need of green chemistry today. In this chapter, the authors point toward the vast needs of green nanotechnology in the progress of science globally. Green processing systems and green process engineering will surely open up new chapters of scientific determination and streamline the new vision of green chemistry and green nanotechnology (Palit and Hussain 2018a, b, c, d, e, f, g, h, i, j).

7 Significant Scientific Advances in Green Nanomaterials and Green Nanotechnology

Green nanotechnology and green nanomaterials are the wonders of science and technology today. Nanotechnology, green engineering and green chemistry are witnessing immense scientific upheavals. In this section, the authors with immense conscience and far-sightedness elucidate on the scientific and engineering success in the domain of green nanotechnology, green nanoengineering and green nanomaterials with the sole vision of progress and emancipation of science, technology and engineering (Palit and Hussain 2018a, b, c, d, e, f, g, h, i, j).

Glaser (2012) discussed with cogent insight green chemistry with nanocatalysts. Nanomaterials are expected to be a promising area for green chemistry catalysis due to its ability to design at the nanoscale. Nanocatalysts are highly esteemed as materials of increased surface area and are found to have catalytic abilities (Glaser 2012). Nanocatalysts are heralded as new and novel process candidates exhibiting higher activity, greater stability, durability or recycling potential, and cost-effectiveness (Glaser 2012). Chemical reaction engineering and chemical catalysis are today in the path of new revamping. These new catalysts are candidates for clean energy applications such as hydrogen generation, hydrogen storage and fuel cell applications (Glaser 2012). The author in this paper elucidates on plastic cycling conversion technology, biofuels for transportation, reaction kinetics in the pharmaceutical industry, future of green chemistry, the importance of green innovation index and ozone chemistry. The author in this paper also discussed the proliferation of literature in the field of nanotechnology, nanomaterials, nano-energy and green chemistry today (Glaser 2012).

Nath et al. (2014) discussed on the domain of green nanomaterials and how green they are as biotherapeutic technique. The technology and the engineering domain of nanomaterials are advancing rapidly overcoming one visionary frontier over another.

The research in the field of green nanoparticles has attracted widespread activity in the scientific and engineering fraternity for decades due to their unique properties and applications in diverse areas including drug delivery and therapy. Green nanotechnology gives opportunities in decreasing the risk of producing nanomaterials and also using the nanomaterials in lowering the formation of chemical intermediates. Nanotechnology is a relatively a new and growing domain of research endeavor. It is the latest innovative trend in modern science and technology. Nanotechnology is the scientific understanding of matter at the nanoscale, at dimensions between 1 and 100 nm, where unique phenomena envision and envisage novel applications. Technology, engineering and science are enshrined and envisioned as nanotechnology surges into new disciplines and opens up newer windows of science in years to come (Nath et al. 2014). In return for making nanomanufacturing techniques less energy and resource-intensive, the inherent principles of green chemistry applied to nanoscience and nanotechnology involve, if at all possible, (a) the use of cost-effective precursors, (b) minimization of carcinogenic reagents and solvents, (c) reduction of experiments which are at deep risk, (d) use of few number of reagents, (e) minimization of reaction steps leading to waste, reagent use and power consumption, (f) development of reactions to generate high purity materials, (g) ambient temperature and pressure synthesis, (h)efficiency of scale-up (Nath et al. 2014). The authors discussed with deep details green nanotechnology and therapeutics, green nanomaterials in therapeutic use, therapeutic nanoparticle platforms, inorganic nanoparticles, and the vast area of green nanoparticles (Nath et al. 2014). The world of technology and engineering of nanoparticles will be veritably reenvisioned as application of nanoparticles in biological sciences, medical science and biomedical engineering enters into new knowledge dimensions.

European Commission Report (2009) discussed in deep details challenges, opportunities and targets of green nanotechnology applications. A critical overview of opportunities and risks is deliberated and discussed in minute details in this report. Materials produced with the help of nanotechnologies are starting to be used and reenvisaged in many areas of human life (cosmetics, clothing, fabrics, paints, packaging, foods, etc.). Mankind's vast knowledge prowess, grit and determination will all lead a visionary way in the true emancipation of the science of nanotechnology today. Nanotechnology, biological sciences and biotechnology are inter-related today. As the applications of nanotechnology expand, many proponents of the science of nanotechnology are positioning this domain toward a greener and a sustainable future. Environmental science and technology and energy sustainability are today in the zenith of scientific enquiry. Nanotechnology and sustainability are related to each other today. In such an important juncture of vision and ingenuity, it is essential for the civil society to gain knowledge on different areas of the emerging nanotechnology development especially in relation to green engineering and green technology. The success of the science of sustainability, the academic rigor of nanotechnology and the needs of the human society are the veritable forerunners toward a new visionary age in technology and engineering (European Commission Report 2009). The authors in this report delineate with immense conscience the following points:

- Challenges, targets and opportunities in application of green nanotechnologies.
- Environmental, health, safety and hazardous management research.
- Regulatory status (European Commission Report 2009).
- NGO guidelines on sustainability assessment of nanotechnologies.

The authors in this report deeply discussed with cogent insight the vision of nanotechnology applications in the ever-growing advances of science and engineering. Nanotechnology and nanoengineering are the wonders of science today. The authors of this report reviewed water purification, renewable energy production, waste management and environmental remediation. This paper also delineated the use of nanotechnologies and materials and its impact on biodiversity, ecosystems and human health (European Commission Report 2009). This report also treads the path toward regulatory regimes and options to secure the safe and responsible future of nanotechnologies, with vast reference to green nanotechnologies. Environmental and health impacts of nanomaterials stand as major pillars of this well-researched treatise (European Commission Report 2009). The challenges of environmental protection in the twenty-first century are vast and versatile. Climate change, overdependence on fossil fuel resources, overexploitation of natural resources as well as the impacts of Western economies are the biggest environmental engineering challenges today. In the similar manner, chemical technology and petroleum engineering are the domains which are highly stressed as science moves forward toward a new scientific age and a new knowledge dimension (European Commission Report 2009). Nanotechnology is the next technological revolution. The authors in this report deeply elucidate the success, the ingenuity and the futuristic technological validation of nanotechnology applications to human society.

Biao et al. (2018) deeply discussed with scientific foresight green synthesis of gold nanoparticles. Green synthesis of gold nanoparticles using plant extracts is one of the promising avenues for obtaining environmentally friendly nanomaterials for biological applications and environmental protection applications. Natural products chemistry is a marvel of science today. In this study, proanthocyanidins—functionalized gold nanoparticles—were synthesized via a hydrothermal method. The obtained gold nanoparticles were characterized by ultraviolet and visible spectrophotometry (UV-vis) (Biao et al. 2018). The research endeavor shows that functionalized gold nanoparticles had high removal rates for the heavy metal ions and dye. Gold nanoparticles are one of the most widely investigated noble metal nanoparticles due to its vast applications in catalysis, electronic devices, medicine and physical chemistry. In this study, the authors investigated the water extract of pomegranate peels as both a reducing agent and a capping agent to prepare gold nanoparticles via a hydrothermal method. Technological vision and the vast world of scientific validation are the veritable pivots toward nanoparticle applications to the human society. The authors in this paper deeply elucidate the immense need of nanoparticles and environmental engineering toward future scientific progress (Biao et al. 2018).

Lu and Ozcan (2015) elucidated with immense scientific insight green nanoparticles and its research and development forays for a sustainable future. Nanotechnology is one of the most important breakthroughs in the twenty-first century. Technological and scientific vision is immense today as regards application of nanoparticles to human society today (Lu and Ozcan 2015). Nanomaterials and engineered nanomaterials are the veritable successes of scientific endeavor in modern civilization today. With vast and varied applications that surpass scientific boundaries—from electronics to biomedical engineering, to advanced manufacturing technology, to cosmetics—nanotechnology has scientific potential to drastically change lifestyles and entire economic status of human race. But nanomaterials and processes currently used are not only dependent on nonrenewable processes but also highly produce dangerous products (Lu and Ozcan 2015). Thus, the need of environmental sustainability and greater realization of the science of environmental engineering. This paper reviews the recent advancements and vision in the field of green nanotechnology and green nanomaterials (Lu and Ozcan 2015). The challenges and the vision of nanomaterials and engineered nanomaterials are groundbreaking and need to be envisioned and readdressed with the passage of science. In the past few years, nanomaterials have immensely shown superior performance in numerous applications, including medicine, energy, environment and manufacturing engineering. Advanced manufacturing tools, human factor engineering and technology management are the needs of science and engineering and technology today (Lu and Ozcan 2015). The authors in this paper elucidate the success of green nanotechnology and the scientific ingenuity of nanomaterials applications in the applications of green nanomaterials and green engineered nanomaterials (Lu and Ozcan 2015). Chemical engineering, process technology and environmental engineering are today linked with nanotechnology in many areas of scientific research pursuit. Current green nanotechnology practices very often involve the use of natural sources, nonhazardous solvents and energy-efficient tools in the preparation of nanomaterials (Lu and Ozcan 2015). These scientific challenges in the world of nanotechnology are described in minute details in this chapter.

Wardencki et al. (2005) deeply elaborated with immense far-sightedness current and future issues in green chemistry. A detailed overview of green chemistry issues in the recent past demonstrated new methodologies that protect human health in an environmentally friendly way. The concept of green chemistry and green engineering has appeared in the USA as a robust research program resulting from an interdisciplinary cooperation of university teams, research groups, scientists, engineers, technologists and educationists. The status of green chemistry thus proliferated with immense vision, might and scientific determination. Green chemistry incorporates a new approach to the synthesis, processing and application of chemical substances whose main aim is to reduce dangers to human health and the environment. The new approach of green chemistry involves: (1) environmentally friendly green chemistry, (2) clean chemistry, (3) atom economy and (4) benign with the help of applied chemistry and chemical engineering. The implementation of green chemistry and green engineering principles into practice in human society assumes immense importance as regards sustainable development today. The authors deeply delved into the unknown depths of green chemistry and green engineering. This paper evolved immense scientific thoughts and is a watershed text in the field of green chemistry and green engineering (Wardencki et al. 2005).

Green nanomaterials and nanotechnology are today in the avenues of immense scientific rejuvenation. The world today stands shocked and mesmerized as environmental disasters and global warming devastate the global scientific firmament. Chemical engineering and environmental biotechnology today stand in the similar way in the middle of profundity, divination and revelation. The authors in this chapter reiterate and focus on the intricate and the vast issue of environmental sustainability and open up new windows of scientific vision in years to come.

8 Recent Scientific Endeavor in the Field of Arsenic and Heavy Metal Groundwater Remediation

Heavy metal poisoning of drinking water is today challenging the vast global scientific fabric. South Asia, particularly India and Bangladesh, is in the threshold of a major disaster. The drinking water catastrophe is challenging science, technology and engineering domains today. Millions of people in Bangladesh and India are affected with health issues such as cancer and many are still struggling to surmount the drinking water contamination issue. Technology and science have few solutions to the growing issues of heavy metal drinking water poisoning. Scientific vision and ingenuity are the utmost needs of the hour in confronting this burning problem. Arsenic and heavy metal groundwater and drinking water remediation and the participation of civil society and engineers are needed if environmental protection science needs to surpass visionary frontiers. The state of West Bengal in India is equally in the threshold of this monstrous issue. The scientific provenance of mankind, the technological motivation and the world of scientific validation in environmental sciences and engineering will truly lead an effective and resounding way in the effective realization of science and engineering.

Hashim et al. (2011) discussed with lucid insight remediation tools for heavy metal contaminated groundwater. The poisoning of drinking water and groundwater by heavy metal, originating either from natural soil resources or from anthropogenic sources, is an ever-growing concern for human civilization and modern science and engineering (Hashim et al. 2011). Remediation of contaminated water should take pivotal role globally as millions of people around the globe depend on drinking water and groundwater. In this research pursuit, thirty-five procedures for groundwater treatment are deeply investigated and are divided under three major classifications which are chemical, biochemical/biological and physicochemical water treatment procedures. Selection of a research area and an effective tool for contamination remediation at a particular site is one of the most daunting tasks due to complex soil chemistry. Keeping the sustainability issues and the environmental ethics in mind, the vast and versatile technologies encompassing natural chemistry, bioremediation, groundwater remediation and bio-sorption are highly recommended in the research and development initiatives in environmental sciences and engineering. Heavy metal is a generally collective term, which applies to the group of metals and metalloids with atomic density greater than 4000 kg m^{-3} and they are natural substances of the earth's crust. In the surrounding environment, the heavy metals are generally more seen than organic compounds such as pesticides or petroleum by-products. Technological vision and the status of scientific validation in environmental engineering will all be the forerunners toward a new era in arsenic groundwater remediation today. The authors (Hashim et al. 2011) discussed in minute details sources, speciation of heavy metals in groundwater, technologies for treatment of heavy metal contaminated groundwater, chemical treatment techniques, reduction with the help of iron-based technologies, biological and biochemical treatment technologies. The state of drinking water provision is immensely dismal today. This paper depicts profoundly the needs of research forays in the field of water pollution control (Hashim et al. 2011).

9 Recent Scientific Research Endeavor in Environmental Protection and Green Engineering

Environmental protection and green engineering are the utmost necessities of human scientific research pursuit today. Sustainability is moving toward a new age of vast engineering ingenuity. Shannon et al. (2008) deeply elucidated on the science of water purification and wastewater treatment. One of the intricate issues globally today is the world of water purification today Shannon et al. (2008). The most important problem affecting people throughout the world is inadequate access to pure water and absolutely improper sanitation. Problems with water purification are expected to grow worse in the coming decades, with water scarcity occurring globally even in drought-prone areas around the globe. In this paper(Shannon et al. 2008), the authors highlighted some of the science and technology being developed to improve the disinfection and decontamination of water, and research and development forays in the safe reuse of wastewater and effective desalination of seawater. 1.2 billion people around the globe lack access to pure drinking water, 2.6 billion have scarce sanitation, millions of people die annually—3900 children a day—from diseases transmitted through unsafe water (Shannon et al. 2008). Innumerable citizens around the world are afflicted with disease and contamination. Water also tremendously affects energy, environment and food production and the quality of environment affects severely economies of developing and developed world (Shannon et al. 2008). Today, science has practically no answers to the arsenic groundwater contamination in South Asia and many nations in industrial world. The world of water pollution control today stands in the midst of scientific introspection and definite scientific redeeming. Conventional techniques of water disinfection can address and envision the immense challenges of water quality and supply (Shannon et al. 2008). Mankind's vast scientific girth and determination, the vast area of technological and engineering validation and the futuristic vision of science and engineering are the pallbearers toward a new era in the field of environmental protection and water pollution control. In this treatise, it was discussed with scientific far-sightedness disinfection, reuse and reclamation and the domain of desalination. Technology of water disinfection is in the state of new rejuvenation today. An overarching aim and objective for providing safe water is a robust approach to decontaminate water from emerging pathogens, without creating more issues due to the disinfection process itself. Thus, the immediate need of environmental engineering research and development initiatives with immense participation of the civil society. Thus also the need for development of new disinfection strategies, coagulation, flocculation, sedimentation and membrane filtration. The much-needed vision for the future of reclamation and reuse of water is to capture water directly from nonconventional water resources such as industrial wastewaters and restore it to drinking quality. Vast technological vision, the immense challenges of water reclamation and the true ingenuity of environmental protection science are the forerunners toward a new era in the domain of science and engineering (Shannon et al. 2008).

Science and engineering are today surpassing vast and difficult boundaries. Technological challenges are immense and groundbreaking as science of environmental protection science trudges forward. Scientific research ingenuity and scientific revelation are the pivots of scientific pursuit today. In this entire chapter, the authors reiterate and pronounce the facts behind water purification, environmental remediation and the vast contribution of nanotechnology in the advancement of science and technology.

10 Environmental Management, Sustainability and the Visionary Future

The vast areas of environmental engineering and environmental management are today highly relevant to the scientific progress and academic rigor. Sustainable development and environmental management are today integrated with each other. The visionary words of Dr. Gro Harlem Brundtland, former Prime Minister of Norway, on "sustainability" need to be reenvisioned and reorganized with the passage of time.

Barrow (2006) deeply discussed with scientific and technological conscience environmental management for sustainable development. Environmental management is a broad and rapidly evolving domain of science and technology.

Human scientific progress today stands at the crossroads of vision and scientific ingenuity. Environmental management and sustainable development are the hallmarks of academic rigor today. This book seeks to offer a comprehensive and understandable introduction (Barrow 2006). Provision of basic human necessities such as water, energy, food and shelter is highly neglected in human civilization today. Environmental engineering and management are evolving rapidly; it is extremely important for more and more sectors of human activity and plays a pivotal role in realization of sustainable development. Provision of basic human necessities such as water and energy is of pivotal importance to the progress of human civilization and human scientific rigor. Water purification and wastewater treatment are today in the middle of immense caution, scientific ingenuity and scientific fortitude. Environmental management generally envisions a multidisciplinary approach and achieving this in an organized manner can be a veritable challenge because suitable supportive frameworks are still in the process of development. The author in this book delineated theory, principles and key concepts of environmental management, environmental management approaches, standards, monitoring, modeling, auditing and coordination of the environmental management framework (Barrow 2006). The author deeply comprehended with scientific conscience the needs of environmental management along with environmental and energy sustainability in the pursuit of science and technology.

Barrow (2005), elucidated on the subject of environmental management and development. The author in this book deeply elucidated theory of environmental management in industrialized and developing countries around the globe. The author also profoundly discusses resource management issues and environmental tools and policies (Barrow 2005). The need of environmental management and sustainable development is immense and thoroughly scientific. The provision of basic human needs such as water and electricity is highly challenged as well as reenvisioned as environmental and energy sustainability assumes immense importance. This book investigates environmental management, with the focus on developing countries and the development process around the world. The vast process of development takes place in the environment, using natural resources, generating waste and causing immense scientific and technological impacts. Both environmental management and development are domains of research pursuit which envision a multidisciplinary view and immensely make it possible to unite together toward a new vision and a new scientific innovation. In the last 15 years, both environmental management and development have had to address global thoughts and global issues; sustainability and sustainable development are the visionary targets for the future. The global scenario has a rich North and a poor South with areas of poverty in the former and some immensely rich people in the latter. Here arises the vast importance of environmental and energy sustainability (Barrow 2005). Providing a vastly and universally acceptable definition of development is impossible, although most would accept that as a vast process of change and transition which can accelerate, slow, stagnate or run backwards. Development management has successfully evolved independently of environmental management and environmental sciences, but the processes of both frequently overlap. Development managers focus on inter-organisational relationships. Sustainable development and energy and environmental sustainability had become a prominent part of the environment and development discourses by the twenty-first century, and visionaries have noted that it is a solid and permanent concept that helps integrate environmental management and development management (Barrow 2005).

The entire treatise explores the key issues and approaches in environmental management, agriculture, food security, urban environmental issues, rural environmental issues, environmental management methods, tools and techniques. The world of science and technology today stands in the midst of vision, scientific fortitude and reawakening. This book explores the vision of sustainable development and its interface with science, technology and engineering.

Kates et al. (2005) discussed in details goals, indicators, and practices in sustainable development. The most vastly accepted definition of sustainability is

> Humanity has the ability to make development and progress sustainable—to ensure that it meets the needs of the present without compromising the ability and urge of future generations to meet their own needs. Science, technology and engineering are in the avenues of new scientific overhauling as civilization faces immense difficulties and barriers. (Kates et al. 2005)

The World Commission on Environment and Development was propounded by the General Assembly of the United Nations in 1982 and its visionary report "Our common future," was published in 1987. It was chaired by then-Prime Minister of Norway, Gro Harlem Brundtland, thus earning the name the Brundtland Commission. The authors in this treatise redefined the idea of sustainable and holistic development. Today is an era of space technology and nuclear science. Nanotechnology, biotechnology and environmental biotechnology are reframing the scientific world. In this paper, the authors focus on the success of sustainable development in developing and developed economies around the globe. The Brundtland Commission's brief definition of sustainability as the "ability to make development sustainable—to ensure that it meets the needs of the present without compromising the ability of future generations to meet their own needs" is surely the most visionary definition. The study of ecosystem services and its engineering has strengthened this definition over time. The entire paper targets the scientific and technological success of energy and environmental sustainability in the progress of civilization today (Kates et al. 2005).

Science and engineering globally are in a dismal state as global warming, loss of ecological biodiversity, environmental catastrophes and loss of fossil fuel resources devastates the scientific landscape. In this chapter, the author elucidates the realization of sustainable development in the furtherance of science and engineering mainly green nanotechnology and green nanomaterials.

11 Nanotechnology and Sustainability

The science of "sustainability" today stands in the midst of technological travails and vast scientific difficulties. Nanoscience and nanotechnology are today's visionary research questions. Scientific articulation and ingenuity are the pallbearers toward a new visionary era in nanotechnology and green engineering. Environmental protection, deep scientific vision of environmental engineering and chemical engineering will all lead a profound way in unraveling the scientific truth of nanoscience and nanotechnology. Provision of basic human needs such as food, water, energy, electricity and shelter is highly stressed as civilization and science surge forward. The domain of nanotechnology is a veritable wonder of science. Water purification science in a similar vision is highly challenged as nations around the world faces the tremendous wrath of heavy metal drinking water and groundwater contamination. Thus, the need of a concerted effort from the scientists, engineers and the civil society. The visionary words of Dr Gro Harlem Brundtland, former Prime Minister of Norway, on sustainability need to be readdressed and reenvisaged with the passage of history and time. Green engineering and environmental engineering science are today linked with each other. Thus, the immense need of green nanomaterials and its sustainable solutions in present-day human society. Nuclear science and space technology are moving at a drastic pace overcoming one frontier over another. In the similar vein, nanotechnology is today in the path of new beginning (Palit 2017a, b, 2018a, b).

12 Heavy Metal Groundwater Remediation and the Future of Science

Heavy metal groundwater and drinking water contamination are veritably challenging the fabric of vision, might, technology and determination. Technology, engineering and science today have practically no solutions to the growing and the monstrous issue of arsenic groundwater contamination. The challenges of scientific endeavor in the field of water science and technology need to be reenvisioned and reorganized with the passage of history and time. The scientific travails, the scientific ingenuity and the vast needs of the science of sustainability are the veritable torchbearers toward a new visionary era in green nanotechnology and green nanomaterials today. Today is the visionary world of green technology and green chemistry. In South Asia, particularly India and Bangladesh are in the threshold of world's largest environmental disaster that is arsenic drinking water poisoning. There should be a concerted effort on the part of the engineers and scientists to deal with this unending environmental crisis of immense proportions. The authors along with applications of green nanomaterials also target the travails, the provenance and the barriers in water purification and industrial wastewater treatment. The domain of water purification today stands in the middle of scientific vision and deep scientific introspection. Heavy metal groundwater remediation and the areas of environmental protection are the utmost need of scientific progress today. In this chapter, the authors vastly pronounce the immense importance of provision of basic human necessities such as water, energy, food and shelter toward the further emancipation of science, engineering and technology (Hashim et al. 2011; Hussain and Palit 2017; Palit and Hussain 2018a, b).

13 Challenges–Utilization and Commercialization

Green nanomaterials and green nanotechnology are the cornerstones of scientific emancipation, realization and technological motivation of human civilization today. The present need of the human society is the greater commercialization of green nanotechnology and green chemistry. The utilization and commercialization of nanotechnology to human society assume importance as science and engineering gear forward. The vision and the targets of scientific advancements in sustainable green nanomaterials are immense and groundbreaking. In this chapter, the authors deeply enumerate the success, the scientific truth and the scientific adjudication in the domain of green nanomaterials applications in society. The visionary words of Dr. Gro Harlem Brundtland, the former Prime Minister of Norway, on the science of "sustainability" need to be revisited with every step of history of science and technology. Today, the world of green engineering and environmental engineering is in the midst of scientific introspection and vision. The veritable challenges of utilization and commercialization are vast and versatile as environmental engineering surges forward toward a new knowledge dimension. In this well-researched paper, the authors pointedly focus on the success of sustainability issues in the application of green nanomaterials in different branches of science and engineering such as environmental protection and groundwater remediation. Energy and environmental sustainability are the important needs of scientific progress in present-day civilization. Application of nanotechnology in sustainability whether it is energy, environmental, social and economic is the pillars of scientific rejuvenation today. The target of scientific endeavor today should be to transform lab-scale research to human society. In such a situation, application of green nanomaterials to human society and its sustainable solutions are the veritable forerunners toward an age of scientific validation. Application of nanotechnology to water purification and water treatment are the needs of research and development initiatives. This chapter will surely lead a visionary way in the true realization of nanotechnology, green engineering, green chemistry and green nanomaterials (Hashim et al. 2011; Hussain and Palit 2017; Palit and Hussain 2018a, b, c, d, e, f).

14 Future Research Directions in Green Nanomaterials, Nanotechnology and Sustainability

Green nanotechnology and the its applications are today undergoing drastic changes. The scientific endeavor and the vast introspection in the field of green nanotechnology need to be reorganized with the march of civilization and academic rigor. Future research pursuit in the field of green nanomaterials should be directed toward more visionary applications of sustainable green nanomaterials and the proper application of environmental sustainability to human society. The domain of environmental protection today is in the juncture of immense catastrophe and deep scientific contemplation. Provision of pure drinking water is an area of science which is immensely neglected. So, the future lies in the areas of green nanotechnology and green nanomaterials applications in water purification (Hashim et al. 2011; Hussain and Palit 2017; Palit and Hussain 2018a, b, c, d, e, f, g, h, i, j).

The status of human environment is highly challenged and in the same vision immensely dismal with the frequent environmental disasters, global warming and loss of fossil fuel resources. Thus, chemical process engineering, environmental engineering and petroleum engineering need to be reorganized with the progress of civilization. Future research directions and future research endeavor in the world of nanotechnology should be directed toward greater emancipation of energy and environmental sustainability. Water purification and industrial wastewater treatment are the needs of human civilization at present or the future. Thus, the immediate need of sustainable and feasible solutions to human scientific and academic rigor. Future recommendations of the study should be directed toward more enhancement of environmental sustainability applications toward the greater emancipation and greater realization of infrastructural development and holistic sustainable development. Pollution control, groundwater heavy metal contamination mitigation and amelioration of environmental catastrophes stand as major scientific imperatives toward the progress of human society. Sustainability and nanotechnology research directions in today's scientific world should be sound and environmentally safe. Research pursuit in green nanomaterials and nanotechnology applications should be linked with human factor engineering, reliability engineering and systems science. Today, the gamut of challenges lies in the environmental science issues of nanomaterials. Here come the issues of environmental sustainability and industrial hazard management. The authors deeply pursue these vital issues of environmental engineering and green nanomaterials applications. Research initiatives should be directed toward commercial-scale production. It is then a larger scientific truth and a deeper scientific adjudication will be realized (Hashim et al. 2011; Hussain and Palit 2017; Palit and Hussain 2018g, f, g, h, i, j).

The status of research endeavor in the field of advanced materials and green materials is bright and far-reaching. Green materials and sustainability will merge together toward new advanced materials and nanomaterials. The research pursuit and its vast scientific imagination need to be reorganized as regards research in advanced green materials. The major thrust area of vision is the application of green nanomaterials in human society. This chapter will veritably usher in a new age in the field of green material science. Scientific understanding will reach a new zenith if sustainability is merged with new areas of green material science.

15 Conclusions and Visionary Scientific Perspectives

Civilization, scientific advancement and research forays today are in the juncture of immense contemplation and scientific difficulties. Green nanotechnology, green nanomaterials and the world of green engineering are in the path of scientific comprehension. The status of global environment is immensely disastrous as science and engineering trudge forward toward new scientific profundity. Technology has today few answers to the monstrous issue of drinking water heavy metal poisoning. Health issues in the developed and disadvantaged nations around the world are immense as arsenic and heavy metal drinking water contamination assumes vast importance. Thus, the need of the application of green nanomaterials. Sustainable development whether it is energy, environmental, social or economic is the utmost need of the moment as human civilization moves forward toward a newer economic order. The target of the science and the civil society should be toward sustainable development and provision of basic human needs such as water, food, energy, electricity, shelter and education. Water science and technology thus assume immense importance. This paper will surely open up new thoughts in the field of nanoscience and nanotechnology in years to come. Human mankind's perseverance, revelation and determination, the vision of nanotechnology and the utmost needs of sustainability will all today lead a profound way in the true redeeming of science and engineering. Human scientific regeneration will then result in a new beginning and a new chapter and a new scientific and engineering age will evolve.

References

Barrow CJ (2005) Environmental management and development. Rzoutledge perspectives on development, 1st edn. Routledge Press, United Kingdom

Barrow CJ (2006) Environmental management for sustainable development, 2nd edn. Routledge Press, United Kingdom

Biao L, Tan S, Meng Q, Gao J, Zhang X, Liu Z, Fu Y (2018) Green synthesis, characterization and application of proanthocyanidins—functionalized gold nanoparticles. Nanomaterials 8(53):1–12. https://doi.org/10.3390/nano8010053

European Commission Report (2009, April) Challenges and opportunities to green nanotechnologies. Nanotechnologies in the 21st century, Issue 1. (Author of the Report-Dr. Rye Senjen)

Glaser JA (2012) Green chemistry with nanocatalysts. Clean Technol Environ Policy 14:513–520. https://doi.org/10.1007/s10098-012-0507-0

Hashim MA, Mukhopadhyay S, Sahu JN, Sengupta B (2011) Remediation technologies for heavy metal contaminated groundwater. J Environ Manage 92:2355–2388. https://doi.org/10.1016/j.jenvman.2011.06.009

Hussain CM, Palit S (2017) Nanomaterials, ecomaterials and wide vision of material science. In: Martinez LMT, Kharissova OV, Kharisov BI (eds) Springer handbook of ecomaterials, 1st edn. Springer International Publishing, Springer Nature Switzerland AG, pp 1–29

Kates RW, Parris TM, Leiserowitz AA (2005) What is sustainable development? Goals, indicators, values and practices. Environ Sci Policy Sustain Dev 47(3):8–21. https://doi.org/10.1080/00139157.2005.10524444

Lu Y, Ozcan S (2015) Green nanomaterials: on track for a sustainable future. Nano Today 10:417–420. https://doi.org/10.1016/j.nantod.2015.04.010

Nath D, Banerjee P, Das B (2014) Green nanomaterial-how green they are as biotherapeutic tool. J Nanomed Biotherapeutic Discov 4:2. https://doi.org/10.4172/2155-983X.1000125

Palit S (2017a) Application of nanotechnology, nanofiltration and drinking and wastewater treatment- a vision for the future. In: Grumezescu AM (ed) Water purification, 1st edn. Academic Press, USA, pp 587–620

Palit S (2017b) Nanomaterials for industrial wastewater treatment and water purification. In: Martinez LMT, Kharissova OV, Kharisov BI (eds) Springer handbook of ecomaterials, 1st edn. Springer International Publishing, Springer Nature Switzerland AG, pp 1–41

Palit S (2018a) Recent advances in corrosion science: a critical overview and a deep comprehension. In: Kharisov B (ed) Direct synthesis of metal complexes, 1st edn. Elsevier, USA, pp 379–410

Palit S (2018b) Recent advances in the application of engineered nanomaterials in the environment industry-a critical overview and a vision for the future (Chap. 47). In: Hussain CM (ed) Handbook of nanomaterials for industrial applications, 1st edn. Elsevier, Amsterdam, pp 883–893

Palit S, Hussain CM (2018a) Recent advances in green nanotechnology and the vision for the future (Chap. 1). In: Kanchi S, Ahmed S (eds) Green metal nanoparticles, Wiley Scrivener Publishing LLC, USA, pp 3–22

Palit S, Hussain CM (2018b) Engineered nanomaterial for industrial use (Chap. 1). In: Hussain CM (ed) Handbook of nanomaterials for industrial applications, 1st edn. Elsevier, Amsterdam, pp-3–12

Palit S, Hussain CM (2018c) Environmental management and sustainable development. In: Hussain CM (ed) Springer handbook of environmental materials management, 1st edn. Springer Nature America, Inc., pp 1–17

Palit S, Hussain CM (2018d) Sustainable biomedical waste management. In: Hussain Cm (ed) Springer handbook of environmental materials management, 1st edn. Springer Nature America, Inc., pp-1–23

Palit S, Hussain CM (2018e) Nanomembranes for environment. In: Hussain CM (ed) Springer handbook of environmental materials management, 1st edn. Springer Nature America, Inc., pp-1–24

Palit S, Hussain CM (2018f) Remediation of industrial and automobile exhausts for environmental management. In: Hussain CM (ed) Springer handbook of environmental materials management, 1st edn. Springer Nature America, Inc., pp 1–17

Palit S, Hussain CM (2018g) Nanomaterials for environmental science: a recent and future perspective. In: Hussain CM, Mishra AK (eds) Nanotechnology in environmental science, 1st edn. Wiley, Weinheim, pp 3–18

Palit S, Hussain CM (2018h) Biopolymers, nanocomposites and environmental protection: a far-reaching review. In: Ahmed S (ed) Bio-based materials for food packaging. Springer Nature Singapore Pte. Ltd., Singapore, pp 217–236

Palit S, Hussain CM (2018i) Nanocomposites in packaging: a groundbreaking review and a vision for the future. In: Ahmed S (ed) Bio-based materials for food packaging. Springer Nature Singapore Pte. Ltd., Singapore, pp 287–303

Palit S, Hussain CM (2018j) Green sustainability, nanotechnology and advanced materials- a critical overview and a vision for the future (Chap. 1). In: Ahmed S, Hussain CM (eds) Green and sustainable advanced materials-applications, vol 2. Wiley Scrivener Publishing, Beverley, pp 1–18

Shannon MA, Bohn PW, Elimelech M, Georgiadis JG, Marinas BJ, Mayes AM (2008) Science and technology for water purification in the coming decades. Nature Publishing Group, USA, pp 301–310. https://doi.org/10.1038/nature06599

Wardencki W, Curylo J, Namiesnik J (2005) Green chemistry-current and future issues. Pol J Environ Stud 14(4):389–395

Important Websites for References

https://books.google.co.in/books?isbn=1498762433

https://impact.ref.ac.uk/CaseStudies/CaseStudy.aspx?Id=38375

https://www.sciencedirect.com/science/article/pii/S1748013215000572

https://en.wikipedia.org/wiki/Green_nanotechnology
https://ascelibrary.org/doi/10.1061/9780784414088.ch23
https://www.azocleantech.com/article.aspx?ArticleID=330
https://pubs.acs.org/doi/abs/10.1021/es900021v
merid.org/en/Content/News_Services/Nanotechnology.../green_nanotech.aspx
https://www.elsevier.com/books/nanomaterials-for-green.../978-0-12-813731-4
https://greennano.org/sites/greennano2.uoregon.edu/files/GCI_WP_GN10.pdf
https://www.ncbi.nlm.nih.gov/pmc/articles/PMC4005277/
https://www.nanowerk.com/nanotechnology-in-green-industries.php
https://en.wikipedia.org/wiki/Nanotechnology
https://www.nano.gov/nanotech-101/what/definition
ieeexplore.ieee.org/document/6045222/
iopscience.iop.org/journal/0957-4484
iopscience.iop.org/article/10.1088/0957-4484/23/29/290201/meta
ostaustria.org/bridges...14.../2294-green-nanotechnology-a-path-to-sustainability
www.nnin.org/news-events/spotlights/what-nanotechnologya
www.insituarsenic.org/
www.insituarsenic.org/origin.html
www.nmlindia.org/pollutionmetigation.html
www.thestandrewsprize.com/finalist/arsenic-removal-from-water
www.worstpolluted.org/projects_reports/display/76

Characterisation of Green Nanomaterials

T. Anukiruthika, S. Priyanka, J. A. Moses and C. Anandharamakrishnan

Abstract Green synthesis of nanoparticles is becoming popular due to its sustainability, eco-friendly and inexpensive approach. Production of submicron-sized particles from the biogenic sources is referred as green-mediated synthesis. Different kinds of green metal nanoparticles available are Ag, Au, Cu, Cd, Pt, Pd, Fe, and other metal oxides include ZnO, CeO_2, TiO_2, ZrO_2, In_2O_3, Cu_2O and CuO, PbS, Fe_3O_4. These nanoparticles are characterised for its surface morphology and its compositional structure. The common techniques employed for characterisation of metal and metal oxide nanoparticles include UV-visible absorption spectroscopy, X-ray diffraction (XRD), Fourier transform infrared (FTIR) spectroscopy, energy dispersive spectroscopy (EDAX), atomic force microscopy (AFM), scanning electron microscope (SEM), transmission electron microscope (TEM), etc. The main characteristics that determine the properties and functionality of green nanomaterials are particle size, size distribution, crystallinity, specific surface area, surface morphology and elemental composition. Characterisation of green nanoparticles is significant to study about the nature, behaviour and functional properties of synthesised green nanoparticles (NPs). These NPs are well known for its antimicrobial and antioxidant effects as they have potential applications in food processing as in food packaging, food quality analysis and food/drug delivery systems. In order to find potential applications of green nanomaterials, characterisation of these nanoparticles is significant. This chapter essentially focusses on various characterisation techniques for the green-synthesised nanomaterials.

Keywords Green synthesis · Nanoparticles · Characterisation

T. Anukiruthika · S. Priyanka · J. A. Moses (✉) · C. Anandharamakrishnan
Indian Institute of Food Processing Technology (IIFPT), Ministry of Food Processing Industries, Government of India, Thanjavur, Tamil Nadu, India
e-mail: moses.ja@iifpt.edu.in

S. Ahmed and W. Ali (eds.), *Green Nanomaterials*,
Advanced Structured Materials 126,
https://doi.org/10.1007/978-981-15-3560-4_3

1 Introduction

Green nanomaterials have gained attention in due course of time for its wide applications in various fields. Nanomaterials have unique properties that make it different from that of the bulk materials (Shameli et al. 2012). Green synthesis is the production of nanomaterials from the biological source that possess minimum toxicity compared to that of the other synthetic methods. The utilisation of the biodegradable materials and biowastes paves a way for sustainable eco-friendly approach for the development of the green-mediated synthesis of the nanomaterials which is cost-effective and can be adopted economically as a valuable alternative to chemical methods (Kouvaris et al. 2012). Different ranges of biological sources for the synthesis of green nanomaterials are plant-based synthesis like biopolymers and biowaste; microbial-based synthesis like bacteria, fungi, yeast and algae. Green synthesis is a biogenic reduction process based on bottom-up approach where the potential natural products act as a reducing agent with inherent stabilising and capping properties as that of the chemical reducing agent (Hussain et al. 2016).

2 Characterisation of Green Nanomaterials

Various characterisation techniques are available for analysing the physiochemical properties of the green nanoparticles. The factors to be considered for green synthesis of NPs include the solvent medium, nature of the reducing and stabilising agent (Tripathi 2018). The distinct physiochemical properties of the synthesised green NPs are size, shape, molecular weight, purity, chemical composition, stability and solubility. These properties are responsible for the unique interactions of NPs that found its potential applications in agriculture and food, drug delivery system, medical applications, biomarkers, biosensors, cosmetics, etc., with improved efficacy and reduced toxic effects (Bai et al. 2018; Jain et al. 2018; Kale and Jagtap 2018). With this context, there is a need for visualising the characteristics of the functionalised green-mediated nanoparticles (García-Betancourt et al. 2018). Various characterisation techniques as shown in Table 1 are summarised in the following sections. Thus, robust and reliable characterisation techniques are essential for safety and quality assurance (Lin et al. 2014).

2.1 UV Visible Spectroscopy

UV-Vis spectroscopy is a simple useful technique employed for the qualitative and quantitative characterisation of biochemical compounds (Tripathi 2018).UV-Vis absorbance spectroscopy is commonly employed for the physiochemical characterisation of the nanomaterials that include size, concentration and aggregation of molecules (Lin et al. 2014).

Table 1 List of characterisation techniques for green nanomaterials

Characterisation techniques	Nanomaterial characteristics	References
UV-Vis Spectroscopy	Structure and conformation of bio-conjugate surface properties and optoelectronic properties	(De Oliveira et al. 2017; Shameli et al. 2012; Stuart 2005; Tripathi 2018; Weckhuysen 2004)
Fourier transform infrared spectroscopy (FTIR)	Surface properties, conformation of conjugates and bonding structure	(Asou et al. 2018; Faix 1992; Jaggi and Vij 2006; Tripathi 2018)
Mass spectroscopy (MS)	Molecular weight, surface properties, compositional structure	(Dixon 2005; Kumar et al. 2018; Tanaka et al. 1988; Taylor et al. 2008)
Raman spectroscopy (RS) Surface enhanced Raman spectroscopy (SERS) and Tip enhanced Raman spectroscopy (TERS)	Hydrodynamic size, vibrational and electronic property, conformation of protein and NPs conjugate	(Devience et al. 2015; Kneipp et al. 2010; Kumar and Thomas 2011; Popovic et al. 2011; Suganya et al. 2015)
Nuclear magnetic resonance spectroscopy (NMR)	Indirect analysis of size, structure, composition, purity and conformational change	(Agarwal et al. 2018; Alam and Jenkins 2012; Lupoi et al. 2014; Ma et al. 2003; Taylor et al. 2008)
X ray photoelectron spectroscopy (XPS)	Elemental surface analysis, chemical state and energy distribution	(Hollander and Jolly 1970; Khan et al. 2018; Konno 2016; Paterson and Swaffield 1994; Pleul and Simon 2008; Yang et al. 2018)
Scanning electron microscopy (SEM)	Size and size distribution, shape, aggregation, dispersion	(Goldstein et al. 2017; Henning and Adhikari 2017)
Transmission electron microscopy (TEM)	Size and size distribution, shape heterogeneity, aggregation, dispersion	(Walther 2017; Hall et al. 2007)
Atomic force microscopy (AFM)	Size and size distribution, shape, structure, sorption, dispersion, aggregation and surface properties	(Song and Bhushan 2007; Khan et al. 2016a, b)
X ray diffraction (XRD)	Size, shape, structure and crystallinity	(Bunaciu et al. 2015; Salame et al. 2018)
Energy dispersive X ray spectroscopy (EDAX)	Elemental surface analysis	(Hafner 2006; Mishra et al. 2017)
Dynamic light scattering (DLS)	Hydrodynamic size distribution and aggregation state	(Lin et al. 2014; Goyal 2017)

2.1.1 Working Principle

UV-Vis spectroscopy is based on the principle of absorption of light in the ultraviolet and visible range of electromagnetic spectrum which undergoes electronic transition of molecules and also differs with the perceived colour of the chemicals involved (Saxena et al. 2010). In UV-Vis spectroscopy, as the energy difference between the two states is equal to the energy absorbed by the molecule, so the electrons will get excited from lower to the higher states, resulting in the absorption spectra specific to the composition of the material (Weckhuysen 2004). UV-Vis spectroscopy obeys the Beer–Lambert law which states that the absorbance and the concentration of the absorbing species are in a linear relation (Stuart 2005).

$$A = \in cl$$

where

A absorbance of the solution,
c concentration,
l path length of the sample.

2.1.2 Instrumentation

Various components of UV-Vis spectrophotometer are light source, optical table, sample and reference ports, monochromator and detector (Tripathi 2018) as shown in Fig. 1. The characterisation of materials by UV-Vis spectroscopy utilises the optical properties of the materials like colour, absorptivity, transmissivity and reflectivity.

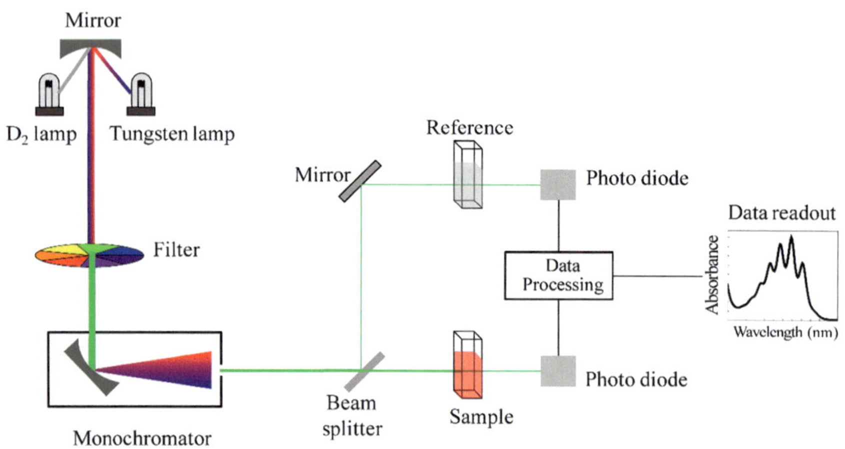

Fig. 1 Schematic diagram of UV spectroscopy. *From* https://en.Wikipedia.org/wiki/Ultraviolet%E2%80%93visible_spectroscopy

2.1.3 Characterisation by UV-Vis Spectroscopy

UV-Vis spectroscopy is helpful for determining the properties of organic materials, polymers and magnetic ceramic nano-based materials. The surface plasmon resonance of the metallic nanopaticles like Au, Ag and Cu is characterised based on its resonance that depends on size and the shape of nanomaterials. These surface plasmon resonance properties are found their applications in catalytic studies, surface-enhanced Raman scattering studies, biosensors, etc. The performance characteristics of the nanoparticles depend on the optical and catalytic properties that are influenced by its composition and morphology (De Oliveira et al. 2017).

UV-Vis absorption spectra were used as a colorimetric sensors as gold nanoparticles functionalised with glutathione being used in heavy metal detection especially Pb^{2+} sensing. Au-NPs are employed in the colorimetric detection of the pigment melamine in milk which was based on the aggregation of particles that result in change of colour from blue to red due to the shift of the absorption band (De Oliveira et al. 2017). UV-Vis spectrum absorption helps to monitor the growth of the nanostructured film. A linear relationship between the growth of layers and absorbance indicates the uniform deposition of the solid substrate, while a nonlinear relation indicates that different adsorption processes are controlling the growth of the film (De Oliveira et al. 2017).

Faix (1992) made a comparative study on characterisation of two different colloidal systems by UV-Vis. The relationship between the particle distribution and its absorbance of ZnO nanocrystals was revealed by using UV-Vis spectroscopic analysis (Pesila et al. 2003). Kouvaris et al. (2012) reported that green synthesis and characterisation of Ag-NPs from *Arbutus unedo* leaf extract. The study showed that free electrons of the metal nanoparticles gave rise to plasmon resonance which is due to combined vibration in resonance with the light. The resulting colour change from colourless to reddish yellow was due to the excitation and the sharp bands of the silver colloids observed at 436 nm. Similar study reported by Singhal and Bhavesh (2011) showed a colour change from colourless to dark yellow upon the addition of *O. sanctum* leaf extract to silver nitrate solution. The UV spectra band centred at 413 nm indicates the reduction of $AgNO_3$ into silver nanoparticles.

Green mediated synthesis of Ag-NPs from *Thymbra spicata* extract and its application as a nanocatalyst was studied by Veisi et al. (2017). It was well known that Ag-NPs can act as a nanocatalyst, which can reduce the dyes like 4-nitrophenol (4-NP), Rhodamin B (RhB) and methylene blue (MB) in the presence of $NaBH_4$ in water at room temperature. From the UV-Vis spectra of Ag-NPs, the dispersity of the colloidal solution can be evaluated and a comparison of the full width at half maximum of the absorbance band was detected at 395–420 nm. The monodispersity of the sample indicated that the colloidal Ag specimens are identical at 85–100 nm. The excess amounts of the reductants used ensured that all the Ag^+ were reduced to Ag (Medina-Ramirez et al. 2009).

Comparing to the other methods for the synthesis of nanomaterials, green synthesis can minimise the toxicity generation (Patel 2015). Biao et al. (2018) adopted a new hydrothermal method for the green-mediated synthesis of the Au-NPs from pomegranate fruit peel. Proanthocyanidins, an antioxidant from the plant source, is well known for its pharmacological and therapeutic activities against free radicals (Crozier et al. 2009; Dixon 2005). The single plasmon resonance band (PRB) of the Au-NPs peaking at 532 nm indicates the spherical shape of it, and the resulted colloidal gold particles remain stable during the storage. Further the funtionalised Au-NPs showed good stability at different concentrations ranging from 0.1 to 0.5 M NaCl and pH solutions of 3.0–9.0. Further the extra stability of the proanthocyanidins-functionalised Au-NPs was due to association of the Au-NPs with the proanthocyanidins (Biao et al. 2018).

Shameli et al. (2012) reported the synthesis and characterisation of polyethylene glycol-mediated green Ag-NPs. The surface plasmon of the Ag-NPs were detected at 412–437 nm which shows the spherical shape of the nanoparticles, and its formation was confirmed. The reduction of Ag^+ ions was monitored by continuous scanning from 280–760 nm with maximum absorbance at 438 nm (Dipankar and Murugan 2012). Another study on the characterisation and evaluation of antibacterial effects of silver nanoparticles from aqueous extract of *Ilex paraguariensis* was reported by Miguel et al. (2018). This study confirmed the formation of Ag-NPs at maximum absorbance of 460 nm with a blue shift to 440 nm was occurred in the extract.

Arunachalam (2013), reported one-step green synthesis method and characterisation of *Memecylon umbellatum* leaf extract-mediated Ag-NPs and Au-NPs nanoparticles. The study showed that the water-soluble fractions of plant extracts aided in the biological reduction of Ag^+ silver ions with the absorption spectra peaked at 440 nm. Similarly, the reduction of the gold ions to Au NPs was detected by a colour change with the appearance of surface plasmon resonance band at about 540 nm. The formation and stability of the iron nanoparticles can be observed from the UV spectrum of the reaction media in the range of 200–700 nm with a sharp peak at 450 nm indicates the synthesised Fe_2O_3 nanoparticles (Asou et al. 2018).

2.2 *Fourier Transform Infrared (FTIR) Spectroscopy*

FTIR spectroscopy describes the analysis of the generated signal by adopting the mathematical transformation. FTIR spectroscopy involves absorption, reflection, emission of the IR light by the sample (Lin et al. 2014). Among the other spectroscopic techniques, FTIR is more sensitive and offers rapid analysis.

2.2.1 Working Principle

In FTIR, as the applied frequency equals the natural resonant frequency of the molecules absorption takes place. The light from the IR source is passed through the interferometer where it is divided by beam splitter. The output of the interferometer is a record of the signal intensity and is referred as interferogram (Jaggi and Vij 2006). The resultant signal intensity was further subjected to Fourier transformation to obtain the spectral information. Hence,

$$S(v) = \int_{-\infty}^{+\infty} I(x)e^{i2\pi vx}\mathrm{d}x = F^{-1}[I(x)]$$

is referred as the inverse Fourier transform.

$$I(x) = \int_{-\infty}^{+\infty} S(v)e^{-i2\pi vx}\mathrm{d}v = F[S(v)]$$

is referred as the Fourier transform.

Thus, the integral given in (1) converts the interferogram $I(x)$, function of path difference x into a spectrum $S(v)$, which is a function of frequency v. All these transformations can be carried out in computer.

2.2.2 Instrumentation

FTIR is different from other dispersive spectroscopy techniques as FTIR uses interferometers. Different interferometers that are generally used in FTIR are Michelson interferometer, lamellar grating interferometer and Fabry—perot interferometer. Most commonly used one is step scan interferometer also referred as Michelson interferometer (Stuart 2005). The output of the interferometer is the interferogram which further transformed using Fourier transform to get more information (Faix 1992). Figure 2 shows the basic components of FTIR are optical system and interferometer.

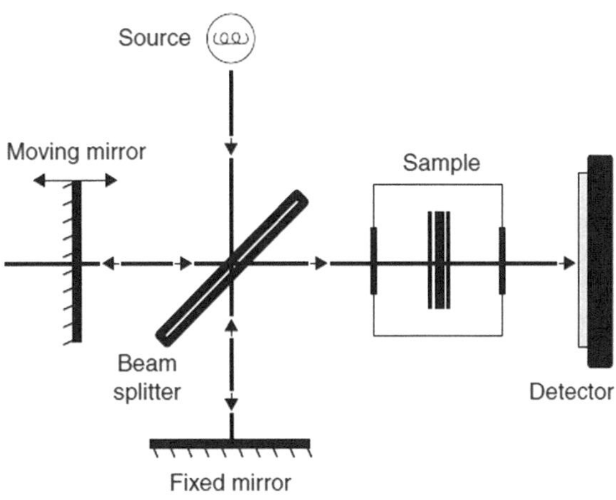

Fig. 2 Schematic diagram of FTIR spectroscopy (Subramanian and Rodriguez-Saona 2009)

2.2.3 Characterisation by FTIR

FTIR found its applications in specialised areas as in microanalysis, analysis of aqueous or solid samples that employ special reflectance techniques, quantitative and qualitative evaluations (Faix 1992). IR spectroscopy is the commonly employed technique for the identification of the organic compounds and also useful in identification and characterisation of the biopolymers (Stuart 2005). Both the mid IR and near IR are being employed for qualitative and quantitative analysis of samples. FTIR spectroscopy was helpful in the measurement of wavelength and the intensity of the absorption of IR radiation. The spectral data provides the information about the molecular vibration pattern specific to the individual molecules that help in the identification of the functional groups (Kong and Yu 2007). Further, FTIR was an effective tool for analysing the structural dynamics, composition, stability, aggregation and other conformational changes.

Arunachalam (2013) explored the size distribution and characterisation of the green nanomaterials using FTIR. From the spectra, the changes in functional groups after encapsulation were observed with a narrow peak. The study concluded that the proteins of the plant extract of *M.umbellatum* stabilised the Ag-NPs due to the presence of quinones and phenolic compounds. The FTIR spectra of the green synthesised Fe_2O_3 indicated the chemical linkage of the functional groups. The study showed that *A.excelsa* leaf extract has a potential ability to bind with iron oxide nanoparticles through the carbonyl groups of the proteins which possess stabilising and dispersing property which prevents the agglomeration of the hematite nanoparticles (Asou et al. 2018). Similar study conducted by Cheng et al. (2005) revealed the biomedical applications of the Fe_3O_4 nanoparticles. FTIR analysis was made to characterise the surface nature of the iron oxide particles. It was observed from FTIR spectra that the magnetite particles were covered with the hydroxide groups through electrostatic interactions.

The preparation and characterisation of magnetite NPs coated by amino silane was studied by Ma et al. (2003). The reduction of the Fe_3O_4 to nanoscale dimension made the absorption band of IR spectra to shift to higher wave numbers with a blue shift of Fe-O bond. FTIR was useful in the deduction of the structural characterisation of the molecules. The spectrum of the micelle-based synthesis of cobalt ferrite nanoparticles confirmed the transfer of the cations between the octahedral and tetrahedral sites (Rana et al. 2010). Dipankar and Murugan (2012) used FTIR for identification of the biomolecules in the *I.herbstii* leaf extract. The FTIR spectrum of purified IhAg-NP powder determined the chemical compounds which showed better antimicrobial activity.

A study reported on biosynthesis of the Ag-NPs using the fungus *Fusarium oxysporum*. The fungus reduces the aqueous silver ions in the solution leading to the formation of the extremely stable silver hydrosol. FTIR analysis for the drop-coated Ag-NPs-based film in fungus solution showed the characteristic bands of amide linkages of the protein and also indicated that the secondary structure of protein was not affected during binding with the Ag-NPs.

FTIR spectrum with attenuated total reflectance (ATR) configuration was performed for *Ilex paraguariensis* leaf extract. The spectrum showed a complex absorption banding due to polyphenol compounds with silver nitrate and a characteristic band at 1310 cm^{-1} because of nitro stretching. Difference in the intensity of the band was due to different bonding pattern of the hydrogen molecules (Miguel et al. 2018). Singhal and Bhavesh (2011) performed FTIR analysis for the biosynthesised Ag-NPs in order to identify the interaction effects of protein with the Ag-NPs. The spectroscopic studies confirmed the binding ability of the nanoparticles which can act as stabilising and capping agent (Sathyavathi et al. 2010).

Veisi et al. (2017) studied FTIR spectrum of the Ag NPs obtained from *Thymbra spicata* extract that found its application as recyclable nanocatalyst showed the presence variety of the organic compounds like ketones, aldehydes, terpenoids, alcohols and carboxylic acid surrounding the Ag-NPs. Jiang et al. (2005) studied the effect of the colloidal gold particles on conformational changes using IR spectroscopy. The study revealed the interaction effects of the 3D protein structure with the nanoparticles. The antibacterial effect of the green synthesised Ag-NPs was revealed by characterising the components of the root extract. The FTIR analysis of the powdered Ag-NPs was recorded at a frequency of 4000–400 cm^{-1} with a resolution at 4 cm^{-1} using KBr pellet method (Arokiyaraj et al. 2016).

Davar et al. (2009) reported the synthesis and characterisation of the Ni-NPs and NiO-NPs. From FTIR analysis, the study showed the reduction mechanism with the presence of the N-H peaks indicating that the amines were bound to Ni-NPs surface. Song et al. (2009) studied the synthesis and characterisation of the biosynthesis of Au-NPs using *Magnolia kobus* and *Diopyros kaki* leaf extracts. The IR spectrum of the extract was used to characterise the hydroxyl functional groups of alcohols and phenolic compounds of the *M.kobus* leaf extract. The study indicated that the synthesised Au-NPs are surrounded by proteins and secondary metabolites like terpenoid which also has functional groups of amines, alcohols, ketones, aldehydes and carboxylic acids.

2.3 *Mass Spectrometry*

Mass spectrometry (MS) is one of the characterising techniques employed for nanoparticles. MS was a powerful analytical tool for examining the mass, elemental composition, chemical structure of the biomolecules and the identification of the purity of the substance (Dixon 2005; Kumar et al. 2018; Taylor et al. 2008). Further, MS was useful for determination of the molecular mass of the unknown compounds in the sample. Mass spectroscopic methods provide a better insight in the structural analysis of the biological and polymer nanomaterials (Khitrov and Strouse 2003).

2.3.1 Working Principle

The basic principle behind mass spectrometry involves the identification of the charged ions based on the molecular mass to charge ratio (Tripathi 2018). MS provides information with higher precision and accuracy for molecular mass determination. Small amount of the sample in the range of 10^{-9}–10^{-12} mol was required for analysis (Lin et al. 2014).

With different ionisation techniques being coupled with the MS, the mass spectrometer was named as follows: matrix-assisted laser desorption/ionisation (MALDI), electrospray ionisation (ESI) and inductively coupled plasma mass spectroscopy (ICP-MS) (Lin et al. 2014). MALDI-MS was used for the determination of the molecular mass of macro- and micro-molecules, polymers and proteins binding with the nanoparticles (Tanaka et al. 1988). ICP-MS was helpful in deducing the structural compositional of and validating conjugation reaction of functionalised metal nanoparticles (Bandyopadhyay 2013). Laser-induced fluorescence and ion trap mass spectrometer are a newly developed mass spectrometer useful in analysing the fluorescent and dye molecules (Tripathi 2018).

Other form of mass spectroscopy technique includes aerosol time of flight mass spectrometer (ATOF-MS) which consists of aerosol interface region, light scattering region and TOF-MS that aids in the characterisation of aerosol and differential mobility analysis (DMA) which is coupled with MS for aerosol sizing and analysis (Taylor et al. 2008).

2.3.2 Instrumentation

The major components of the mass spectroscopy instrument are ion source, an analyser and a detector system. The ion source will be responsible for the generation of the ions. Analyser aids in determining mass to charge ratio of the molecules, while the detector will detect the various ionic species present in the sample. First, the biomolecules are ionised and converted to volatile components without thermal decomposition. For the identification of molecules of the unknown compounds, the conventional mass spectrometry was employed (Yu et al. 2017).

2.3.3 Characterisation by MS

Pace et al. (2011) reported a study on determining the transport efficiency for counting and sizing nanoparticles by single particle inductively coupled plasma mass spectroscopy (spICP-MS). The study briefed the method of counting and sizing nanoparticles using spICP-MS. ICP-MS detects metal ions based on the amount of metal particles detected during a user-specified time interval (Pace et al. 2011). This analysis further helps in differentiating the particles which are having same elemental component as the target nanoparticle like Ag and AgS.

A quantitative characterisation of the gold nanoparticles using ICP-MS was reported by (Schmidt et al. 2011). ICP-MS was established as an effective technique for separation and quantification of size and mass of NPs at relatively low level (Djuris et al. 2014; Kumar et al. 2018; López-sanz et al. 2016). The study showed that the ICP-MS quantification of the separated Au-NPs corresponded to the mean recoveries of 50, 95 and 67% at 10, 20 and 60 nm Au-NPs respectively (Schmidt et al. 2011). Silver nanoparticles were synthesised through sunlight-mediated process using *A.arborescens*. The spICP-MS study determined the mean particle size of Ag-NPs in the range of 38 $\pm$ 2 nm with a stability that can withstand over 24 h (Kumar et al. 2018).

ZnS nanomaterial was characterised using MALDI-TOF MS. Single high-intensity Gaussian-shaped peak was exhibited by ZnS nanocrystal with m/z values of 12,000 16,000, 26,500 and 37,500 Da, respectively. It was observed that there was an increase in the mass with increase in size and the peak shifted to the higher m/z values. Thus, the study showed use of MALDI-MS data for analysing the size measurement and size distribution (Khitrov and Strouse 2003). One-pot green synthesis of Ag-NPs using *Acmella oleracea* flower extract and its dopamine (DA) sensing properties were studied. A fibre optic sensor based on surface plasmon resonance (SPR) has been constructed by green synthesis by keeping Ag-NPs as sensing material. Compared with other techniques for DA analysis, fibre optic spectroscopic method was a simple and efficient method. Localised surface plasmon resonance (LSPR)-based sensor showed a liner response towards DA in 0.2–3.0 μM concentration range. The result showed that the time response of the sensor was found to about 6 min with detection limit of 0.2 μM and sensitivity of 3 nm/μM (Ritheshraj et al. 2015).

Green synthesised Ag-NPs from *Rheum palmatum* root extract were characterised for their antibacterial activity against *Staphylococcus aureus* and *Pseudomonas aeruginosa*. A gas chromatography was coupled with mass spectrometer which was used for the identification of unknown compounds from the root extract. Thus, the GCMS analysis revealed the presence of 35 chemical compounds which were solely responsible for the antibacterial activity of the root extract (Arokiyaraj et al. 2016). A similar study on anti-inflammatory activity of green synthesised Ag-NPs from European black elderberry extract was performed. The ICP-MS analysis of Ag-NPs showed a corresponding molar concentration of 4.10 nM for effective anti-inflammatory properties (David et al. 2014). Sun et al. (2014) reported a study on the green synthesis of Ag-NPs using tea leaf extract and evaluation of its antibacterial activity. The amount of the Ag present in the aqueous solution and the time dependent release of Ag^+ ions were determined by ICP-MS. On comparing the release rates of silver ions, the lower ion release indicates a good stability and prolonged preservation time (Sun et al. 2014).

Malachite green (MG) is a cationic compound being used in the aquaculture industry for its antimicrobial effect. Different protocols for malachite green determination were based on mass spectrometry (MS) or liquid chromatography (LC) combined with triple quadrupole mass spectrometry. The explored a new dispersive NP solid phase microextraction method based on stationary phase NPs combined with spectrometry for the quantification of MG in the water samples (Asfaram et al. 2015).

Similar study was reported on the application of iron and iron oxide nanoparticles synthesised from green tea extract for examining the ecotoxicological profile and to determine the ability of Fe-NPs to degrade malachite green (Plachtova et al. 2018). Another study on synthesis of bimetallic Fe/Ni NPs from eucalyptus leaf extract was reported by Weng et al. (2017), where eight major chemical compounds which serve as reducing and capping agents were characterised by GC-MS.

ICP-MS was performed for examining the uptake of the nanomaterials by the organisms and tissue samples as well as for the quantification of metal ion release from metal oxide (MO) nanoparticles. Thus, the chemical analysis using ICP-MS was found to be effective method for determining the ion release (Djuris et al. 2014). (López-sanz et al. 2016) reported a methodology for monitoring the Au-NPs for nanotoxicity using ICP-MS. The results of ICP-MS confirmed the size distribution and ion release of the gold nanoparticles.

Aqueous solutions of the *Salvadora persica* root extract was used for the green synthesis of palladium nanoparticles. The study explored the application of green synthesised Pd-NPs as an effective catalyst for coupling reactions. The formation of the coupled product was confirmed by subjecting the mixture to crystallisation using ethanol. The resulted crystallised material was analysed by GC-MS that revealed the presence of biphenyl compound (Khan et al. 2016a, b).

The study on biological synthesis and characterisation of the silver/copper nanopaticles for catalytic degradation of the toxic compounds in water was reported by (Rosbero and Camacho 2017). Mass spectroscopic studies were performed for identification of the degradation products like bimetallic Ag/Cu nanopaticles. These bimetallic NPs showed an efficient degradation of chlorpyrifos in water samples (Rosbero and Camacho 2017). A similar study on application of graphene quantum dots as green nanomaterials for photocatalytic degradation of cationic dye was reported by Roushani et al. (2015).

2.4 Raman Spectroscopy

Raman spectroscopy is the most widely employed spectroscopic technique for the characterisation and identification of vibrational pattern and electronic structures of nanomaterials (Popovic et al. 2011).

2.4.1 Working Principle

The principle behind the Raman spectroscopy is inelastic scattering of light. When the sample is exposed to the laser light source, the particles may absorb, transmit and reflect the light. Raman spectroscopy is based on the measuring of the scattering of light by the molecules, when the monochromatic light interacts with the sample (Lin et al. 2014). Based on the difference in the energy and frequency of the incident photons, there are two types of the scattering either stokes (lower than incident photons) or anti-stokes (higher than incident photons) associated with its characteristic emission (Tripathi 2018). These light scattering patterns were detected and analysed by means of photodetector.

The major advantage of Raman scattering is its ability to investigate the concentration and composition of the biological molecules that helps in detecting the abnormalities in the tissues (Kumar and Thomas 2011). As no sample preparation is required, Raman scattering can be used for the in situ analysis with high spatial resolution (Tripathi 2018). The resulting spectra which will be showing the characteristic of the nanomaterials known as Raman shift can be obtained when there is a difference in the energy levels of incident and scattered photons due to the transfer of energy from lower to higher energy states. The spectra can be expressed in terms of intensity of the Raman shift (Tripathi 2018).

2.4.2 Different Forms of Raman Spectroscopy

The conventional Raman spectroscopy provides indirect characterisation with negligible spatial resolution. Thus, Raman spectroscopy of nanomaterials results in low frequency of about 1 in 10^7 photons due to nanosize of the particles (Tripathi 2018). In order to overcome the drawback of the conventional Raman spectroscopy, surface-enhanced Raman scattering (SERS) and tip-enhanced Raman spectroscopy were employed for higher sensitive characterisation and provides high spectral specificity, multiplex capability and photostability (Kneipp et al. 2010; Zhang et al. 2015). Thus, SERS can enhance signal intensity and increase the spatial resolution of the metallic NPs like Ag, Au and Cu which have been characterised based on particle size distribution, electron–photon coupling and photon dispersion.

SERS a direct and non destructive spectroscopic technique is most widely used for high sensitive characterisation of the NPs (Liu et al. 2018). SERS was used for study and monitor the conformational changes in proteins conjugated with metal NPs. It was also useful in tracking intracellular drug release and in pH measurement of the aqueous medium (Ando et al. 2013).The other form of RS that is tip-enhanced Raman spectroscopy (TERS) was employed to overcome the diffraction limited resolution of conventional technique. The optical fibre in the conventional RS was replaced by the metallic tip in TERS in order to enhance the Raman signals (Wang et al. 2012). Both SERS and TERS were being used for obtaining the topological information of the NPs (Popovic et al. 2011).

2.4.3 Characterisation by RS

Biofunctionalised nanoparticles play a significant role in nanomedicine (Nagajyothi et al. 2017). Raman spectroscopy was generally performed for the confirmation of the functional groups which were responsible for reduction reaction and stabilisation of green synthesised Au-NPs (Suganya et al. 2015) and to determine the anti-cancerous property of yttrium oxide nanoparticles (Nagajyothi et al. 2017). Similar study on the characterisation and evaluation of the antimicrobial activity of green Au NPs from *Euphorbia hirta* (Arunachalam 2013) and Ag-NPs from *Nyctanthes arbortristis* (Gogoi et al. 2015) was reported. The presence of the powdered NPs was confirmed by Raman spectroscopy. Further, the Raman spectra aided in the identification of the organic and inorganic compounds which were responsible for the antibacterial activity of synthesised Au-NPs and Ag-NPs, respectively.

Hyllested et al. (2015) reported work on the green synthesis of the Ag-NPs by using fruit extracts using apple and oranges as reducing agents. The study also investigated the ability of Ag-NPs for enhancement of Raman signals in SERS using NIR excitation and non-resonant para mercaptobenzoic acid (pMBA). Muñetón et al. (2018) reported a study on the green synthesis and stability of the Ag-NPs. Micro-Raman spectroscopy was performed for the analysing the presence of silver oxide species, and their possible interactions with its adjacent molecules with Raman shift at 230 and 340 cm^{-1}. A biocompatible probe was developed for targeted vibrational characterisation of the biological systems based on SERS signal of the indocyanine green (ICG) on colloidal silver and gold nanoparticles (Kneipp et al. 2010).

Green nanoparticles from flower extract of *Hibiscus sabdariffa* were synthesised and characterised using several techniques. The purity of the green synthesised CeO_2 NPs was investigated by Raman spectroscopy (Thovhogi et al. 2015). Microbial synthesis of the nanocrystalline silver particles from fungus *T.asperellum* was reported by (Mukherjee et al. 2008). Raman spectroscopic studies showed enhancement of bond vibrations of the NPs due to the excitation of particles when exposed to laser light. SERS was used to detect the presence of malachite green and leucomalachite green in the fish muscles. Thus, SERS with gold nanospheres enabled the detection of these pigments stored in the muscles bound to proteins or lipids at lower levels from 1 to 10 ng g^{-1} (Zhang et al. 2015). Apart from the spectral analysis, Raman spectroscopy was also used for area mapping to examine the homogeneity of the sample and in determining the depth profile in order to select the changes in Raman signals with respect to the depth (Costa 2008).

2.5 Nuclear Magnetic Resonance (NMR) Spectroscopy

Nuclear magnetic resonance (NMR) spectroscopy is an efficient non-invasive method for analysing the chemical constituents of the sample. The features of the NMR-based characterisation are high throughput, high reproducibility and easily quantifiable data (Tripathi 2018). NMR spectroscopy provides information about the molecular weight and structure, purity, dynamics and diffusion properties of the nanomaterials (Devience et al. 2015; Lin et al. 2014; Marbella and Millstone 2015; Roming et al. 2008). Different forms of NMR were one-dimensional (1D), two-dimensional (2D) and multi-dimensional (nD) useful either in liquid state or solid state (Agarwal et al. 2018). On comparing with other imaging and diffraction techniques of long-range order, NMR aids in the characterisation of amorphous materials, polymers and biomolecules.

2.5.1 Working Principle

The principle of NMR characterisation is that the nuclei of certain atoms possess magnetic properties which aid in obtaining the information about the biochemical compounds. All the subatomic particles like protons, electrons and neutrons will spin around its own axes. Only certain nuclei are NMR active containing either odd number of proton or neutron which possesses a property called spin. When an external magnetic field is applied on magnetic nuclei, it absorbs and re-emits electromagnetic radiations. This energy that is present is actually at a specific resonance frequency known as Larmor frequency ω_0 that depends on the gyromagnetic ratio (γ) and applied magnetic field B_o represented as follows:

$$\omega_0 = \gamma \mathrm{B_o}$$

When the external magnetic field is applied, the nuclei will get aligned in either parallel or anti-parallel direction. As the frequency of the radiation matches with the frequency of the precession, the absorption of the energy takes place. As the applied magnetic field diminishes, the nucleus returns back to ground state from the excited state by a process of relaxation (Agarwal et al. 2018). The emission of the energy during relaxation will be recorded in order to obtain the characteristics of the molecule.

In addition of evaluating the chemical structure and composition, NMR was useful in determining the dynamic interactions of the species. Based on these dynamic interactions, the molecular relaxation, conformation and mobility can be evaluated with respect to the gradient sequence of radio frequency pulses (Taylor et al. 2008).

For evaluation of diffusion properties of the green nanoparticles pulsed field gradient NMR was used (Muñetón et al. 2018). High-resolution magic angle spinning (HR-MAS), a type of NMR technique, was designed for the nonsolid materials which are generally used for analysing the biological samples. The advantage of this technique was the generation of the high-resolution NMR spectra with accurate characterisation of heterogeneous samples (Alam and Jenkins 2012).

One of the major advantages of the NMR technique was it requires less sample preparation, but due to low detection sensitivity, sometimes large amount of sample was required for analysis. The other constrains of NMR were more time consuming for spectral analysis (Lupoi et al. 2014).

2.5.2 Types of NMR Techniques

Some recent advancements and technological developments made the application of NMR for solid-state samples (Alam and Jenkins 2012). Various NMR techniques useful in characterisation are magic angle spinning (MAS), high-resolution magic angle spinning (HR-MAS), cross polarisation (CP) and cross-polarisation magic angle spinning (CP-MAS). Some other 1D NMR techniques are correlation spectroscopy (COSY), heteronuclear single quantum correlation spectroscopy (HSQC), heteronuclear multiple bond quantum correlation spectroscopy(HMBC), total quantum coherence spectroscopy (TOCSY), exchange spectroscopy (EXSY), diffusion spectroscopy (DOSY) which can be used in combination with several other techniques like magnetic resonance imaging (MRI) and magnetic resonance scanning (MRS) for solution-state and solid-state samples (Agarwal et al. 2018; Alam and Jenkins 2012; Zhang et al. 2012). Proton relaxation of the synthesised magnetic Fe_3O_4 nanoparticles was determined by the solid-state NMR spectrophotometer (Cheng et al. 2005; Ma et al. 2003; Yamaura et al. 2004).

2.5.3 Characterisation by NMR

A biopolymer levan is used as a reducing and capping agent for the green synthesis of Ag-NPs and Au-NPs. The obtained NPs were characterised by H NMR to determine the presence of the sugar protons from the peaks of NMR spectra (Ahmed et al. 2014; Gopi et al. 2014). Gopi et al. (2014) reported pectin-mediated green synthesis of the hydroxyapatite NPs from the banana peel. The NMR spectral characterisation of H NMR confirms the synthesis of NPs from pectin by means of D-galacturonic acid, and also C^{13} NMR spectrum generates a signal in the range of 20.2–20.7 ppm that is specific to O-acetylated residues of galacturonic units (Gopi et al. 2014). H NMR spectra of silver ions from the fig leaf extract revealed the presence of different functional groups like alkenes, alkanes, alkynes, alcohols and phenols that were responsible for photo-induced reduction of silver ions (Ulug et al. 2015).

2.6 X-Ray Photoelectron Spectroscopy

X-ray photoelectron spectroscopy (XPS) is widely used for surface properties of the nanomaterials that provide quantitative, chemical state and energy distribution information (Li and Zhang 2007; Pleul and Simon 2008). This technique is also referred as electron spectroscopy for chemical analysis (ESCA) as it was helpful in examining the physiochemical characteristics of the nanomaterials (Yang et al. 2018). Photoelectron spectroscopy is based on photoionisation principle which was used to study the compositional and electronic state of the sample surface (Hollander and Jolly 1970).

2.6.1 Working Principle

When the sample was excited by electromagnetic radiation of a particular wavelength, photoelectrons are emitted from the surface of the sample. This emitted energy was detected by photomultiplier device, and the output was resulted in the form of electron yield against electron energy (Paterson and Swaffield 1994). Thus, in XPS, the exciting radiation will be X-ray beam and the photoelectrons are emitted front the core and valance shell of the atom represented by:

$$M + h\nu \rightarrow M^{+} + e^{-}$$

where

M is the ground state species,
$h\nu$ is X-ray photon
M^{+} is excited ionised species
e^{-} is the emitted photoelectron
Applying the conservation of energy,

$$KE = h\nu - BE$$

where

KE kinetic energy of the photoelectron
BE binding energy of the electron
h planck's constant
ν wavenumber of incident radiation

The emission spectrum will be different for each element, and the characteristic binding energy of the atom gives the characteristic set of peaks. There are two types of peaks, namely auger peak which results from the auger process that forms the basis of another surface analytical technique, auger electron spectroscopy (AES) and satellite peak which results from the photoionisation process that involves X-rays of slightly different energy as that of primary beam (Chen et al. 2000). The energy of transition for filling the place of ejected core electron by outer electron was balanced by emission of the characteristic X-ray, also referred as auger electrons. The energy of the emitted electrons was determined as a function of their velocity at which it enters the detector. Further, the intensity of the XPS peaks was related to the concentration of the element for the quantitative analysis of the sample (Hollander and Jolly 1970).

2.6.2 Instrumentation

The instrumentation part of XPS comprises mainly of radiation (X-ray) source, separation system which can disperse the emitted electrons and analysing system which measures the flux of emitted electrons as shown in Fig. 3. The other requirements

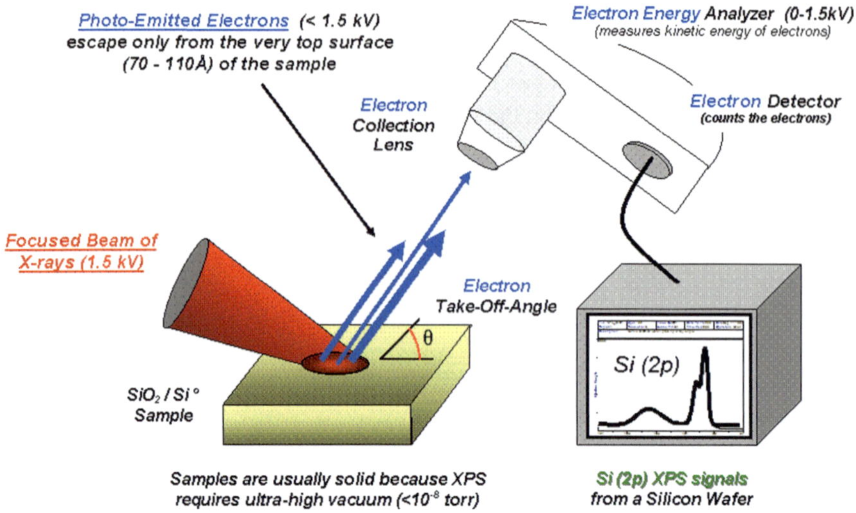

Fig. 3 Schematic diagram of XPS. https://en.wikipedia.org/wiki/X-ray_photoelectron_spectroscopy

are provision of a vacuum condition at 10^{-7} Pa that enables the analysis of the emitted photoelectrons without any interference (Konno 2016). The incident X-rays strike the sample surface, and photons release the electrons from the atom known as photoelectrons. The spectrometer helps in determining the velocity of the escaping photoelectrons (Chen et al. 2000). The ejected photoelectrons allowed passing through a slit between two hemispheres of optics. An applied electric field directs the photoelectrons through a circular path, and those photoelectrons with sufficient energy could reach the count unit at the end of hemispherical analyser. Thus, based on difference in the applied field strength, the photoelectrons were distinguished (Pleul and Simon 2008). An XPS spectrum was obtained by counting the amount of the photoelectrons and auger electrons which represent the surface composition of the nanomaterials. The energy corresponding to each peak denotes specific element, and area under the curve is a measure of relative amount of chemical element (Hollander and Jolly 1970).

2.6.3 Charcterisation by XPS

XPS was used for examining the interaction effects of proteins from *Capsicum annuum* in the formation of silver nanoparticles. The whole process was carried under pressure of above 10^{-6} Pa. The general spectra of different elements were characterised using ESCA (Li and Zhang 2007). The surface chemistry of the green synthesised silver nanoparticles was studied using XPS in order to explore its antimicrobial and antioxidant effects. Further, the study was extended to determine the toxicity of the phenolic azo dye against fish, Daphnia and green algae (Khan et al. 2018).

2.7 *Scanning Electron Microscopy*

Scanning electron microscopy (SEM) is commonly employed technique for characterisation of green nanomaterials. The SEM technique provides data regarding the morphological features such as shape and size (Luo et al. 2016), topography and surface composition (Henning and Adhikari 2017). Moreover, at present, SEM is the most prevalent microscopic technique due to its simplicity in preparation of samples, effortless interpretation of images and user-friendly nature (Michler and Goerg 2008). Figure 4 shows different resolution ranges of different microscopic techniques.

2.7.1 Principle

The scanning electron microscopy is customised to operate under high vacuum conditions to obtain a magnified image of the surface of the specimen with the aid of beam of electrons which scan the surface of the specimen in raster motion; electromagnetic lenses are employed for controlling the resolution of the image (Lin et al. 2014). Furthermore, the image is generated as result of interaction of beam of electrons with the sample specimen. The moment the electron beam is incident on sample surface, there is emission of secondary electron (inelastic scattering) and

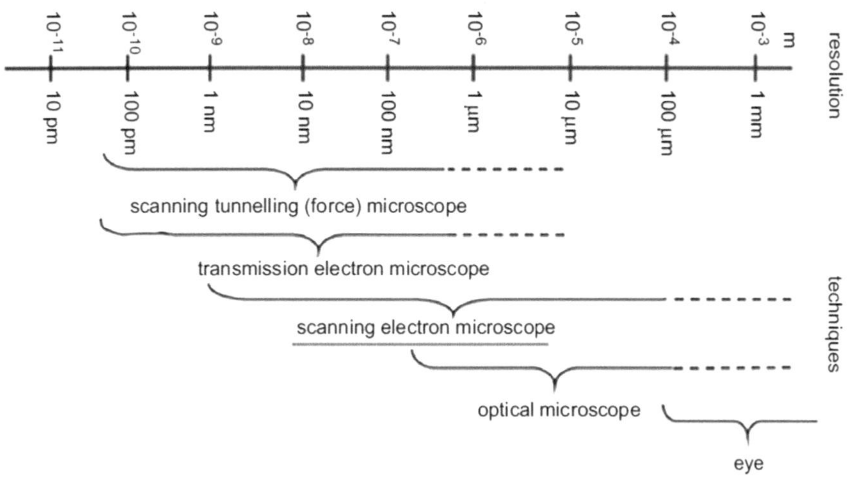

Fig. 4 Resolution ranges of various microscopic techniques (Michler 2008)

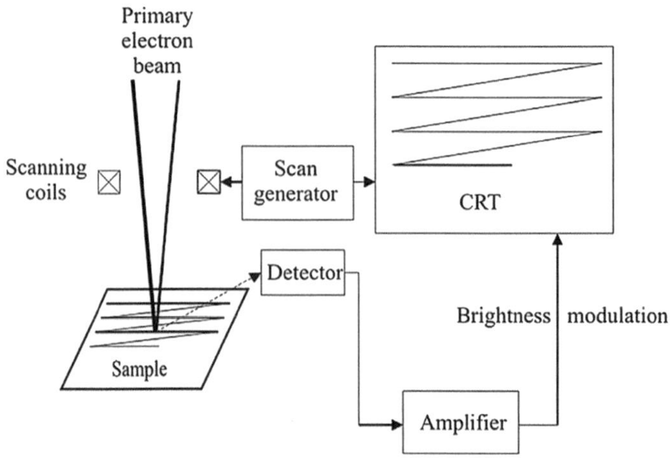

Fig. 5 Schematic diagram scanning electron microscopy (Michler 2008)

backscattered electron (elastic scattering) (Johal 2012) which are of prime importance in SEM imaging, and they can be detected using their respective detectors. The detected signals are amplified and then converted to image consisting of spots of low or high intensity based on the signal strength, in the same raster motion on the screen which is usually a cathode ray tube (Henning and Adhikari 2017). Figure 5 illustrates different components of scanning electron microscope.

2.7.2 Electron and Sample Interaction

The electrons are accelerated by electrical energy in the column, and as result of this they possess kinetic energy which is dissipated as they impinge the specimen surface and simultaneously generates numerous signals for investigation as shown in Fig. 6 (Lee et al. 2011). Some of the basic features of SEM are listed in Table 2.

Some of the most prevalent electron-sample interactions are

(a) **Secondary electrons**: they are produced as a result of primary electron ejection from the atomic orbital. They usually have lower energy compared to the primary electron (below 50 eV) since they are produced due to inelastic scattering. The secondary electrons generated close to the surface carry information about the topographical features of the sample and inherit high spatial resolution which makes them the prime choice for SEM imaging.
(b) **Backscattered electrons**: they are produced when the primary electrons are elastically scattered due the positive charge of the nucleus. These are electrons that take back a U-turn with losing much energy. Furthermore, BSEs usually have higher energy compared to SEs because of which they can travel to greater depth within the sample. The BSEs carry details regarding the composition of the sample.

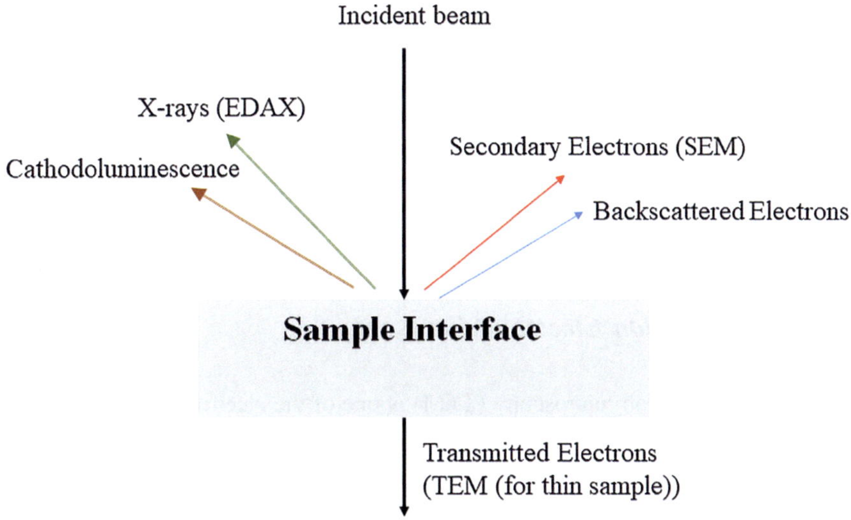

Fig. 6 Schematic illustration of electron- sample interactions (Salame et al. 2018)

Table 2 Basic features of SEM (Henning and Adhikari 2017)

Basic features of SEM	
Electron beam generator	Thermionic emission guns (Tungsten, LaB_6) Field-emission guns (FEGs)
Vacuum system	Thermionic LaB_6 guns (10^{-6}–10^{-7} torr) Field emission guns (10^{-9}–10^{-10} torr)
Signals	Secondary electrons (SEs) (low energy of 3–50 eV) Backscattered electrons (BSEs) (higher energy than SEs)
Detectors	SEs—Everhart-Thornley detector BSEs—Semiconductor detectors Characteristic X-rays—WDS/EDS detector

2.7.3 Characterisation

SEM is mainly employed for analysing the morphological features like size, shape and size distribution of the green nanomaterials. However, there may be some alterations such as shrinkage in the nanomaterial due to various sample preparation procedures such as drying and contrasting, which will modify the nanomaterials characteristics (Hall et al. 2007).

In addition, an ultra-thin coating of electrically conductive material like Au is mandatory for characterising the non-conductive biomolecules because they do not produce enough signals for proper imaging (Suzuki 2002).

The morphology of the nanoparticles like shape and particle aggregation was studied by SEM analysis (Sarmento et al. 2006). Green synthesis of Ag-NPs from

Ocimum sanctum was reported by Rao et al. (2013). SEM analysis was performed for studying the dispersion of the silver NPs which reveal the size and size distribution of the green NPs. Similarly, gold NPs was produced by green method from *Magnolia kobus* and *Diopyros kaki* plants, respectively. The size and structure of the synthesised NPs at different reaction temperatures was studied using field emission spectroscopy (FE-SEM) (Song et al. 2009). Further, the TEM images of these Au NPs were in accordance with the TEM images.

2.8 *Transmission Electron Microscopy*

Transmission electron microscopy (TEM) is one of the electron microscopic techniques which are frequently utilised for characterisation of nanomaterials. Unlike SEM, TEM is a much more sophisticated technique which will generate a direct image with higher spatial resolution in atomic range (<1 nm), and apart from this, it also provides details regarding the chemical composition of the specimen (Wang 1999). Thus, TEM is an indispensable technique for the interpretation and characterisation of morphological and compositional features of nanomaterial(Mayeen et al. 2018).

The significant difference between SEM and TEM is thickness of the sample, its position and the way of data collection from the specimen. Furthermore, SEM also have scan and descan coils above and below the specimen, respectively, to perform raster scan of the sample surface (Walther 2017). Apart from the above-mentioned differences, all other features of TEM like position of electron gun, electromagnetic lenses and condenser are same as that of SEM (Salame et al. 2018). Figure 7 illustrates the schematic of optical microscopy, SEM and TEM.

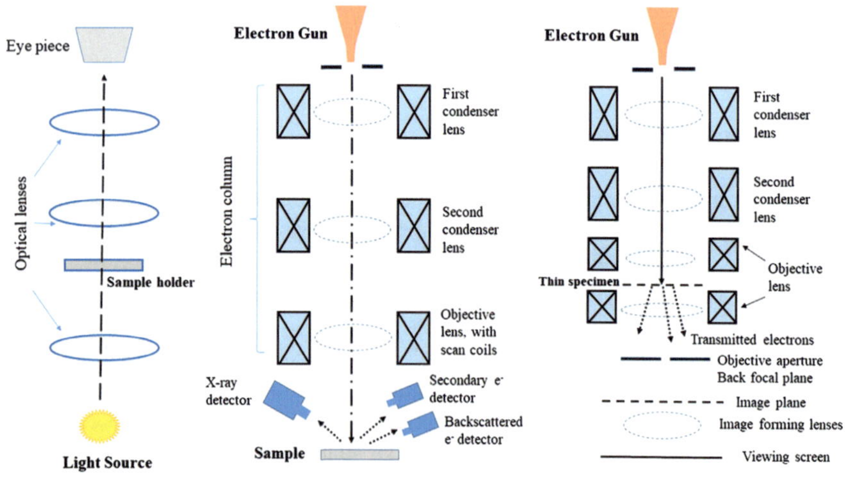

Fig. 7 Schematic diagram of optical, SEM and TEM (Salame et al. 2018)

Table 3 Basic features of TEM (Walther 2017)

Basic features of SEM	
Electron transparency	Soft bulk matter (polymers) can be cut by ultramicrotomes, down to 30 nm in thickness
	Conductive bulk material (metals, alloys) can be electropolished or chemically etched to perforation
	Semiconductor and ceramic bulk specimens or thin films can be cut, ground, polished, and finally ion milled
Electron beam generator	Tungsten (W) hair-pin cathodes heated to >2000 K
	Lanthanum hexaboride (LaB_6)/Cerium hexaboride (CeB_6) cathodes heated to about 1800 K
	Field emission guns: Zirconia (ZrO_2)-covered W tips heated to 1500 K (so-called Schottky electron guns) and cold (i.e., room temperature)
Vacuum system	Thermionic LaB_6/CeB_6 guns (10^{-6}–10^{-7} torr) Field emission guns (10^{-9}–10^{-10} torr)
Signals	Transmitted electrons
Detectors	Two-dimensional detectors charge-coupled device (CCD) detector Direct electron detectors

2.8.1 Principle

In the conventional TEM method, when the high-energy electron beam impinges on the sample of very less thickness (<100 nm) (Walther 2017), the electrons beam gets transmitted through the specimen. The electron–electron interaction occurs as the beam of electron is transmitted through the specimen. Moreover, the primary electrons are transformed to various interaction products such as unscattered, elastically scattered or inelastically scattered electrons which are formed due to the interaction of electron beam with the sample (Salame et al. 2018). Furthermore, series electromagnetic lenses are used for focusing the scattered or unscattered electrons to generate various types of images like phase-contrast, amplitude-contrast or shadow image of varying darkness based on the density of unscattered electrons (Williams and Carter 2009). The basic features of TEM are listed in Table 3.

2.8.2 Characterisation

Energy dispersive X-ray spectroscopy can be integrated with both SEM and TEM to study about the chemical composition of the nanomaterials (Tiede et al. 2008). In general, both SEM and TEM can be used for determining the size and shape of the nanomaterials, along with the degree of dispersion and aggregation. But, TEM has greater advantage compared to SEM by providing images of better spatial resolution and details regarding the internal features of the sample (Hall et al. 2007).

TEM analysis of the green silver nanoparticles showed an average mean size of nanoparticles and revealed the spherical morphology of the Ag NPs. Kouvaris et al. (2012) reported the green synthesis of the Ag-NPs from *Arbutus unedo* leaf extract and studied the morphology of the formed NPs by TEM. Similar study was reported by Shrivastava et al. (2007), for characterisation of the Ag-NPs and its potential application as antibacterial effects. TEM micrograph was also used for studying the degree of aggregation of the NPs (Medina-Ramirez et al. 2009). Further, the high-resolution TEM was used for analysing the crystalline nature of the Ni and NiO NPs (Davar et al. 2009).

2.9 *Atomic Force Microscopy*

Atomic force microscopy (AFM) is an advanced technology for generating images of surface features in range of nanoscale. It is a very promising non-invasive analytical technique which can be performed in various surrounding conditions such as air, liquid or vacuum (Ferri et al. 2012; Song and Bhushan 2007). The AFM has an immense potential to produce topographic images of very high-resolution down to atomic resolution. In numerous fields, AFM is employed to gather three-dimensional data regarding the surface defects, scratches, gouges, corrosion, scribes, pits, etc. The potential AFM to perform surface imaging from nano- to atomic scale, in any possible surrounding conditions and ease in handling, makes AFM proficient tool for the characterisation of the surface properties of the specimen (Khan et al. 2016a, b).

2.9.1 Principle

The AFM consists of a sharp tip attached to a flexible cantilever usually made of silicon (Si) or silicon nitride (Si_3N_4) which are employed to scan over the sample surface. The force (van der Waals force) between the sample surface and tip is sensed by the cantilever as it scans the surface (Khan et al. 2016a, b). The piezoelectric scanner is attached with the sharp tip flexible cantilever set-up which has the ability to move in all three dimensions. While examining the sample, the laser beam is continuously emitted by the laser diode on the backside of the cantilever attached to the tip.

The cantilever gets deflected based on the forces acting between the sample and the tip; these deflections of the cantilever are precisely detected using a position-sensitive photodiode by sensing the angular deflection of laser beam. There will be detectable changes in the scale of the beam deflection based on the force of interaction between the sample and the tip. Thus, the AFM technique is used for mapping the surface topography by sensing the interaction force between the tip and the sample (Liu and Wang 2011) as shown in Fig. 8.

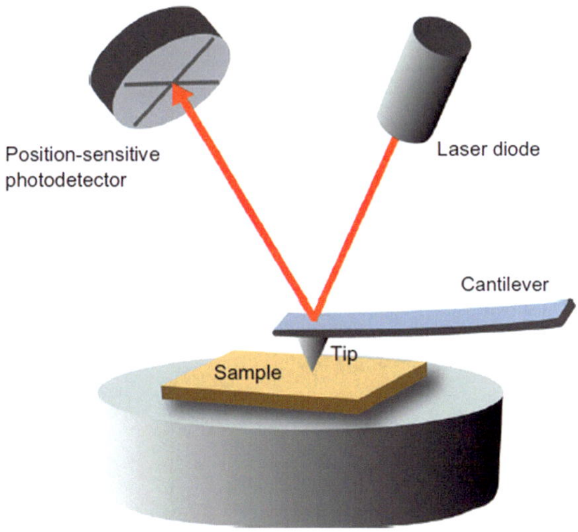

Fig. 8 Schematic diagram of AFM (Liu and Wang 2010)

2.9.2 Working Modes of AFM

The AFM technique provides three different modes of imaging, and all three modes work differently based on differences in the force of interaction between the tip and the sample as listed in Table 4.

Contact Mode

In the contact mode, the sharp tip is employed to raster scan the surface of the sample to obtain an image of very high resolution. So in this method, as the name suggests, there is always a contact between the tip and the sample as illustrated in Fig. 9 (Liu and Wang 2011). It is essential that the stiffness of the cantilever should be less than that of the sample stiffness in order to prevent the breakage of cantilever tip. However, due to significant lateral contact force, there are chances that surface of the specimen will get distorted or damaged. So, contact mode is preferred only for samples that are moderately robust in nature and scanning should be done quickly (Khan et al. 2016a, b).

Table 4 Different modes of AFM (Khan et al. 2016a, b)

S. No.	Type of AFM	Force experienced by the probe
1	Contact mode	Repulsive forces
2	Non-contact mode	Attractive forces
3	Tapping or oscillating mode	Both attractive and repulsive force

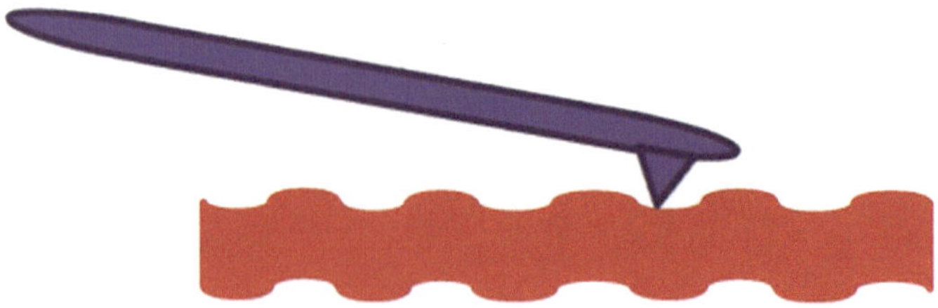

Fig. 9 Schematic diagram of contact mode of AFM (Khan et al. 2016a, b)

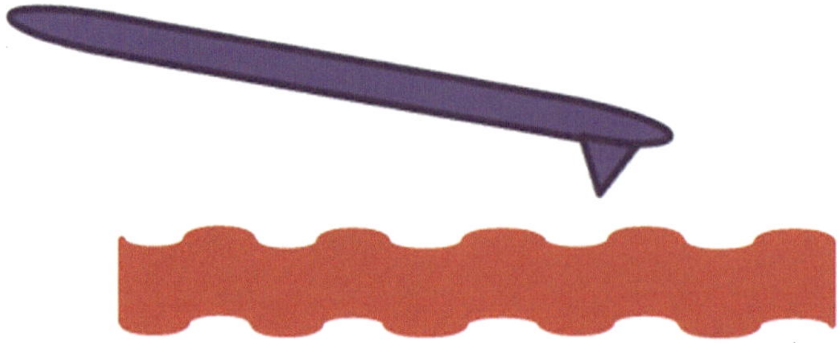

Fig. 10 Schematic diagram of non-contact mode of AFM (Khan et al. 2016a, b)

Non-Contact Mode

In this mode, the will be distance of 0.1–10 nm between the tip and sample surface. The cantilever will vibrate at a frequency higher than its resonant frequency in close proximity to the surface of the sample as shown in Fig. 10. As the distance between the tip and the sample surface increases, there will be changes in the attractive van der Waals forces which will in turn decrease the resonant frequency and the amplitude of vibration. The non-contact mode can be employed to analyse the surface of the sample in various environments. Furthermore, this mode provides flexibility to image those surfaces that were not possible to image by other electron microscopic techniques (Khan et al. 2016a, b).

Tapping or Oscillating Mode

In tapping mode, the tip of the vibrating cantilever occasionally contacts the surface of the sample as shown in Fig. 11. Furthermore, in this mode, we will obtain images of high resolution, and there will be comparatively less damage to both sample and the tip than contact mode because only meagre force is exerted by the tip on the sample.Thus, this is the most preferred mode for characterising nanomaterials in food and life science fields (Liu and Wang 2011).

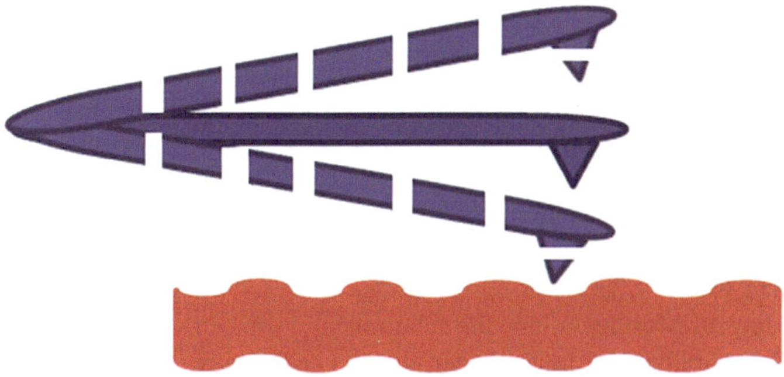

Fig. 11 Schematic diagram of tapping mode of AFM (Khan et al. 2016a, b)

2.9.3 Characterisation

Atomic force microscopy (AFM) was useful in analysing the surface properties of the biogenic NPs. Biosynthesis of the metal NPs like Ag, Au was confirmed by AFM. AFM image of the Au NPs synthesised with 5% Magnolia leaf broth showed the nano-plate structure of the gold NPs with width in the range of 250–300 nm and thickness of 5–7 nm (Song et al. 2009). The potential of antioxidant tagged Au NPs was studied for its potent antimicrobial activity. EDAX analysis was performed to characterise the specific functional groups of formulated Au-NPs and antioxidant tagged gold NPs for its antimicrobial activity (Bhattacherjee et al. 2018).

2.10 Energy Dispersive X-Ray Analysis (EDX)

Energy dispersive X-ray (EDX) is a sophisticated X-ray-based non-destructive technique, useful in characterising the elementary composition of materials by X-ray mapping (In). Various studies by different scientists have inferred that combining EDX with electron microscopic techniques such as SEM and TEM can be effectively used to find the underlying relationship between the elemental spectra, composition and microstructures (Anake et al. 2016; Hafner 2006).

EDX analysis can be of qualitative, semiquantitative or quantitative that provides information on the spatial distribution through mapping of the components. EDX analysis also deals with characterising the composition and properties of nano-metal oxides (Mishra et al. 2017). This is one of the most powerful tools for analysing the physical, chemical and compositional properties of micro- to nanomaterials in the material science and technology field.

2.10.1 Principle

The characteristic X-rays produced during electronic transition when high-energy electrons are bombarded with sample knock out an electron from the interior shell, which leaves a hole there, i.e. vacant space for electron. Eventually, the electrons will get transferred from the higher energy level to the lower energy level (Mishra et al. 2017). The difference in the energy between two transition levels will be emitted as the characteristic X-rays.

According to Moseley's law, the relationship between the energy of the characteristic X-rays and the atomic number is given as follows:

$$E = c_1(Z - c_2)^2$$

where E is the energy of the characteristic X-ray, Z is the atomic number, and c_1 and c_2 are constants for a given line type (Instruments 1999).

2.10.2 Characterisation

Energy dispersive X-ray spectroscopy is useful for elemental characterisation of NPs and biomaterials (Gangadoo and Chapman 2015). Silver and gold NPs are synthesised from *Memecylon umbellatum* leaf extract, and the synthesised NPs were characterised by energy dispersive X-ray spectroscopy that gives information about the surface chemistry of the NPs. This study confirmed the presence of Ag and Au NPs with strong signal peaks that correspond to silver and gold atoms, respectively (Arunachalam 2013). EDAX was used for establishing the chemical identity of the ZnO nanocrystals produced using *Nephelium lappaceum* peel (Yuvakkumar et al. 2014).

2.11 X-Ray Diffraction

The XRD is the eminent non-destructive analytical technique that helps in studying the crystal structures and atomic spacing of the crystalline material (Bunaciu et al. 2015). Furthermore, they can provide details regarding the crystal structures, phases, orientations and other structural parameters of crystal such as strain, crystal defects and crystallinity. Since, the wavelength of the X-rays is in the range of 0.2 and 10 nm which is similar to that of the interatomic spacing, they are used for the study of crystallinity of nanomaterials (Salame et al. 2018). The different applications of XRD are shown in Table 5.

Table 5 Applications of XRD (Bishnoi et al. 2017)

Application of XRD	Peak position	Pattern indexing
		Phase identification
		Unit cell parameters
	Intensity	Reaction kinetics
		Phase abundance
		Texture analysis
	Shape	Crystallite size
		Size distribution
		Crystalline growth kinetics

2.11.1 Principle

X-ray diffraction is grounded on constructive interference of monochromatic X-rays and the crystalline sample (Bunaciu et al. 2015). For diffraction to take place, the spacing in the grating should be the same order of the electromagnetic radiation. When an incident beam of X-rays comes in contact with crystalline material, a lot of interactions take place such as elastic or inelastic scattering by electrons of material as shown in Fig. 12. The resultant radiations may either undergo constructive or destructive interference. Those interactions of the incident rays with the sample produce constructive interference when they satisfy Bragg's law, whereas the destructive interference does not obey Bragg's law (Bishnoi et al. 2017).

Bragg's law:

$$n\lambda = 2d \sin \theta$$

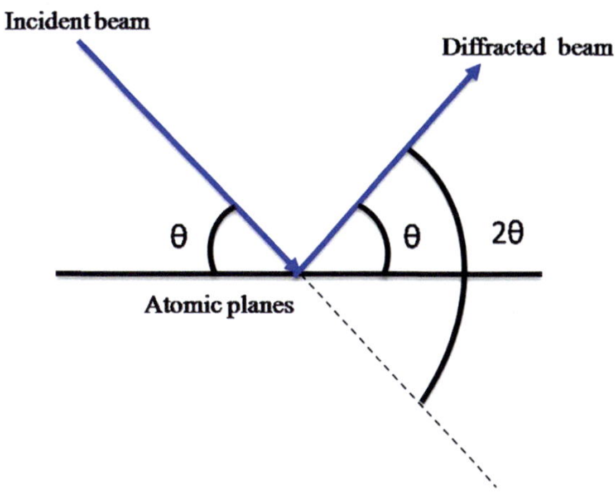

Fig. 12 Schematic illustration of Bragg's law (Bishnoi et al. 2017)

where n is an integer, λ is the wavelength of the X-rays, d is the interplanar spacing generating the diffraction, and θ is the diffraction angle (Bunaciu et al. 2015).

2.11.2 Types of XRD

There are two common types:

i. Laue method: In this method, the wavelength of the X-ray is varied keeping the diffraction angle constant.
ii. Powder diffraction method: The diffraction angle is varied keeping wavelength of the X-ray constant.

The intensity of the diffracted X-ray beam is measured against the diffraction angle 2θ using both the methods that provide information about the diffraction pattern of the crystalline material (Bishnoi et al. 2017). Furthermore, the crystallite size of the materials can be calculated by using Scherrer formula and full width at half of the maximum intensity (FWHM), i.e. peak broadening (Bunaciu et al. 2015).

$$t = \frac{k\lambda}{\beta_c \cos\theta}$$

where t is the crystallite size, β_c is the corrected FWHM, K is Scherrer's constant for diffractions, and λ is the wavelength of the X-ray (Goyal 2017).

2.11.3 Characterisation

The crystalline nature of the Ag NPs was evaluated using X-ray diffraction (XRD). From XRD peaks, it was confirmed that the main composition of the formed NPs was silver and no other impurities were present (Shameli et al. 2012). XRD patterns of precipitated iron oxide powder were studied for evaluating its crystalline nature for its potential biomedical applications (Cheng et al. 2005). XRD pattern was helpful in determining the conversion of the metal nickel NPs into metal oxide nickel NPs, and also the obtained XRD pattern was close in agreement with TEM studies (Davar et al. 2009). A study was reported on the extracellular biosynthesis of Ag-NPs using fungus *Fusarium oxysporum,* and synthesised NPs were characterised by XRD (Ahmad et al. 2003; Khan et al. 2017).

3 Conclusion

The green-mediated synthesis of nanoparticles is basically produced through plants, microorganisms, algae, etc. In order to study the significance of the green nanomaterials, it is necessary to characterise various morphological features like size, shape, crystallinity, surface and other functional properties like optical properties, elemental composition of the synthesised green nanomaterials. For characterisation purpose, various high-end analytical techniques such as electron microscopy techniques like SEM, TEM and AFM to analyse the topographical features, internal features and 3D imaging of the surface topography are being employed. Furthermore, X-ray-based techniques such as EDAX, XRD, XPS are used to determine the elemental composition, crystal structure, chemical state and energy distribution, respectively. Apart from the above-mentioned techniques, various spectroscopic techniques such as UV, FTIR, MS and NMR are also being utilised for characterising the green nanomaterials. Thus, the above-mentioned techniques are inevitable in the characterising the nanomaterials.

References

Agarwal N, Nair MS, Mazumder A (2018) Characterization of nanomaterials using nuclear magnetic resonance spectroscopy. In: Characterization of nanomaterials. Elsevier Ltd. https://doi.org/10.1016/B978-0-08-101973-3.00003-1

Ahmad A, Mukherjee P, Mandal D, Senapati S, Khan MI, Kumar R, Sastry M (2003) Extracellular biosynthesis of silver nanoparticles using the fungus *Fusarium oxysporum* composite metal particles, and the atom to metal. Colloids Surf B: Biointerfaces 28:313–318

Ahmed KBA, Kalla D, Uppuluri KB, Anbazhagan V (2014) Green synthesis of silver and gold nanoparticles employing levan, a biopolymer from Acetobacter xylinum NCIM 2526, as a reducing agent and capping agent. Carbohyd Polym 112:539–545. https://doi.org/10.1016/j.carbpol.2014.06.033

Alam TM, Jenkins JE (2012) HR-MAS NMR spectroscopy in material science. Intech Open 2:64. https://doi.org/10.5772/32009

Anake WU, Ana GR, Benson NU (2016) Study of surface morphology, elemental composition and sources of airborne fine particulate matter in Agbara industrial estate, Nigeria. Int J Appl Environ Sci 11(4):881–890

Ando J, Yano T, Fujita K, Kawata S (2013) Metal nanoparticles for nano-imaging and nano-analysis. Phys Chem Chem Phys 15(33):13713–13722. https://doi.org/10.1039/c3cp51806j

Arokiyaraj S, Vincent S, Saravanan M, Oh YK, Kim KH (2016) Green synthesis of silver nanoparticles using Rheum palmatum root extract and their antibacterial activity against Staphylococcus aureus and Pseudomonas aeruginosa. 1401 (March). https://doi.org/10.3109/21691401.2016.1160403

Arunachalam KD, Annamalai SK, Hari S (2013) One-step green synthesis and characterization of leaf extract-mediated biocompatible silver and gold nanoparticles from Memecylon umbellatum. Int J Nanomedicine 8:1307–1315

Asfaram A, Ghaedi M, Goudarzi A (2015) Solid-phase extraction for the determination of. New J Chem 39:9813–9823. https://doi.org/10.1039/C5NJ01730K

Asou HM, Al-antary M, Awwad AM (2018) Environmental nanotechnology, monitoring & management green route for synthesis hematite (α-Fe_2O_3) nanoparticles: toxicity e ffect on the green peach aphid. Myzus persicae (Sulzer) 9:107–111. https://doi.org/10.1016/j.enmm.2018.01.004

Bai RG, Sabouni R, Husseini G (2018) Green Nanotechnology—A Road Map to Safer Nanomaterials. In: Applications of Nanomaterials, Elsevier Ltd. https://doi.org/10.1016/B978-0-08-101971-9.00006-5

Bandyopadhyay S (2013) Advanced analytical techniques for the measurement of nanomaterials in food and agricultural samples: a review 1, 30(3): 118–125. https://doi.org/10.1089/ees.2012.0325
Bhattacherjee A, Ghosh T, Datta A (2018) Green synthesis and characterisation of antioxidant-tagged gold nanoparticle (X-GNP) and studies on its potent antimicrobial activity. J Exp Nanosci 13(1):50–61. https://doi.org/10.1080/17458080.2017.1407042
Biao L, Tan S, Meng Q, Gao J, Zhang X, Liu Z, Fu Y (2018) Green synthesis, characterization and application of proanthocyanidins-functionalized gold nanoparticles. Nanomaterials 8(1):53. https://doi.org/10.3390/nano8010053
Bishnoi A, Kumar S, Joshi N (2017) Wide-angle X-ray diffraction (WXRD): technique for characterization of nanomaterials and polymer nanocomposites. In: Microscopy Methods in Nanomaterials Characterization, Elsevier, pp 313–337
Bunaciu AA, UdriŞTioiu EG, Aboul-Enein HY (2015) X-ray diffraction: instrumentation and applications. Crit Rev Anal Chem 45(4):289–299
Chen M, Wang X, Yu YH, Pei ZL, Bai XD, Sun C, Wen LS (2000) X-ray photoelectron spectroscopy and auger electron spectroscopy studies of Al-doped ZnO films. Appl Surf Sci 158(1):134–140. https://doi.org/10.1016/S0169-4332(99)00601-7
Cheng F, Su C, Yang Y, Yeh C (2005) Characterization of aqueous dispersions of Fe_3O_4 nanoparticles and their biomedical applications 26: 729–738. https://doi.org/10.1016/j.biomaterials.2004.03.016
Costa S, Borowiak-Palen E, Kruszynska M, Bachmatiuk A, Kalenczuk R (2008) Characterization of carbon nanotubes by Raman spectroscopy. Mater Sci Poland 26(2):433–441
Crozier A, Jaganath IB, Clifford MN, Crozier A, Jaganath B, Clifford MN (2009) Dietary phenolics: chemistry, bioavailability and effects on health. 26(8). https://doi.org/10.1039/b802662a
Davar F, Fereshteh Z, Salavati-niasari M (2009) Nanoparticles Ni and NiO: synthesis, characterization and magnetic properties 476:797–801. https://doi.org/10.1016/j.jallcom.2008.09.121
David L, Moldovan B, Vulcu A, Olenic L, Perde-schrepler M, Fischer-fodor E, Adriana G (2014) Green synthesis, characterization and anti-inflammatory activity of silver nanoparticles using European black elderberry fruits extract. Colloids Surfaces B Biointerfaces. https://doi.org/10.1016/j.colsurfb.2014.08.018
De Oliveira O, Marystela F, de Lima Leite F, Da Róz AL (2017) Nanocharacterization Techniques. Matthew Deans. https://doi.org/10.1016/B978-0-323-49778-7.00001-1
Devience SJ, Pham LM, Lovchinsky I, Sushkov AO, Bar-gill N, Belthangady C, Walsworth RL (2015) Nanoscale NMR spectroscopy and imaging of multiple nuclear species. Nature Nanotechnol, 2–7. https://doi.org/10.1038/nnano.2014.313
Dipankar C, Murugan S (2012) Colloids and Surfaces B: Biointerfaces The green synthesis, characterization and evaluation of the biological activities of silver nanoparticles synthesized from Iresine herbstii leaf aqueous extracts. Colloids Surf B 98:112–119. https://doi.org/10.1016/j.colsurfb.2012.04.006
Dixon R (2005). Proanthocyanidins—a final frontier in
Djuris AB, Leung YH, Ng AMC, Xu XY, Lee PK, Degger N, Wu RS (2014) Toxicity of metal oxide nanoparticles: mechanisms, characterization, and avoiding experimental artefacts. pp 1–19. https://doi.org/10.1002/smll.201303947
Faix O (1992) Fourier transform infrared spectroscopy. Methods Lignin Chem pp 83–109. https://doi.org/10.1007/978-3-642-74065-7_16
Ferri FA, Perreira-da-Silva MDA, Marega E (2012) Atomic force microscopy-imaging, measuring and manipulating surfaces at the atomic scale. In: Bellitto V (ed) InTech, p 268
Gangadoo S, Chapman J (2015) Nanoparticle and biomaterial characterisation techniques, 30. https://doi.org/10.1179/1753555714Y.0000000201
García-Betancourt ML, Rahier H, Van Assche G (2018)Physicochemical characterization of nanomaterials: polymorph, composition, wettability, and thermal stability. In: Emerging applications of nanoparticles and architecture nanostructures, Elsevier Inc. https://doi.org/10.1016/B978-0-323-51254-1.00009-9

Gogoi N, Jayasekhar P, Mahanta C, Bora U (2015) Green synthesis and characterization of silver nanoparticles using alcoholic fl ower extract of Nyctanthes arbortristis and in vitro investigation of their antibacterial and cytotoxic activities. Mater Sci Eng C 46:463–469. https://doi.org/10.1016/j.msec.2014.10.069
Goldstein JI, Newbury DE, Michael JR, Ritchie NW, Scott JH, Joy DC (2017) Scanning electron microscopy and X-ray microanalysis, Springer
Gopi D, Kanimozhi K, Bhuvaneshwari N, Indira J, Kavitha L (2014) Novel banana peel pectin mediated green route for the synthesis of hydroxyapatite nanoparticles and their spectral characterization. Spectrochim Acta Part A Mol Biomol Spectrosc 118:589–597. https://doi.org/10.1016/j.saa.2013.09.034
Goyal RK (2017) Nanomaterials and nanocomposites: synthesis, properties, characterization techniques, and applications, CRC Press
Hafner B (2006) Energy dispersive spectroscopy on the SEM: a primer. In: Characterization Facility, University of Minnesota, pp 1–26
Hall JB, Dobrovolskaia MA, Patri AK, McNeil SE (2007) Characterization of nanoparticles for therapeutics
Henning S, Adhikari R (2017) Scanning electron microscopy, ESEM, and X-ray microanalysis. In: Microscopy Methods in Nanomaterials Characterization, Elsevier, pp. 1–30
Hollander JM, Jolly WL (1970) X-ray photoelectron spectroscopy. Acc Chem Res 3(6):193–200
https://en.Wikipedia.org/wiki/Ultraviolet%E2%80%93visible_spectroscopy
https://en.wikipedia.org/wiki/X-ray_photoelectron_spectroscopy#/media/File:System2.gif
Hussain I, Singh NB, Singh A, Singh H, Singh SC (2016) Green synthesis of nanoparticles and its potential application. Biotech Lett 38(4):545–560. https://doi.org/10.1007/s10529-015-2026-7
Hyllested JÆ, Palanco ME, Hagen N, Mogensen KB (2015) Green preparation and spectroscopic characterization of plasmonic silver nanoparticles using fruits as reducing agents, pp 293–299. https://doi.org/10.3762/bjnano.6.27
Instruments N (1999) Energy-dispersive x-ray microanalysis an introduction. Noran Instruments Inc., Middleton
Jaggi N, Vij DR (2006) Fourier transform infrared, pp 411–450
Jain A, Ranjan S, Dasgupta N, Ramalingam C (2018) Nanomaterials in food and agriculture: an overview on their safety concerns and regulatory issues. Crit Rev Food Sci Nutr 58(2):297–317. https://doi.org/10.1080/10408398.2016.1160363
Jiang X, Jiang J, Jin Y, Wang E, Dong S (2005). Effect of colloidal gold size on the conformational changes of adsorbed cytochrome c: probing by circular dichroism, UV—Visible, and Infrared Spectroscopy, pp 46–53
Johal MS (2012) Understanding nanomaterials, CRC Press
Kale RD, Jagtap P (2018) Biogenic synthesis of silver nanoparticles using Citrus Limon leaves and its structural investigation, pp 11–20. https://doi.org/10.1007/978-981-10-7122-5_2
Khan AU, Malik N, Khan M, Hwan M (2017) Fungi-assisted silver nanoparticle synthesis and their applications. Bioprocess Biosyst Eng. https://doi.org/10.1007/s00449-017-1846-3
Khan MK, Wang QY, Fitzpatrick ME (2016) Atomic force microscopy (AFM) for materials characterization. In: Materials Characterization Using Nondestructive Evaluation (NDE) Methods, Elsevier, pp 1–16
Khan M, Albalawi GH, Shaik MR, Khan M, Adil SF, Kuniyil M, Alkhathlan HZ, Al-Warthan A, Siddiqui MRH (2017) Miswak mediated green synthesized palladium nanoparticles as effective catalysts for the Suzuki coupling reactions in aqueous media. J Saudi Chem Soc 21(4):450–457. https://doi.org/10.1039/b615357g.
Khan ZUH, Khan A, Chen YM, Shah NS, Khan AU, Muhammad N, Wan P (2018) Enhanced antimicrobial, anti-oxidant applications of green synthesized AgNPs- an acute chronic toxicity study of phenolic azo dyes & study of materials surface using X-ray photoelectron spectroscopy. J Photochem Photobiol B 180:208–217. https://doi.org/10.1016/j.jphotobiol.2018.02.015
Khitrov GA, Strouse GF (2003) ZnS nanomaterial characterization by MALDI-TOF mass spectrometry 10: 10465–10469

Kneipp J, Kneipp H, Wittig B, Kneipp K (2010) Novel optical nanosensors for probing and imaging live cells. Nanomed Nanotechnol Biol Med 6(2): 214–226. https://doi.org/10.1016/j.nano.2009.07.009

Kong J, Yu S (2007) Fourier transform infrared spectroscopic analysis of protein secondary structures protein FTIR data analysis and band assign 39(20745001):549–559. https://doi.org/10.1111/j.1745-7270.2007.00320.x

Konno H (2016) X-ray Photoelectron Spectroscopy. In: Materials Science and Engineering of Carbon, Tsinghua University Press Limited. https://doi.org/10.1016/B978-0-12-805256-3.00008-8

Kouvaris P, Delimitis A, Zaspalis V, Papadopoulos D, Tsipas SA, Michailidis N (2012) Green synthesis and characterization of silver nanoparticles produced using Arbutus Unedo leaf extract. Mater Lett 76:18–20. https://doi.org/10.1016/j.matlet.2012.02.025

Kumar J, Thomas KG (2011) Nanorod edges and dimer junctions

Kumar U, Ankamwar B, Karmakar S, Halder A (2018) Science direct green synthesis of silver nanoparticles using the plant extract of Shikakai and Reetha. Mater Today Proc 5(1):2321–2329. https://doi.org/10.1016/j.matpr.2017.09.236

Li X, Zhang W (2007) Sequestration of metal cations with Zerovalent Iron nanoparticles: a study with high resolution X-ray Photoelectron Spectroscopy (HR-XPS). J Phys Chem C Saudi Chem Soc 111(19): 6939–6946. https://doi.org/10.1021/jp0702189. https://doi.org/10.1016/j.jscs.2016.03.008

Lee SH, Heng Desmond, Ng WK, Chan H-K, Tan RBH (2011) Nano spray drying: a novel method for preparing protein nanoparticles for protein therapy. Int J Pharm 403(1–2):192–200

Lin PC, Lin S, Wang PC, Sridhar R (2014) Techniques for physicochemical characterization of nanomaterials. Biotechnol Adv 32(4):711–726. https://doi.org/10.1016/j.biotechadv.2013.11.006

Liu S, Wang Y (2010) Application of AFM in microbiology: a review. Scanning 32(2):61–73

Liu S, Wang Y (2011) A review of the application of atomic force microscopy (AFM) in food science and technology. In: Advances in food and nutrition research, Elsevier, pp 201–240

Liu J, Jiang P, He C, Jiang X, Lu L (2018) Preparation and characterization of Ag @ C composites with hexagon silver core and controllable carbon shell for surface-enhanced Raman scattering, pp 3443–3454. https://doi.org/10.1007/s10853-017-1801-3

López-Sanz S, Fariñas NR, Vargas RS, Martín-Doimeadios RDCR, Ríos Á (2017) Methodology for monitoring gold nanoparticles and dissolved gold species in culture medium and cells used for nanotoxicity tests by liquid chromatography hyphenated to inductively coupled plasma-mass spectrometry. Talanta 164:451–457. https://doi.org/10.1016/j.talanta.2016.11.060

Luo F, Yang D, Chen Z, Megharaj M, Naidu R (2016) One-step green synthesis of bimetallic Fe/Pd nanoparticles used to degrade Orange II. J Hazard Mater 303:145–153

Lupoi JS, Singh S, Simmons BA, Henry RJ (2014) Assessment of Lignocellulosic biomass using analytical spectroscopy: an evolution to high-throughput techniques. Bioenergy Res 7(1):1–23. https://doi.org/10.1007/s12155-013-9352-1

Ma M, Zhang Y, Yu W, Shen H, Zhang H, Gu N (2003) Preparation and characterization of magnetite nanoparticles coated by amino silane 212:219–226

Marbella LE, Millstone JE (2015) NMR techniques for noble metal nanoparticles. https://doi.org/10.1021/cm504809c

Mayeen A, Shaji LK, Nair AK, Kalarikkal N (2018) Morphological characterization of nanomaterials. In: Characterization of Nanomaterials, Elsevier, pp 335–364

Medina-Ramirez I, Bashir S, Luo Z, Liu JL (2009) Green synthesis and characterization of polymer-stabilized silver nanoparticles. Colloids Surf B 73(2):185–191. https://doi.org/10.1016/j.colsurfb.2009.05.015

Miguel L, Rivera R, Silva LP (2018) https://doi.org/10.1016/j.matchemphys.2018.05.068 (Accepted Manuscript)

Michler GH (2008) Electron microscopy of polymers, Springer Science & Business Media

Mishra RK, Zachariah AK, Thomas S (2017) Energy-dispersive X-ray spectroscopy techniques for nanomaterial. In: Microscopy Methods in Nanomaterials Characterization, Elsevier, pp 383–405

Mukherjee P, Roy M, Mandal BP, Dey GK, Mukherjee PK, Ghatak J, Kale SP (2008) Green synthesis of highly stabilized nanocrystalline silver particles by a non-pathogenic and agriculturally important fungus T. asperellum, p 075103. https://doi.org/10.1088/0957-4484/19/7/075103

Muñetón D, Santillán JMJ, Arce VB, Fernández MB , Raap V, Muraca D, Sca LB (2018) Materials characterization a simple and "green" technique to synthesize long-term stability colloidal Ag nanoparticles: Fs laser ablation in a biocompatible aqueous medium, vol 140, pp 320–332. https://doi.org/10.1016/j.matchar.2018.04.021

Nagajyothi PC, Pandurangan M, Veerappan M, Kim DH, Sreekanth TVM, Shim J (2017) Green synthesis, characterization and anticancer activity of yttrium oxide nanoparticles. Mater Lett. https://doi.org/10.1016/j.matlet.2017.12.081

Pesila NS, Kathleen JS, Searson PC (2003) Relationship between absorbance spectra and particle size distributions for quantum-sized nanocrystals, pp 10412–10415. https://doi.org/10.1021/jp0303218

Pace HE, Rogers NJ, Jarolimek C, Coleman VA, Higgins CP, Ranville JF (2011) Determining transport efficiency for the purpose of counting and sizing nanoparticles via single particle inductively coupled plasma mass spectrometry, pp 9361–9369

Patel NH (2015). Basic principle, working and instrumentation of experimental techniques

Paterson E, Swaffield R (1994) X-ray photoelectron spectroscopy

Plachtova P, Med Z, Zbo R, Varma RS, Maršálek B (2018) Iron and iron oxide nanoparticles synthesized using green tea extract: differences in ecotoxicological profile and ability to degrade malachite green. https://doi.org/10.1021/acssuschemeng.8b00986

Pleul D, Simon F (2008) X ray photoelectron spectroscopy. Polymer Surf Interfaces, pp 71–89. https://doi.org/10.1007/978-3-540-73864-0

Popovic ZV, Mitrovic ZD, Scepanoviv M, Brojčinm M, Askrabic S (2011) Raman scattering on nanomaterials and nanostructures, 74(1): 62–74. https://doi.org/10.1002/andp.201000094

Rana S, Philip J, Raj B (2010) Micelle based synthesis of cobalt ferrite nanoparticles and its characterization using Fourier transform infrared transmission spectrometry and thermogravimetry. Mater Chem Phys 124(1):264–269. https://doi.org/10.1016/j.matchemphys.2010.06.029

Rao YS, Kotakadi VS, Prasad TNVKV, Reddy AV, Gopal DS (2013) Green synthesis and spectral characterization of silver nanoparticles from Lakshmi tulasi (*Ocimum sanctum*) leaf extract. Spectrochim Acta Part A: Mol Biomol Spectrosc 103:156–159

Ritheshraj D, Prasanth S, Vineeshkumar TV, Sudarsanakumar C (2015) Sens Actuators B Chem. https://doi.org/10.1016/j.snb.2015.10.106

Roming M, Feldmann C, Avadhut YS (2008) Characterization of noncrystalline nanomaterials: NMR of Zinc Phosphate as a case study. Chem Mater 4:5787–5795

Rosbero TMS, Camacho DH (2017) Green preparation and characterization of tentacle-like silver/copper. Biochem Pharm. https://doi.org/10.1016/j.jece.2017.05.009

Roushani M, Mavaei M, Rajabi HR (2015) Graphene quantum dots as novel and green nanomaterials for the visible-light-driven photocatalytic degradation of cationic dye, Elsevier B.V. https://doi.org/10.1016/j.molcata.2015.08.011

Salame PH, Pawade VB, Bhanvase BA (2018) Characterization tools and techniques for nanomaterials. In: Nanomaterials for green energy, Elsevier, pp 83–111

Sarmento B, Ferreira D, Veiga F, Ribeiro A (2006) Characterization of insulin-loaded alginate nanoparticles produced by ionotropic pre-gelation through DSC and FTIR studies. Carbohydr Polym 66(1):1–7

Sathyavathi R, Krishna MB, Rao SV, Saritha R, Rao DN (2010) Biosynthesis of silver nanoparticles using *Coriandrum Sativum* leaf extract and their application in nonlinear optics. Adv Sci Lett 3(2):1–6. https://doi.org/10.1166/asl.2010.1099

Saxena A, Tripathi RM, Singh RP (2010) Biological synthesis of silver nanoparticles by using onion (Allium cepa) extract and their antibacterial activity. Digest J Nanomaterials Biostructures 5(2):427–432

Schmidt B, Loeschner K, Hadrup N, Mortensen A, Sloth JJ, Koch CB, Larsen EH (2011) Quantitative characterization of gold nanoparticles by field-flow fractionation coupled online with light scattering detection and inductively coupled plasma mass spectrometry. Anal Chem 83:2461–2468

Shameli K, Ahmad MB, Jazayeri SD, Sedaghat S, Shabanzadeh P, Jahangirian H, Abdollahi Y (2012) Synthesis and characterization of polyethylene glycol mediated silver nanoparticles by the green method. Int J Mol Sci 13(6):6639–6650. https://doi.org/10.3390/ijms13066639

Shrivastava S, Bera T, Roy A, Singh G, Ramachandrarao P, Dash D (2007) Characterization of enhanced antibacterial effects of novel silver nanoparticles. Nanotechnology 18(22):225103.

Singhal G, Bhavesh R (2011) Biosynthesis of silver nanoparticles using *Ocimum sanctum* (Tulsi) leaf extract and screening its antimicrobial activity, 2981–2988. https://doi.org/10.1007/s11051-010-0193-y

Song Y, Bhushan B (2007) Finite-element vibration analysis of tapping-mode atomic force microscopy in liquid. Ultramicroscopy 107(10–11):1095–1104

Song JY, Jang H, Kim BS (2009) Biological synthesis of gold nanoparticles using *Magnolia kobus* and *Diopyros kaki* leaf extracts. Process Biochem 44:1133–1138. https://doi.org/10.1016/j.procbio.2009.06.005

Stuart B (2005) Infrared spectroscopy 1(1): 1–20

Subramanian A, Rodriguez-Saona L (2009) Fourier Transform Infrared (FTIR) spectroscopy. In: Sun DW (ed) Infrared spectroscopy for food quality analysis and control. Academic Press, pp 145–178.

Suganya KSU, Govindaraju K, Kumar VG, Dhas TS, Karthick V, Singaravelu G, Elanchezhiyan M (2015) Blue green alga mediated synthesis of gold nanoparticles and its antibacterial efficacy against Gram positive organisms. Mater Sci Eng, C 47:351–356. https://doi.org/10.1016/j.msec.2014.11.043

Sun Q, Cai X, Li J, Zheng M, Chen Z, Yu C (2014) Colloids and surfaces a: physicochemical and engineering aspects green synthesis of silver nanoparticles using tea leaf extract and evaluation of their stability and antibacterial activity. Colloids Surf A Physicochem Eng Aspects 444:226–231. https://doi.org/10.1016/j.colsurfa.2013.12.065

Suzuki E (2002) High-resolution scanning electron microscopy of immunogold-labelled cells by the use of thin plasma coating of osmium. J Microsc 208(3):153–157

Tanaka K, Waki H, Ido Y, Akita S, Yoshida Y (1988) Protein and polymer analyses up to mlz 100 000 by laser ionization time-of-flight mass spectrometry. Rapid Commun Mass Spectrom 2(8):151–153

Taylor P, Tiede K, Boxall ABA, Tear SP, Lewis J, David H, Hassellöv M (2008) Food additives & contaminants: part a detection and characterization of engineered nanoparticles in food and the environment, pp 37–41. https://doi.org/10.1080/02652030802007553. Oct 2012

Thovhogi N, Diallo A, Gurib-fakim A, Maaza M (2015) Nanoparticles green synthesis by Hibiscus Sabdariffa flower extract: main physical properties. J Alloys Compd 647:392–396. https://doi.org/10.1016/j.jallcom.2015.06.076

Tiede K, Boxall ABA, Tear SP, Lewis J, David H, Hassellöv M (2008) Detection and characterization of engineered nanoparticles in food and the environment. Food Addit Contam 25(7):795–821

Tripathi DK et al (2018) Nanomaterials in plants, algae, and microorganisms: concepts and controversies, vol 1. https://doi.org/10.1016/C2016-0-00176-6

Ulug B, Haluk Turkdemir M, Cicek A, Mete A (2015) Role of irradiation in the green synthesis of silver nanoparticles mediated by fig (Ficus carica) leaf extract. Spectrochim Acta Part A Mol Biomol Spectrosc 135:153–161. https://doi.org/10.1016/j.saa.2014.06.142

Veisi H, Azizi S, Mohammadi P (2017) Green synthesis of the silver nanoparticles mediated by Thymbra spicata extract and its application as a heterogeneous and recyclable nanocatalyst for catalytic reduction of a variety of dyes in water. J Cleaner Prod. https://doi.org/10.1016/j.jclepro.2017.09.265

Walther T (2017) Transmission electron microscopy of nanostructures. In: Microscopy Methods in Nanomaterials Characterization, Elsevier, pp 105–134

Wang ZL (1999) Transmission electron microscopy and spectroscopy of nanoparticles. Charact Nanophase Mater 3:37–80

Wang Y, Irudayaraj J, Irudayaraj J (2012) Surface-enhanced Raman spectroscopy at single-molecule scale and its implications in biology

Weckhuysen BM (2004) Ultraviolet-visible spectroscopy, pp 255–270

Weng X, Guo M, Luo F, Chen Z (2017) One-step green synthesis of bimetallic Fe/Ni nanoparticles by eucalyptus leaf extract: Biomolecules identification, characterization and catalytic activity. Chem Eng J 308:904–911. https://doi.org/10.1016/j.cej.2016.09.134

Williams DB, Carter CB (2009) The transmission electron microscope. In: Transmission electron microscopy: a textbook for materials science, Springer, Boston, MA, US, pp 3–22

Yamaura M, Camilo RL, Sampaio LC, Macêdo MA, Nakamura M, Toma HE (2004) Preparation and characterization of (3-aminopropyl)triethoxysilane-coated magnetite nanoparticles. J Magn Magn Mater 279(2–3):210–217. https://doi.org/10.1016/j.jmmm.2004.01.094

Yang Z, Xing Z, Xu J, Liu X, Lei Y (2018) A new synthetic strategy of Ag-TiO_2 Nanocomposites based on Ligand-to-Metal charge transfer under visible light irradiation: characterizations and photocatalytic activity. Phys Status Solidi (A) 215(14):1800026. https://doi.org/10.1002/pssa.201800026

Yu J, Wang H, Zhan J, Huang W (2017) Review of recent UV-Vis and infrared spectroscopy researches on wine detection and discrimination. https://doi.org/10.1080/05704928.2017.1352511

Yuvakkumar R, Suresh J, Nathanael AJ, Sundrarajan M, Hong SI (2014) Novel green synthetic strategy to prepare ZnO nanocrystals using rambutan (Nephelium lappaceum L.) peel extract and its antibacterial applications. Mater Sci Eng, C 41:17–27. https://doi.org/10.1016/j.msec.2014.04.025

Zhang J, Higashi K, Limwikrant W, Moribe K, Yamamoto K (2012) Molecular-level characterization of probucol nanocrystal in water by in situ solid-state NMR spectroscopy. Int J Pharm 423(2):571–576. https://doi.org/10.1016/j.ijpharm.2011.11.028

Zhang Y, Yu W, Pei L, Lai K, Rasco BA, Huang Y (2015) Rapid analysis of malachite green and leucomalachite green in fish muscles with surface-enhanced resonance Raman scattering. Food Chem 169:80–84. https://doi.org/10.1016/j.foodchem.2014.07.129

Green Synthesis of Metal Nanoparticles for Electronic Textiles

Ashish Kapoor, Pramod Shankar and Wazed Ali

Abstract Nanoparticles (NPs) are one of the primary elements of nanotechnology and nanoengineering. Metal nanoparticles (MNPs) particularly are of great interest because metals in the form of nanoparticles, due to their size, display properties that are different from their bulk counterpart. MNPs are widely used in the form of printable conductive inks and conducting polymer composites (CPCs) to impart electrical functionality to textiles to create electronic (E)-textiles. Metal nanoparticles have long been synthesized by conventional physical methods and using chemical agents which pose various environmental hazards due to their toxic nature. Due to this, green synthesis of metal nanoparticles using plants and microbes has attracted attention. In this chapter, we focus on silver (Ag) and gold (Au) nanoparticles synthesized using plants and their potential use in fabricating E-textiles for sensing applications.

Keywords Nanoparticles · Nanotechnology · Green synthesis · Electronic textiles

1 Introduction

Metal particles of sizes 1–100 nm are considered as NPs. Their size is responsible for different physical and chemical properties than their bulk form (refer Table 1). MNPs find multiple uses in the fields of health care, electronics, energy science, catalysis, biomedical, optics, and chemicals. Nanoparticles can be seen as bridging the size gap between substances engineered by synthetic chemistry (Angstrom scale) and photolithographic techniques as shown in Fig. 1.

A. Kapoor
Department of Textile Engineering, Chemistry and Science, Wilson College of Textiles, North Carolina State University, Raleigh, NC 27695, USA

P. Shankar · W. Ali (✉)
Department of Textile Technology, Indian Institute of Technology Delhi, New Delhi, India
e-mail: wazed@textile.iitd.ac.in

S. Ahmed and W. Ali (eds.), *Green Nanomaterials*,
Advanced Structured Materials 126,
https://doi.org/10.1007/978-981-15-3560-4_4

Table 1 Size dependency of nanomaterials physical properties (Abdullaeva 2017)

Properties	Material response to size decrease
Phase changes	Decrease in phase change temperature including melting point
Electrical	Increase in electrical resistance
Magnetic	Increase in coercive force, magnetic resistance, appearance of supermagnetism
Mechanical	Increase in yield strength, hardness, toughness and manifestation of superplasticity at higher temperatures
Kinetic	Higher anomalous coefficient values for diffusion, heat capacity and decrease in thermal conductivity

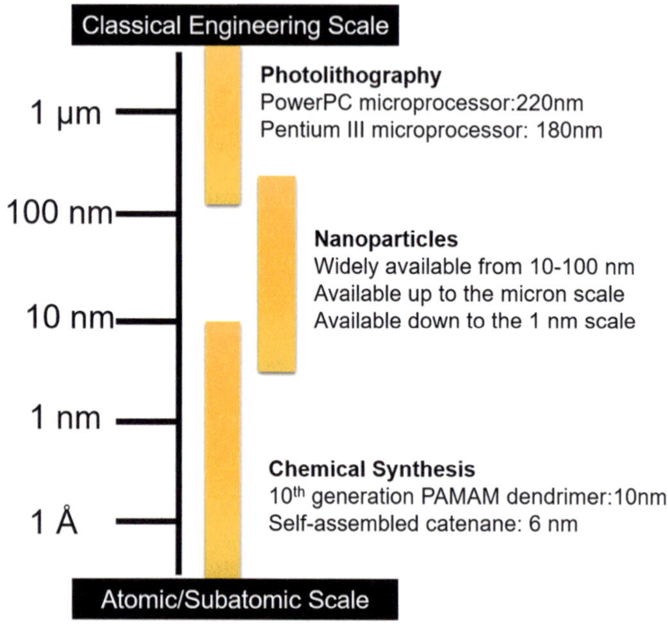

Fig. 1 Nanoparticles bridge the gap between photolithography and synthetic chemical engineering (Shipway et al. 2000)

Metal nanoparticles are mostly synthesized by synthetic methods, but the current focus is shifting toward green synthesis because synthetic techniques produce environmentally toxic and hazardous by-products and are expensive. Green synthesis includes synthesis of NPs using biological routes such as plants and microorganisms (Parveen et al. 2016). Green synthesis is considered a safe, eco-friendly, and cost-effective approach for synthesis of MNPs. A green chemistry approach for NP synthesis should comprise of a harmless solvent, nontoxic reducing agent, and environmentally safe stabilization agent (Raveendran et al. 2003; Annu and Ahmed 2018).The advantage of biological synthesis is slow reaction kinetics, because the growth and stabilization of NPs can be controlled.

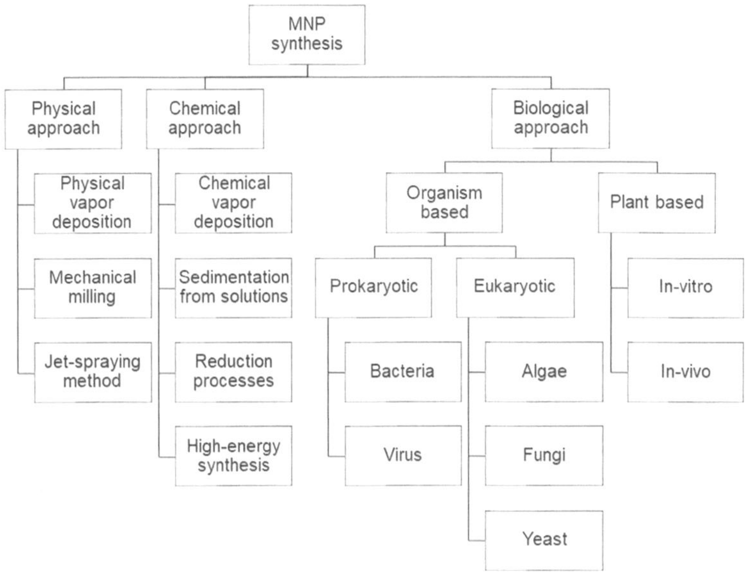

Fig. 2 Various routes for synthesis of nanoparticles

Plant-based synthesis is preferable over microorganism-based synthesis because:

- Plants have an innate ability to hyperaccumulate and reduce metal ions and can be used for large-scale synthesis of NPs (Iravani 2011; Ahmed et al. 2016).
- Plants are also the best choice because there is no cost of maintaining microorganism culture media (Annu and Ahmed 2018).
- Besides, the rate of NP synthesis by plants is faster than microorganisms and NPs are more stable (Iravani 2011).

Figure 2 summarizes physical, chemical, and biological routes available for NP synthesis. In this chapter, only plant-based NP synthesis is focused upon.

2 Nanoparticle Synthesis

There are two primary approaches for synthesis of NPs: bottom-top and top-down (Fig. 3). In *bottom-up approach*, NPs are grown from individual elements like atoms, molecules, or biological cells using chemical and biological methods where atoms self-assemble to form nanosized particle. This approach offers advantage of large-scale and fast synthesis of NPs. In *top-down approach*, NPs are prepared by breaking down bulk material into fine particles through mechanical and chemical processes like milling and electrolysis, respectively (Ahmed et al. 2016; Awasthi et al. 2017; Abdullaeva 2017).

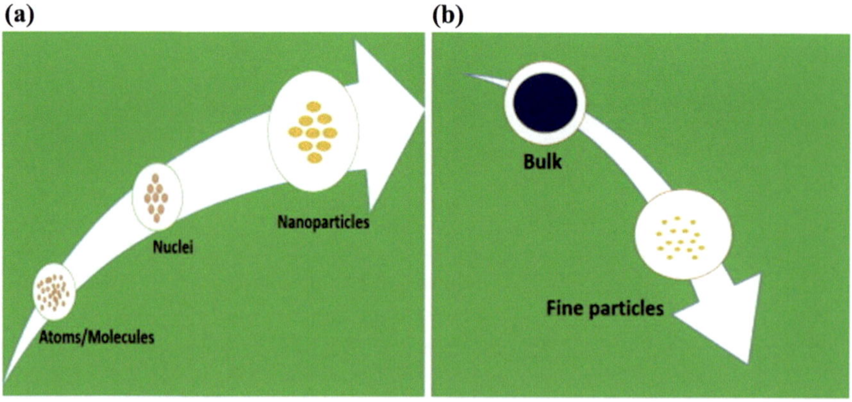

Fig. 3 Protocols for synthesis of nanoparticles. **a** Bottom-up approach. **b** Top-down approach (Ahmed et al. 2016)

The important aspects while synthesizing NPs using green pathways include selection of best organism/plant species depending on their enzyme activities or biochemical pathways and using optimal reaction conditions like pH, temperature, and light (Iravani 2011). Plant-based synthesis can be in vitro and in vivo (Awasthi et al. 2017) where in vitro refers to synthesis of NPs in a laboratory using plant or plant extract and in vivo refers to synthesis of nanoparticles within plants (refer Fig. 4).

In vitro

Different parts of a plant like stem, leaf, and seed have different metabolic and biochemical processing capabilities to synthesize a specific type of NP. In vitro techniques use plant extracts as reducing agents (Castro et al. 2011). The shape and size of NPs can be controlled by reaction conditions like pH, temperature, and time.

In vivo

Plants have the capability to absorb bioavailable metal through roots which is stored as elemental form in their tissues. The absorbed metal is transported via cells and subsequently reduced and finally deposited in form of NPs of varied morphologies. This mode of NP production by plants is still under investigation (Haverkamp and Marshall 2009).

3 Green Synthesis of Silver and Gold Nanoparticles

Noble metals, namely silver and gold have been the main focus of plant-based synthesis (Iravani 2011). Conventional NP synthesis methods use chemicals as reducing agents which is not eco-friendly, whereas green techniques make use of plants/plant extracts as reductants and stabilizers (Annu and Ahmed 2018). Gardea-Torresdey et al. first reported the synthesis of silver (Ag) and gold (Au) nanoparticles within plants (Gardea-Torresdey et al. 2002, 2003).Green synthesis using plants generally

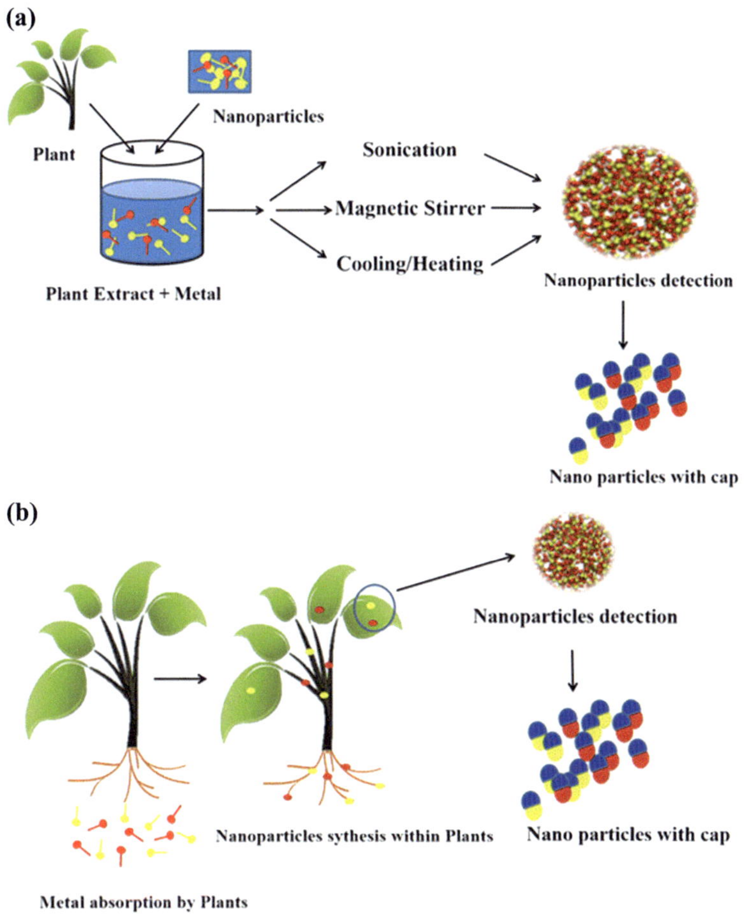

Fig. 4 Synthesis of NPs by plants **a** in vitro and **b** in vivo (Awasthi et al. 2017)

involves an in vitro approach. A general protocol for plant-based synthesis of silver nanoparticles (refer Fig. 5) consists of procuring a part of plant followed by washing, squeezing, and filtering. A metal salt like silver nitrate ($AgNO_3$) is then added to the filtrate, which causes a visible color change of solution, indicating formation of NPs (Ahmed et al. 2016). The synthesized NPs are characterized by UV–Vis spectroscopy.

A lot of research in green synthesis of NPs using plant-based materials focuses on biosynthesis using plant leaf extracts. Shankar et al. used neem (*Azadirachtaindica*) leaf extract to synthesize pure Ag and Au NPs along with bimetallic Au/Ag nanoparticles (Shankar et al. 2004a). The source of Ag and Au was silver *nitrate* ($AgNO_3$) and chloroauric acid ($HAuCl_4$), respectively, which when reacted with neem leaf extract, caused formation of nanoparticles. Triangular-shaped AuNPs as shown in Fig. 6a were produced by the reduction of chloroaurate ions ($AuCl_4^-$) by lemon-

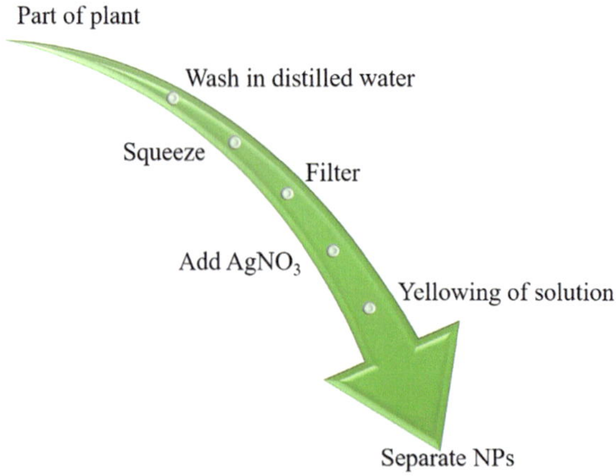

Fig. 5 Synthesis of silver nanoparticles using plant extract (Ahmed et al. 2016)

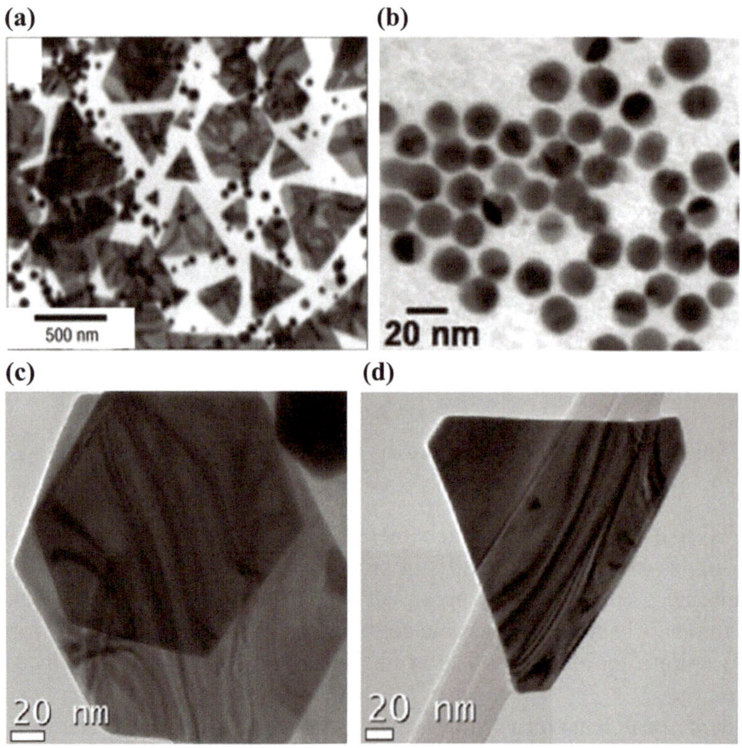

Fig. 6 Representative TEM micrograph of **a** Triangular gold nanotriangles by reduction of aqueous $AuCl_4^-$ by lemongrass extract (Shankar et al. 2004b) and **b** Silver nanoparticles synthesized using aloe vera extract (Chandran et al. 2006). HR-TEM micrographs of **c** Gold nanohexagon and **d** Gold nanotriangle (Ghodake et al. 2010) synthesized from pear extract

grass (*Cymbopogon flexuosus*) plant extract in one-step at room temperature (Shankar et al. 2004b). The change in color of solution from pale yellow to a vivid ruby red indicated the formation of AuNPs. The AuNPs were characterized as thin, flat, and single-crystalline. In another work too, flat and thin gold nanotriangles of thickness 20–40 nm were synthesized using tamarind leaf extract (Ankamwar et al. 2005a). Chandran et al. used aloe vera leaf extract as reducing agent to produce spherical silver NPs as shown in Fig. 6b along with triangular Au NPs (Chandran et al. 2006). AgNPs were also synthesized from reaction mixture of extracts of Eucalyptus hybrida (safeda) leaf and $AgNO_3$ (Dubey et al. 2009).

Fruit extracts have also attracted attention as reducing agents for NP synthesis. Bankar et al. used banana peel (*Musa paradisiaca*) extract in powder form as reducing agent to prepare AgNPs (Bankar et al. 2010). *Emblica officinalis* (amla, Indian gooseberry) fruit extract was added to silver sulfate and chloroauric acid solutions to generate stable silver and gold nanoparticles, respectively (Ankamwar et al. 2005b). It was also found that pear fruit extract can be used to synthesize gold nanoparticles of various shapes like hexagonal and triangular as shown in Fig. 6c, d. NPs were produced by varying the pH of extract where Ghodake et al. found that alkaline conditions of extract produced plate-like morphologies (Ghodake et al. 2010). Other common plant sources for synthesis of silver and gold nanoparticles with their shape and size are summarized in Table 2.

4 Characterization of Nanoparticles

NP characterization includes determining the type, size, shape, agglomeration state, and surface area of NPs. The common characterization techniques are discussed here.

UV–Visible spectroscopy is carried out in wavelength ranges of 200–700 nm to characterize metal NPs. The surface plasmon resonance (SPR) band is the most critical feature in a UV–Vis spectrum. SPR is an interaction between conduction electrons of metal NPs and incident photons relying on size and shape of NP and type of dispersion medium (Noguez 2007; Jana et al. 2016). When recorded as function of reaction time, this time-dependent optical response is a characteristic of formation of NPs and their aggregation as well as anisotropy in particles (Shankar et al. 2004b). *Scanning electron microscopy* (*SEM*) and *transmission electron microscopy* (*TEM*) are used to measure the physical size, morphology, and dispersion state of NPs on micrometer to nanometer length scale. Inclusion of *energy dispersive X-ray spectroscopy* (*EDS*) in an electron microscope like SEM can be used to determine the elemental composition of MNPs. *Dynamic light scattering* (*DLS*) provides particle-size distribution of NPs suspended in a liquid. Fourier transform infrared (FTIR) spectroscopy is the most common technique for surface characterization providing information about various functional groups attached to surface of NPs. Surface characterization on atomic scale is carried out by atomic force microscopy (AFM). X-ray diffraction (XRD) is used to characterize the crystal structure and phase of NPs (Nagajyothi and Sreekanth 2015).

Table 2 Summary of plant based synthesis of silver and gold nanoparticles

Name of plant	Type of NP	Size of NP (nm)	Shape
Acorus calamus (Nakkala et al. 2014) (sweet flag)	Ag	31.83	Spherical
Vitis vinifera (Gnanajobitha et al. 2013) (*common grape wine*)	Ag	30–40	Spherical
Allium sativum (Ahamed et al. 2011) (*garlic*)	Ag	4–22	Spherical
Trachyspermum ammi (Vijayaraghavan et al. 2012) (*ajwain*)	Ag	87(majority)	Triangular
Cocous nucifera (Mariselvam et al. 2014) (coconut palm)	Ag	22	Spherical
Brassica juncea (Haverkamp and Marshall 2009) (mustard)	Ag	2–35	Spherical
Carica papaya (Mude et al. 2009)	Ag	60–80	Spherical
Cinnamon zeylanicum (Sathishkumar et al. 2009)	Ag	31–40	Quasi-spherical, rod
Citrus limon (Prathna et al. 2011) (lemon)	Ag	Below 50	Nearly spherical
Ficus bengalensis (Ravindra et al. 2010) (banyan)	Ag	20	Spherical
Menthapiperita (MubarakAli et al. 2011) (mint)	Ag	90	Spherical
Nelumbo nucifera (Santhoshkumar et al. 2011) (water lilies)	Ag	45	Spherical, triangular, truncated triangular, decahedral
Ocimum sanctum (Philip and Unni 2011) (Krishna tulsi)	Ag	10–20	Spherical
Anacardium occidentale (Sheny et al. 2011) (cashew)	Ag & Au	15(Ag) & 6.5,17(Au)	Spherical
Swietenia mahogani (Mondal et al. 2011)	Ag & Au	–	Spheroidal, triangular, hexagonal
Black tea (Begum et al. 2009)	Au & Ag	20	Spherical, trapezoidal, prismatic, rod

(continued)

Table 2 (continued)

Name of plant	Type of NP	Size of NP (nm)	Shape
Coriandrum sativum (Narayanan and Sakthivel 2008)	Au	6.75–57.91	spherical, triangular, truncated triangular, decahedral
Psidium guajava (Raghunandan et al. 2009) (guava)	Au	27 ± 3	Spherical (mostly), triangular, hexagonal
Tamarindus indica (Ankamwar et al. 2005a) (tamarind)	Au	20–40	Triangular
Terminalia catappa (Ankamwar 2010) (almond)	Au	10–35	Spherical

5 Application of Metal Nanoparticles for Electronic (E)–Textiles

NPs integrated into textiles at fiber, yarn, or fabric level have numerous applications such as antimicrobial, antistatic, electronic, and photonic as shown in Fig. 7. The current focus is on integration of sensing functionality at fiber level into fabrics such that the comfort and flexibility of fabric are not compromised. In this chapter, we focus on special type of textiles, called electronic (E) textiles capable of sensing. E-textiles are fabrics that have electrical functionality or electronics embedded into them. E-textiles have numerous applications, for example, personalized healthcare, sports and fitness, military, and fashion (Coyle et al. 2007). Metal NPs are ideal candidates for imparting electrical properties to fabrics due to their inherent high conductivity (Zeng et al. 2014). Besides, excellent electrical conductivity, NPs are promising for sensor fabrication due to the various reasons (Segev-Bar and Haick 2013). Firstly, NP size and shape can be varied, and as a result, the surface to volume ratio which means that their surface properties can be precisely tuned, and consequently, their response to stimuli like pressure, temperature, etc. (Yin et al. 2011). Secondly, the NPs can be capped with a variety of functional groups which implies that NPs with hybrid combination of physical and chemical properties can be synthesized (Haick 2007).Lastly, NP films with different porosities or interparticle distances can be prepared which is directly related to sensitivity of sensor (Krasteva et al. 2002).

Electrically conducting textiles can be fabricated by two major methods; (a) printing or coating conductive materials on fibers and fabrics and (b) integrating conductive fibers or yarns into textiles. Metal NPs can be used in the form of printing inks or conductive polymer composites to create E-textiles for sensing applications. Figure 8 shows the use of NPs to create electrically functional fabrics. From our literature search, it was found that there are very a few cases where NPs synthesized

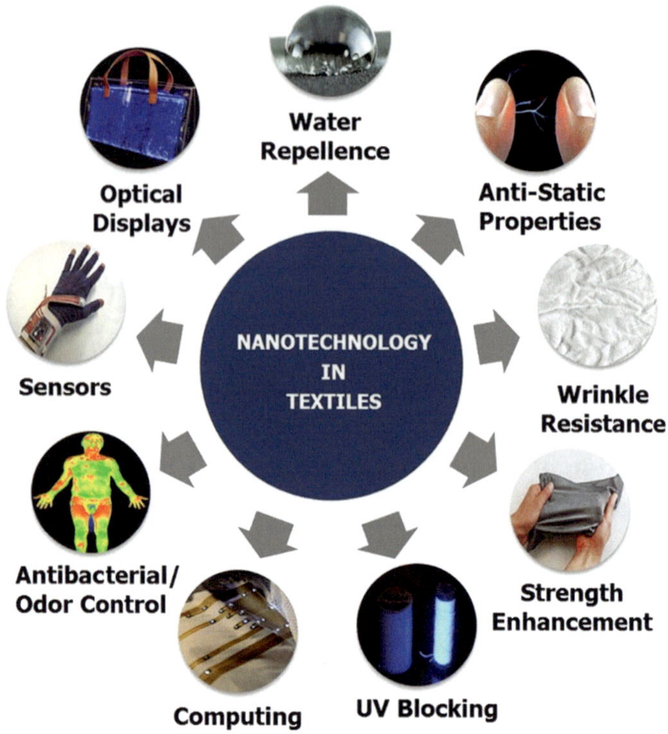

Fig. 7 Nanotechnology in textiles. Reprinted with permission from (Yetisen et al. 2016). Copyright (2016) American Chemical Society

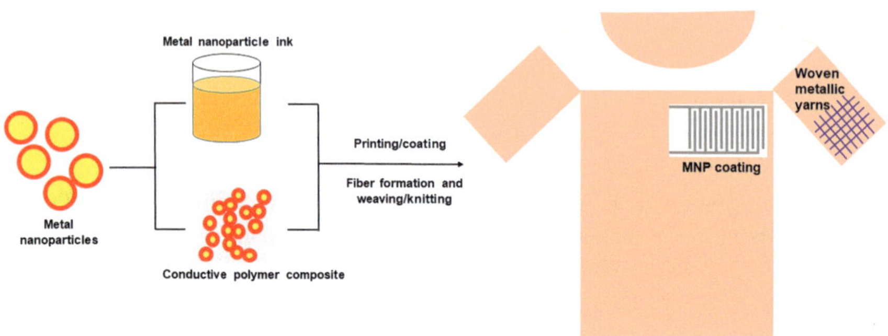

Fig. 8 Schematic showing the use of nanoparticles to create E-textiles

by green techniques have been used to create sensors within textiles. This section discusses a few representative applications of metal NPs irrespective of their mode of synthesis (includes chemically synthesized NPs too) to make conductive textiles and their use in sensing various physical stimuli.

In general, there are two methods for creating conducting fibers from metal NPs (Dong and Hinestroza 2009). In the first method, fibers are immersed into a NP solution prepared by reducing metal salts using reducing agents and second route is an in situ technique where metal ions are first adsorbed onto fiber surface and later reduced to metal NPs via posttreatment using heat, UV radiation, or chemicals (Song et al. 2012). Printing conductive materials onto fibrous substrates is a versatile approach to create sensors. MNPs can be used to form printable conductive inks. MNP printing inks contain a metal precursor-like Ag or Au NPs and a carrier medium such as water. In order to be a good quality ink with reproducible performance, the metal dispersion should not lead to agglomeration or precipitation, which is why a stabilizing agent, mostly a polymer is added to the mixture (Kamyshny 2011). The ink can be printed onto fabrics via inkjet printing or screen printing to create conductive patterns (Stoppa and Chiolerio 2014). The conductivity arises from sintering process where print patterns are exposed to temperature below the melting temperature of metal leading to creation of electrically connected or percolating network (Lee et al. 2005). Gold and silver are widely used metals in printing conductive patterns on fabrics due to their superior conductivity and stability (Gao et al. 2017).

MNPs are also mixed with inherently electrically insulating polymers to create flexible conductive composites (Balazs et al. 2006; Hanemann et al. 2010). These composites can be prepared in two ways: ex situ and in situ (refer Fig. 9). Ex situ is a common approach where pre-synthesized NPs are dispersed in polymer matrix, whereas in situ method consists of synthesis of NPs by reduction of metal salts within polymer matrix (Anyaogu 2008; Palza and Palza 2015). These composites can also be deposited on various substrates to create conductive patterns using commercial printing techniques (Moscicki et al. 2005) and even extruded to form conductive fibers directly.

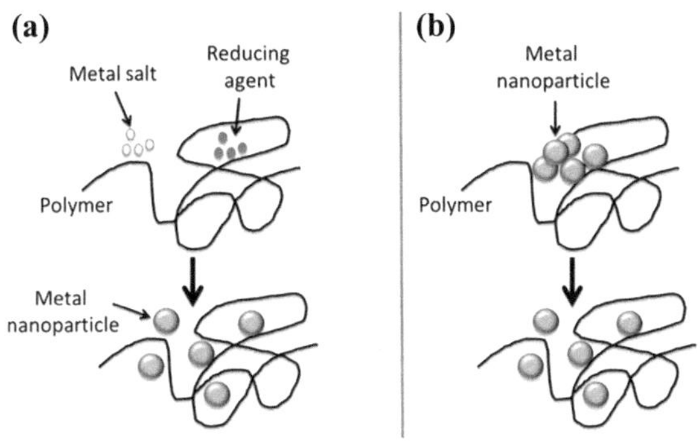

Fig. 9 Methods of fabrication of metal nanoparticle-polymer composites **a** In situ with polymer as reaction medium. **b** Ex situ with polymer as a dispersion medium (Palza and Palza 2015)

Lee et al. fabricated conductive fibers by coating Kevlar fiber with polymer/silver nanoparticle (AgNP) composite. Kevlar fibers were first coated with a layer of poly(styrene-butadiene-styrene) or SBS rubber followed by absorption of Ag precursors which later were reduced to form Ag nanoparticles, confirmed by color change of fiber to dark gray. The conductive fibers were encapsulated with polydimethylsiloxane (PDMS) and assembled orthogonally, like in woven fabric to create a capacitive pressure sensor. A capacitive sensor consists of two conducting materials separated by a dielectric (insulating) medium and the capacitance is a function of overlapping area of conductors, separation between them, and the dielectric constant of the separating medium. There was a variation in capacitance of fiber-based pressure sensor when a load was applied due to change in the thickness of dielectric elastomer layer (refer Fig. 10). The fibers were also woven into a glove to create a smart glove (E-textile) used to control a drone (Lee et al. 2015a).

Lee, Seulah et al. also fabricated stretchable conductive fibers consisting of silver nanowires (AgNWs) and silver nanoparticles (AgNPs) as conductive materials. AgNW-mixed SBS fiber was first produced by wet-spinning technique and then silver precursor was absorbed and reduced repeatedly to create AgNPs on surface and inner region of fiber (refer Fig. 11). AgNWs preserve the electrical conductivity under high-mechanical strain by acting as conducting bridges between AgNPs. The

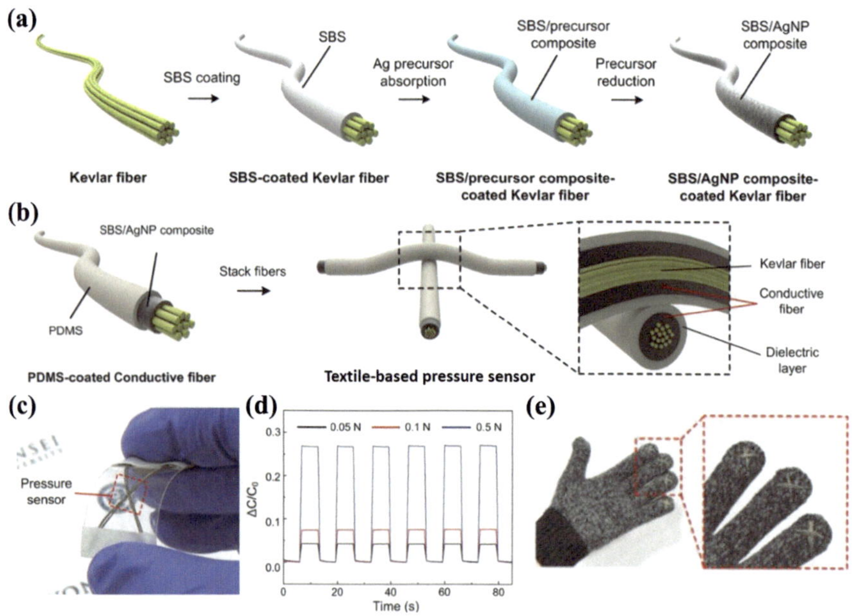

Fig. 10 **a** Schematic of fabrication of conducting fiber with in situ synthesis of NPs. **b** Schematic of fabrication of textile-based capacitive pressure sensor. **c** Pressure sensor on poly (ethyleneterephthalate) (PET) substrate. **d** Capacitive response of pressure sensor to application of different mechanical loads. **e** Integration of conductive fibers in a glove (Lee et al. 2015a)

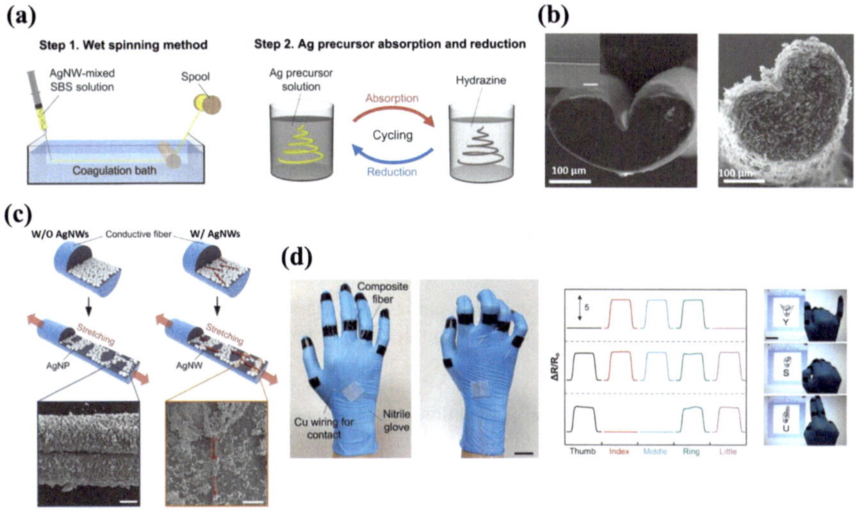

Fig. 11 **a** Fabrication of conductive fiber. **b** Cross-sectional SEM image of fiber without AgNP (left) and with AgNP (right). **c** Modification of AgNW and AgNP network in fiber at 50% strain. **d** Fibers embedded smart glove for motion detection for sign language (Lee et al. 2015b)

fibers were used for detecting sign language by detecting bending motion of fingers as that caused a change in their electrical resistance (Lee et al. 2015b). The change in resistance was due to the modification of conductive network.

6 Conclusion

In this chapter, we focused on green synthesis of silver and gold nanoparticles using plant sources. Plant-based synthesis is preferable over microorganism-based synthesis due to the possibility of large-scale synthesis and cost effectiveness. Two primary approaches for synthesis of NPs: bottom-up and top-down are introduced. In vitro and in vivo methods for preparing NPs using plant sources are highlighted. The primary aspect of green synthesis is that these techniques make use of plants or plant extracts as reducing agents and stabilizers. A lot of research using plant-based materials focuses on biosynthesis using plant leaf and fruit extracts. Major NP characterization techniques are mentioned. Lastly, use of MNPs in the form of conductive printing inks and conductive polymer composites to impart electrical conductivity to textiles in order to create sensing E-textiles has been discussed.

References

Abdullaeva Z (2017) Synthesis of nanoparticles and nanomaterials: biological approaches. Springer, Heidelberg

Ahamed M, Majeed Khan MA, Siddiqui MKJ et al (2011) Green synthesis, characterization and evaluation of biocompatibility of silver nanoparticles. Physica E: Low-dimensional Syst Nanostruct 43:1266–1271. https://doi.org/10.1016/j.physe.2011.02.014

Ahmed S, Ahmad M, Swami BL, Ikram S (2016) A review on plants extract mediated synthesis of silver nanoparticles for antimicrobial applications: a green expertise. J Adv Res 7:17–28. https://doi.org/10.1016/j.jare.2015.02.007

Ankamwar B (2010) Biosynthesis of gold nanoparticles (Green-gold) using leaf extract of *Terminalia catappa*. J Chem. https://www.hindawi.com/journals/jchem/2010/745120/abs/. Accessed 22 Sept 2018

Ankamwar B, Chaudhary M, Sastry M (2005a) Gold nanotriangles biologically synthesized using tamarind leaf extract and potential application in vapor sensing. Synth React Inorg, Met-Org, Nano-Met Chem 35:19–26. https://doi.org/10.1081/SIM-200047527

Ankamwar B, Damle C, Ahmad A, Sastry M (2005b) Biosynthesis of Gold and Silver nanoparticles using *Emblica officinalis* fruit extract, their phase transfer and transmetallation in an organic solution. https://www.ingentaconnect.com/content/asp/jnn/2005/00000005/00000010/art00010. Accessed 2 Sept 2018

Annu AA, Ahmed S (2018) Green synthesis of metal, metal oxide nanoparticles, and their various applications. In: Martínez LMT, Kharissova OV, Kharisov BI (eds) Handbook of ecomaterials. Springer International Publishing, Cham, pp 1–45

Anyaogu KC (2008) Stabilized metal nanoparticle-polymer composites: preparation, characterization and potential applications, 119

Awasthi G, Kumar A, Awasthi KK et al (2017) Green synthesis of nanoparticles: an emerging phytotechnology. In: Singh R, Kumar S (eds) Green technologies and environmental sustainability. Springer International Publishing, Cham, pp 339–363

Balazs AC, Emrick T, Russell TP (2006) Nanoparticle polymer composites: where two small worlds meet. Science 314:1107–1110. https://doi.org/10.1126/science.1130557

Bankar A, Joshi B, Kumar AR, Zinjarde S (2010) Banana peel extract mediated novel route for the synthesis of silver nanoparticles. Colloids Surf A: Physicochemical Eng Aspects 368:58–63. https://doi.org/10.1016/j.colsurfa.2010.07.024

Begum NA, Mondal S, Basu S et al (2009) Biogenic synthesis of Au and Ag nanoparticles using aqueous solutions of Black Tea leaf extracts. Colloids Surf B: Biointerfaces 71:113–118. https://doi.org/10.1016/j.colsurfb.2009.01.012

Castro L, Blázquez ML, Muñoz JA et al (2011) Biosynthesis of gold nanowires using sugar beet pulp. Process Biochem 46:1076–1082. https://doi.org/10.1016/j.procbio.2011.01.025

Chandran SP, Chaudhary M, Pasricha R et al (2006) Synthesis of Gold nanotriangles and Silver nanoparticles using Aloevera plant extract. Biotechnol Prog 22:577–583. https://doi.org/10.1021/bp0501423

Coyle S, Wu Y, Lau K-T et al (2007) Smart nanotextiles: a review of materials and applications. MRS Bull 32:434–442. https://doi.org/10.1557/mrs2007.67

Dong BH, Hinestroza JP (2009) Metal nanoparticles on natural cellulose fibers: electrostatic assembly and In Situ synthesis. ACS Appl Mater Interfaces 1:797–803. https://doi.org/10.1021/am800225j

Dubey M, Bhadauria S, Kushwah BS (2009) Green synthesis of nanosilver particles from extract of *Eucalyptus hybrida* (safeda) leaf, 8

Gao M, Li L, Song Y (2017) Inkjet printing wearable electronic devices. J Mater Chem C 5:2971–2993. https://doi.org/10.1039/C7TC00038C

Gardea-Torresdey JL, Gomez E, Peralta-Videa JR et al (2003) Alfalfa sprouts: a natural source for the synthesis of Silver nanoparticles. Langmuir 19:1357–1361. https://doi.org/10.1021/la020835i

Gardea-Torresdey JL, Parsons JG, Gomez E et al (2002) Formation and growth of Au nanoparticles inside live alfalfa plants. Nano Lett 2:397–401. https://doi.org/10.1021/nl015673+

Ghodake GS, Deshpande NG, Lee YP, Jin ES (2010) Pear fruit extract-assisted room-temperature biosynthesis of gold nanoplates. Colloids Surf B: Biointerfaces 75:584–589. https://doi.org/10.1016/j.colsurfb.2009.09.040

Gnanajobitha G, Paulkumar K, Vanaja M et al (2013) Fruit-mediated synthesis of silver nanoparticles using *Vitis vinifera* and evaluation of their antimicrobial efficacy. J Nanostruct Chem 3:67. https://doi.org/10.1186/2193-8865-3-67

Haick H (2007) Chemical sensors based on molecularly modified metallic nanoparticles. J Phys D Appl Phys 40:7173. https://doi.org/10.1088/0022-3727/40/23/S01

Hanemann T, Szabó DV, Hanemann T, Szabó DV (2010) Polymer-nanoparticle composites: from synthesis to modern applications. Materials 3:3468–3517. https://doi.org/10.3390/ma3063468

Haverkamp RG, Marshall AT (2009) The mechanism of metal nanoparticle formation in plants: limits on accumulation. J Nanopart Res 11:1453–1463. https://doi.org/10.1007/s11051-008-9533-6

Iravani S (2011) Green synthesis of metal nanoparticles using plants. Green Chem 13:2638–2650. https://doi.org/10.1039/C1GC15386B

Jana J, Ganguly M, Pal T (2016) Enlightening surface plasmon resonance effect of metal nanoparticles for practical spectroscopic application. RSC Advances 6:86174–86211. https://doi.org/10.1039/C6RA14173K

Kamyshny A (2011) Metal-based inkjet inks for printed electronics. Open Appl Phys J 4:19–36. https://doi.org/10.2174/1874183501104010019

Krasteva N, Besnard I, Guse B et al (2002) Self-assembled Gold nanoparticle/dendrimer composite films for vapor sensing applications. Nano Lett 2:551–555. https://doi.org/10.1021/nl020242s

Lee H-H, Chou K-S, Huang K-C (2005) Inkjet printing of nanosized silver colloids. Nanotechnology 16:2436. https://doi.org/10.1088/0957-4484/16/10/074

Lee J, Kwon H, Seo J et al (2015a) Conductive fiber-based ultrasensitive textile pressure sensor for wearable electronics. Adv Mater 27:2433–2439

Lee S, Shin S, Lee S et al (2015b) Ag nanowire reinforced highly stretchable conductive fibers for wearable electronics. Adv Func Mater 25:3114–3121. https://doi.org/10.1002/adfm.201500628

Mariselvam R, Ranjitsingh AJA, Usha Raja Nanthini A et al (2014) Green synthesis of silver nanoparticles from the extract of the inflorescence of Cocos nucifera (Family: Arecaceae) for enhanced antibacterial activity. Spectrochim Acta Part A Mol Biomol Spectrosc 129:537–541. https://doi.org/10.1016/j.saa.2014.03.066

Mondal S, Roy N, Laskar RA et al (2011) Biogenic synthesis of Ag, Au and bimetallic Au/Ag alloy nanoparticles using aqueous extract of mahogany (Swietenia mahogani JACQ.) leaves. Colloids Surf, B 82:497–504. https://doi.org/10.1016/j.colsurfb.2010.10.007

Moscicki A, Felba J, Sobierajski T et al (2005) Electrically Conductive Formulations Filled Nano Size Silver Filler for Ink-Jet Technology. In: Polytronic 2005—5th International Conference on Polymers and Adhesives in Microelectronics and Photonics. pp 40–44

MubarakAli D, Thajuddin N, Jeganathan K, Gunasekaran M (2011) Plant extract mediated synthesis of silver and gold nanoparticles and its antibacterial activity against clinically isolated pathogens. Colloids Surf, B 85:360–365. https://doi.org/10.1016/j.colsurfb.2011.03.009

Mude N, Ingle A, Gade A, Rai M (2009) Synthesis of Silver nanoparticles using Callus extract of Carica papaya—a first report. J Plant Biochem Biotechnol 18:83–86. https://doi.org/10.1007/BF03263300

Nagajyothi PC, Sreekanth TVM (2015) Green synthesis of metallic and metal oxide nanoparticles and their antibacterial activities. In: Basiuk VA, Basiuk EV (eds) Green processes for nanotechnology: from inorganic to bioinspired nanomaterials. Springer International Publishing, Cham, pp 99–117

Nakkala JR, Mata R, Gupta AK, Sadras SR (2014) Biological activities of green silver nanoparticles synthesized with *Acorous calamus* rhizome extract. Eur J Med Chem 85:784–794. https://doi.org/10.1016/j.ejmech.2014.08.024

Narayanan KB, Sakthivel N (2008) Coriander leaf mediated biosynthesis of gold nanoparticles. Mater Lett 62:4588–4590. https://doi.org/10.1016/j.matlet.2008.08.044

Noguez C (2007) Surface plasmons on metal nanoparticles: the influence of shape and physical environment. J Phys Chem C 111:3806–3819. https://doi.org/10.1021/jp066539m

Palza H, Palza H (2015) Antimicrobial polymers with metal nanoparticles. Int J Mol Sci 16:2099–2116. https://doi.org/10.3390/ijms16012099

Parveen K, Banse V, Ledwani L (2016) Green synthesis of nanoparticles: their advantages and disadvantages. AIP Conf Proc 1724:020048. https://doi.org/10.1063/1.4945168

Philip D, Unni C (2011) Extracellular biosynthesis of gold and silver nanoparticles using Krishna tulsi (*Ocimum sanctum*) leaf. Physica E: Low-dimensional Syst Nanostruct 43:1318–1322. https://doi.org/10.1016/j.physe.2010.10.006

Prathna TC, Chandrasekaran N, Raichur AM, Mukherjee A (2011) Biomimetic synthesis of silver nanoparticles by *Citrus limon* (lemon) aqueous extract and theoretical prediction of particle size. Colloids Surf B: Biointerfaces 82:152–159. https://doi.org/10.1016/j.colsurfb.2010.08.036

Raghunandan D, Basavaraja S, Mahesh B et al (2009) Biosynthesis of stable polyshaped Gold nanoparticles from microwave-exposed aqueous extracellular anti-malignant Guava (*Psidium guajava*) leaf extract. Nanobiotechnol 5:34–41. https://doi.org/10.1007/s12030-009-9030-8

Raveendran P, Fu J, Wallen SL (2003) Completely "Green" synthesis and stabilization of metal nanoparticles. J Am Chem Soc 125:13940–13941. https://doi.org/10.1021/ja029267j

Ravindra S, Murali Mohan Y, Narayana Reddy N, Mohana Raju K (2010) Fabrication of antibacterial cotton fibres loaded with silver nanoparticles via "Green Approach". Colloids Surf A: Physicochem Eng Aspects 367:31–40. https://doi.org/10.1016/j.colsurfa.2010.06.013

Santhoshkumar T, Rahuman AA, Rajakumar G et al (2011) Synthesis of silver nanoparticles using *Nelumbo nucifera* leaf extract and its larvicidal activity against malaria and filariasis vectors. Parasitol Res 108:693–702. https://doi.org/10.1007/s00436-010-2115-4

Sathishkumar M, Sneha K, Won SW et al (2009) Cinnamon zeylanicum bark extract and powder mediated green synthesis of nano-crystalline silver particles and its bactericidal activity. Colloids Surf B: Biointerfaces 73:332–338. https://doi.org/10.1016/j.colsurfb.2009.06.005

Segev-Bar M, Haick H (2013) Flexible sensors based on nanoparticles. ACS Nano 7:8366–8378. https://doi.org/10.1021/nn402728g

Shankar SS, Rai A, Ahmad A, Sastry M (2004a) Rapid synthesis of Au, Ag, and bimetallic Au core–Ag shell nanoparticles using Neem (*Azadirachta indica*) leaf broth. J Colloid Interface Sci 275:496–502. https://doi.org/10.1016/j.jcis.2004.03.003

Shankar SS, Rai A, Ankamwar B et al (2004b) Biological synthesis of triangular gold nanoprisms. Nat Mater 3:482–488. https://doi.org/10.1038/nmat1152

Sheny DS, Mathew J, Philip D (2011) Phytosynthesis of Au, Ag and Au–Ag bimetallic nanoparticles using aqueous extract and dried leaf of *Anacardium occidentale*. Spectrochimica Acta Part A: Mol Biomol Spectrosc 79:254–262. https://doi.org/10.1016/j.saa.2011.02.051

Shipway AN, Katz E, Willner I (2000) Nanoparticle arrays on surfaces for electronic, optical, and sensor applications. ChemPhysChem 1:18–52. https://doi.org/10.1002/1439-7641(20000804)1:1%3c18:AID-CPHC18%3e3.0.CO;2-L

Song J, Birbach NL, Hinestroza JP (2012) Deposition of silver nanoparticles on cellulosic fibers via stabilization of carboxymethyl groups. Cellulose 19:411–424. https://doi.org/10.1007/s10570-011-9647-3

Stoppa M, Chiolerio A (2014) Wearable electronics and smart textiles: a critical review. Sensors 14:11957–11992

Vijayaraghavan K, Nalini SPK, Prakash NU, Madhankumar D (2012) One step green synthesis of silver nano/microparticles using extracts of *Trachyspermum ammi* and *Papaver somniferum*. Colloids Surf B: Biointerfaces 94:114–117. https://doi.org/10.1016/j.colsurfb.2012.01.026

Yetisen AK, Qu H, Manbachi A et al (2016) Nanotechnology in textiles. ACS Nano 10:3042–3068. https://doi.org/10.1021/acsnano.5b08176

Yin J, Hu P, Luo J et al (2011) Molecularly mediated thin film assembly of nanoparticles on flexible devices: electrical conductivity versus device strains in different gas/vapor environment. ACS Nano 5:6516–6526. https://doi.org/10.1021/nn201858c

Zeng W, Shu L, Li Q et al (2014) Fiber-based wearable electronics: a review of materials, fabrication, devices, and applications. Adv Mater 26:5310–5336

Green Nanofillers for Polymeric Materials

T. P. Mohan and K. Kanny

Abstract Filler materials are commonly added to the polymer in the form of particles for various purposes, such as improving properties, to control processing and for economic benefits of the resulting polymer composite material. During the past five decades, much attention was given on synthetic fillers, such as $CaCO_3$, TiO_2, Al_2O_3, CNT and SiC. However, due to the concerns of environmental and depleting natural resources, researchers were started considering alternative fillers to the polymeric materials. They are moving towards green fillers and much advancement occurred in the past two decades. In this report, we focus on the review on the outline of various types of green nanofillers and their effect on the polymeric system. Green nanofillers are categorized based on their source, namely animal, plant and natural source. The latest progress on green nanofillers was also discussed. Their processing methods were discussed, and their effect on thermo-mechanical and physical properties was evaluated. The latest application of these green fillers was also studied.

Keyword Green nanofillers · Shells · Cellulose · Biomass · Mechanical properties · Thermal properties

1 Introduction

1.1 Composites

A composite is a material made up of two or more material with different properties to make up a material with better properties than each of the constituents. A composite consists of a matrix and fibres. Composites can be categorized into three categories, namely the polymer matrix, ceramics matrix and metal matrix.

T. P. Mohan (✉) · K. Kanny
Composites Research Group, Department of Mechanical Engineering, Durban University of Technology, Durban, South Africa
e-mail: mohanp@dut.ac.za

S. Ahmed and W. Ali (eds.), *Green Nanomaterials*,
Advanced Structured Materials 126,
https://doi.org/10.1007/978-981-15-3560-4_5

1.2 Nanocomposites

Nanocomposites have one or more phases which have dimensions or structures of less than 100 nanometres (nm), primarily the reinforcement or matrix material. The reason for nanocomposites is to design and create new materials with better flexibility and to improve their physical and mechanical properties (Kamigaito 1991). The change in strength and mechanical properties in composites depends highly on the interaction of the nanoparticle and polymer (Visakh et al. 2016; Gusev 2007).

1.3 Nanofillers

Nanofillers are particles at a nanoscale added to a composite to either to lower the consumption of a more expensive binder or to enhance the properties of a composite material. Nanoparticles incorporated in fillers have created reinforced materials and new materials with novel properties, such as stronger mechanics, enhanced optical properties and improved conductivity at an atomic level.

There are three dimensions in the nano range—the first being nanoplates, the second being nanofibres and the third as nanoparticles.

Nanofillers can reduce both interfacial tension and interfacial slip at strong filler–polymer interaction, increasing the modulus. Poor adhesion between the filler and polymer can form strongly bonded nanoparticle aggregates or may cause early failure at the interface, resulting in a change of physical properties in the composite (Bedanokov AJ 2011; Brody 2003; Šupová et al. 2011).

2 Types of Green Fillers

Green fillers are organic materials that can be added to composites to reduce the cost by decreasing the amount of binding materials or to make improvements on some of the mechanical properties.

2.1 Animal Source

These sources are usually by-products or waste products of industries related to the production of animal products.

In this section, the materials under consideration are:

- Eggshells
- Seashells
- Silkworms.

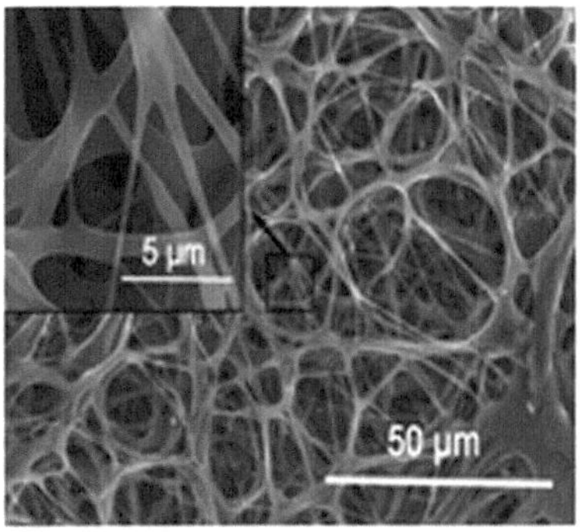

Fig. 1 Eggshell structure at a nanolevel (Vincent and Wegst 2004)

2.1.1 Eggshells

An eggshell has an outer hard shell with 94% calcium carbonate crystals. Eggshells have been used as filler for polymer composites and presented as an alternative to expensive activated carbon as a low-cost absorbent (Bootklad and Kaewtatip 2013). Figure 1 is a microscopic image of an eggshell structure at a nanoscale.

2.1.2 Seashells

Crustaceans are animals such as crabs, lobsters, shrimp and many more. Chitin is a key integrant and can be sourced by these animals which are used as filler for polymer composites. Figure 2 shows the extraction of nanocrab shell fibres into chitin fibres (Wijesena et al. 2012).

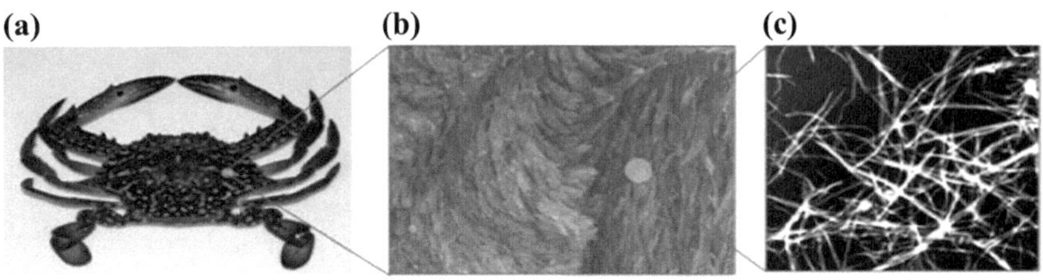

Fig. 2 Extraction of nanocrab shell fibres into chitin fibres—**a** crab shell at a large scale, **b** crab shell at microscale, **c** crab shell fibres at nanoscale (Wijesena et al. 2012)

2.1.3 Silkworms

Nature's strongest fibres are produced from the silk in spiders and silkworms. The silk fibres from spiders and silkworms can prove to be effective in reinforcing bio-composites (Khanam et al. 2015). When you have two silkworm cocoons spun close together, the thread is tougher and rougher than normal silk fibres (Li et al. 2014). Figure 3 shows a comparison of silk fibres from silkworms and spiders (Hardy et al. 2008).

2.2 *Plant Source*

Plant source fillers are fillers that are plant based, meaning they have been extracted from plants and chemically treated to improve the properties in composite structures. In this research, we will be looking at four plant source fillers naming which are:

- Plant fibres
- Bagasse
- Wood
- Starch.

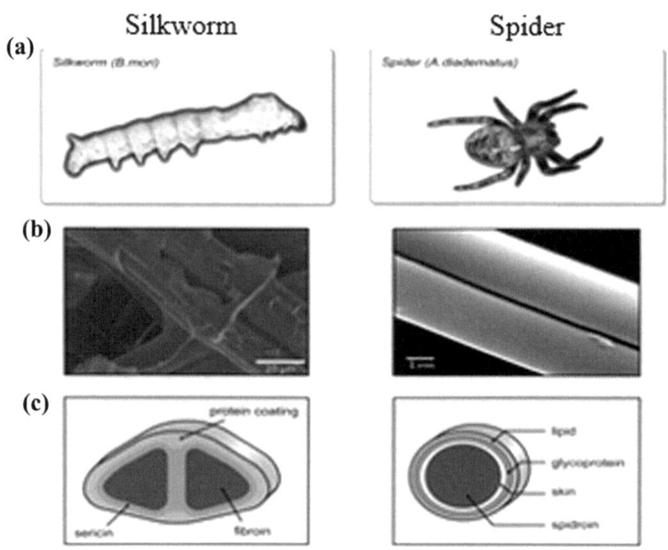

Fig. 3 Nanofibres extraction from silkworms and spiders—**a** species, **b** electron micrograph of silk fibres, **c** top view of silk fibres (Hardy et al. 2008)

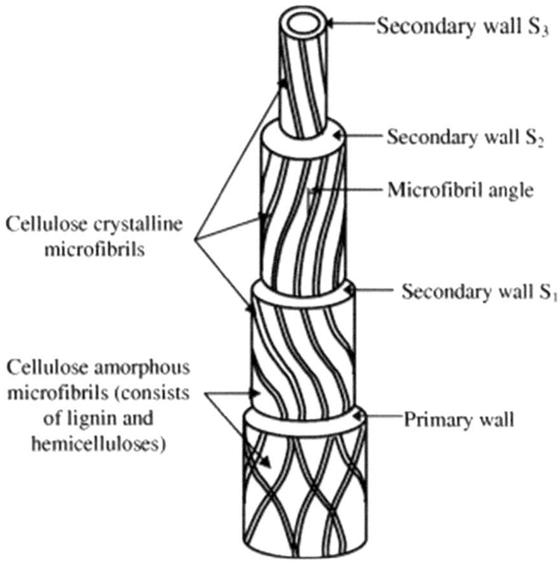

Fig. 4 Compositions structure of a plant fibre (Wi et al. 2018)

2.2.1 Plant Fibres

Plant fibre fillers are fibre based, meaning they have been extracted from fibres and oils and have been chemically treated to improve the properties in composite structures. A plant fibre can be broken down into different constituents such as cellulose, lignin, lignocellulose, nanocellulose and palm oils/wax. In this research, we will be looking at three plant source fillers naming which are:

- Lignin
- Nanocellulose
- Palm oil.

Lignin

Lignin is a shapeless, hyperbranched macromolecule that forms essential materials in the supporting tissues of plants and some algae. Lignin is made up of three different types of phenylpropane units that arrange a stiff, three-dimensional network. Lignocellulose is mainly composed of cellulose, hemicellulose and lignin as well as modicum of pectins, polar compounds and waxes (Wi et al. 2018). The lignocellulosic fibres are used as either as reinforcing or filling materials, with the incorporation of polymeric materials. Figure 4 shows the microstructure of the compositions of plant fibre.

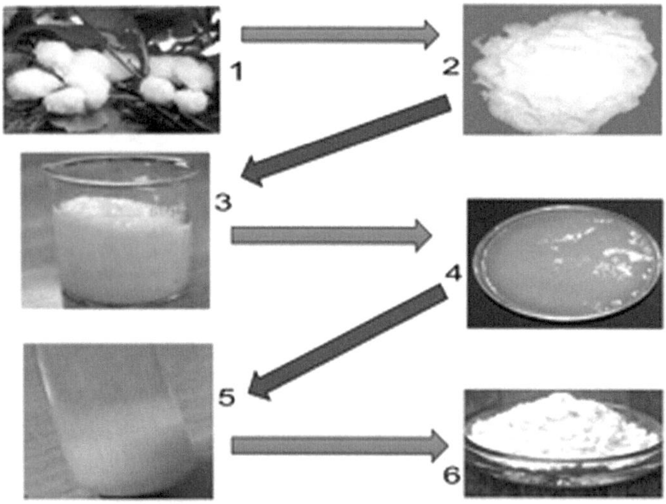

Fig. 5 Synthesis of nanocellulose. (1) Raw cotton; (2) cotton powder; (3) after alkali treatment; (4) after acid hydrolysis; (5) nanocellulose; and (6) nanocellulose powder (Theivasanthi et al. 2018)

Nanocellulose

Cellulose nanofibres are extracted from wheat straw by a chemical treatment, resulting to purified cellulose. To accomplish this, the first step is alkylation treatment of the fibres and what this does is to remove the lignin and hemicellulose in the fibres and is very crucial on the form of your product. The fibres must then be bleached to be as white as they can be for the final step of acid hydrolysis, and the remaining component is nanocellulose. Figure 5 shows the process explained above in picture form of each phase of the process.

Palm Oil

Palm oil fuel ash (POFA) is a by-product formed from combusting leftover materials such as palm kernel shell, palm oil fibre and oil husk. Nanopalm oil fuel ash mainly consists of SiO_2 and that there was an increase in compressive strength of concrete with incorporation of POFA (Henry 1991). The palm oil fuel ash is used as a partial replacement for cement in concrete production, and the oil palm fibres are also incorporated as an additive to reduce thermal conductivity of wall systems (Cho and Park 2011). Figure 6 shows the processing of POFA.

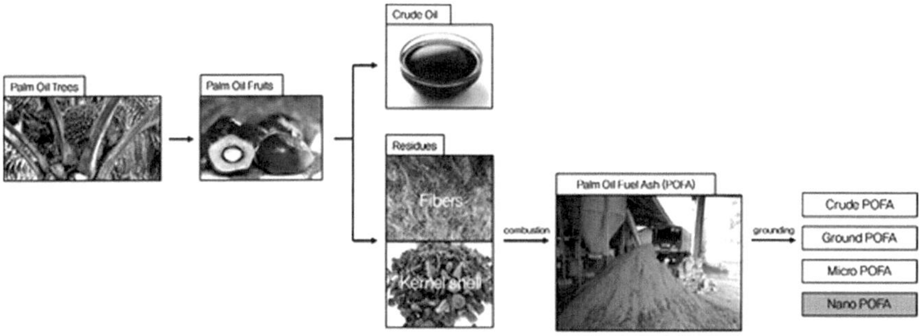

Fig. 6 Palm oil ash (POFA) production (Wi et al. 2018)

2.2.2 Bagasse

Bagasse is a fibrous substance that is a by-product after the extraction of juice from sugar cane. It has been used as filler for cement and polymer materials.

Reduction of Microfibre to Nanobagasse

The reduction in microfibre to nanobagasse was carried out by using a high-energy ball mill made from tungsten carbide. The dried sample of microbagasse fibre is milled in high-energy ball milling at 300 rpm for 5 h to prepare nano-sized bagasse (Cho and Park 2011).

2.2.3 Wood

The usage of natural wood fibre extracts (wood flour) as filler in polymer composites in the production of wood–plastic composites was a major interest to numerous researchers. Overall wood flour is used as filler for plastic, which tends to enhance composite hardness.

Extraction of Nanofibres from Wood

In this study, we show the extraction of nanofibres from wood by de-acidifying wood using a dispersion in 2-propanol of calcium hydroxide nanoparticles. Alkaline nanoparticles are converted into calcium sulphate without mechanical stress to the wood's lumens. Additional applications produce an excess of alkaline nanoparticles that are converted into carbonate (Dana 2012).

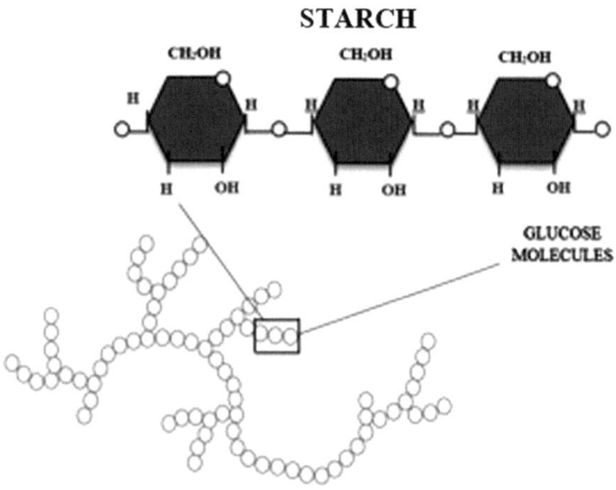

Fig. 7 Chemical structure of starch (Starch Europe 2019)

2.2.4 Starch

Starch also known as amylum is a polymeric carbohydrate which has many glucose units which are joined by glycosidic bonds (Kaisangsri et al. 2012). Starch is mainly produced by majority of green plants as energy. Figure 7 shows the chemical structure of starch as well as glucose molecules bonded together (Saji et al. 2015).

2.2.5 *Moringa oleifera*

Moringa oleifera is part of the moringaceae plant family, and it is the only genus plant. It is also called the drumstick or the miracle tree. It is a rapid growing plant and is resistive to drought, and it grows in the cultivated tropical and subtropical regions.

2.3 *Natural Source*

Natural sources are any source found in the environment which includes plant sources and inorganic materials.

The following sources are discussed:

- Carbon/graphite
- Natural clay
- Biomass
- Waste organic/inorganic.

2.3.1 Carbon/Graphite

Many forms of carbon such as graphite, amorphous carbon and diamond and carbon nanotubes are used as nanofillers (Saji et al. 2015).

The molecular structure of graphite is stacked. These stacked sheets slide over each other creating a greasy texture used for lubrication. Sheets of graphene are shaped into the cylinder with combined edge forming carbon nanotubes. Interchanging graphene layers, bonded by weak van der Waals forces, increase interplanar spacing and with rapid thermal treatment the molecules disintegrate, hence extending the graphite (Saji et al. 2015).

2.3.2 Natural Clay

Natural clay improves plasticity when combined with a specific quantity of water. Clay substances are organized into five classes: smectite; illite; sepiolite; chlorite; and kaolinite. These minerals are widely available, cheaper than most fillers and add benefit to polymeric matrices (Singh and Swami 2015).

Halloysite nanotubes (HNTs) have a hollow tubular structure and are chemically structured like kaolin. HNTs as filler for polymers, such as polyvinyl alcohol, polypropylene and epoxy, greatly improve thermal and mechanical properties of composites. The huge diameter of the halloysite lumen can lodge some molecules which offer specific properties, like high degradation temperature, etc. (Guozhu et al. 2015).

2.3.3 Biomass

Biomass is a general term for acquiring burning wood and other organic matter (Leceta et al. 2014). Composites, made of natural, renewable filler and a polymer matrix, have lightweight, renewable characteristics and mechanical properties (Sue et al. 1997). These raw materials could also solve the problem of increasing fossil fuel prices and environmental issues.

The strategies for introducing biomass-based nano-bioplastics to address the dependence of non-degradable plastics on petroleum and environmental pollution are currently being evaluated. Corn leaf biomass can be used to create biodegradable nano-bioplastics for biomedical and medical and various other industrial applications (Giorgi et al. 2005).

2.3.4 Waste Organic/Inorganic

Waste materials such as nano-oil palm fibres, from empty fruit peels filler-based epoxy nanocomposites, should be used for polymer composites for various uses.

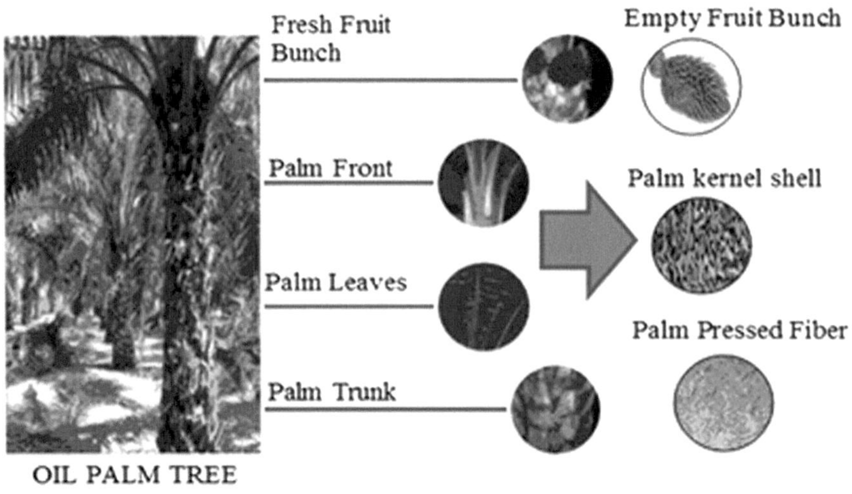

Fig. 8 Various types of oil palm waste in Malaysia (Ng et al. 2015)

Epoxy resins exhibit restrictions due to brittleness, low fracture strength, low-dimensional stability and poor thermal stability (Ng et al. 2015). Waste material used as nanofiller has benefits in comparison to artificial fibre composites, including low cost and an eco-friendly nature (Ng et al. 2015). Figure 8 shows the various types of palm oil waste that is used as nanofillers.

3 Mechanical Properties

3.1 Animal Source

3.1.1 Eggshells

Investigation of the mechanical properties of polymeric composites with eggshell powder is used as filler (Asha and Chandra Sekhar 2014).

The study was based on two polymeric composites polyimide and five specimens of carbon black. The specimens are filled with 5, 10, 15, 20 and 25 weight per cent of eggshell powder with an average size of 100 μm. The study consists of three tests, namely the tensile test, flexural test and impact test. The result of the investigation is shown in Figs. 9, 10 and 11.

Tensile Test

See Fig. 9.

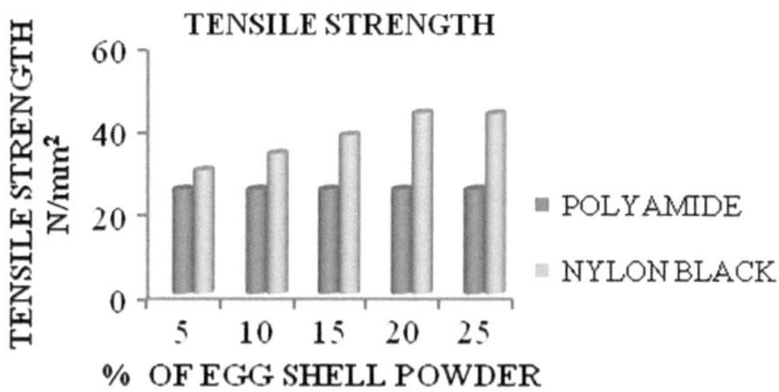

Fig. 9 Change in tensile strength in polyamide and nylon black when filled with eggshell powder (Hiremath et al. 2018)

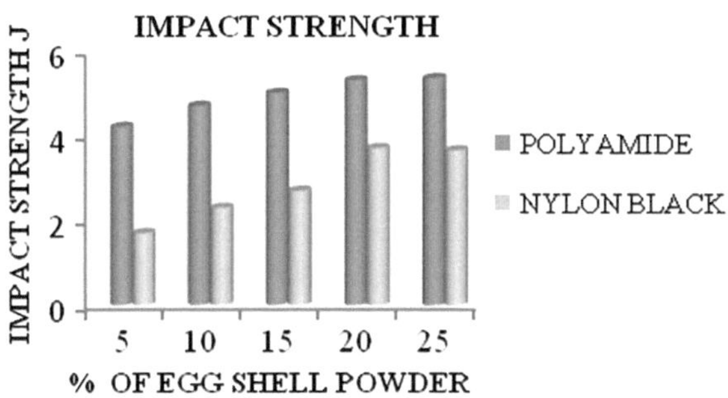

Fig. 10 Change in impact strength in polyamide and nylon black when filled with eggshell powder (Hiremath et al. 2018)

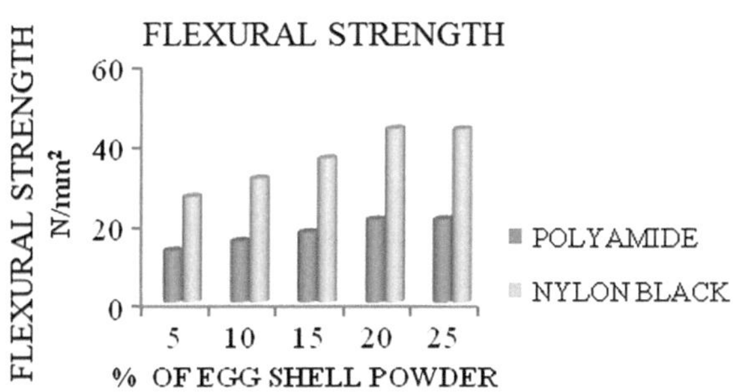

Fig. 11 Change in flexural strength in polyamide and nylon black when filled with eggshell powder (Hiremath et al. 2018)

Impact Test

See Fig. 10.

Flexural Test

See Fig. 11.

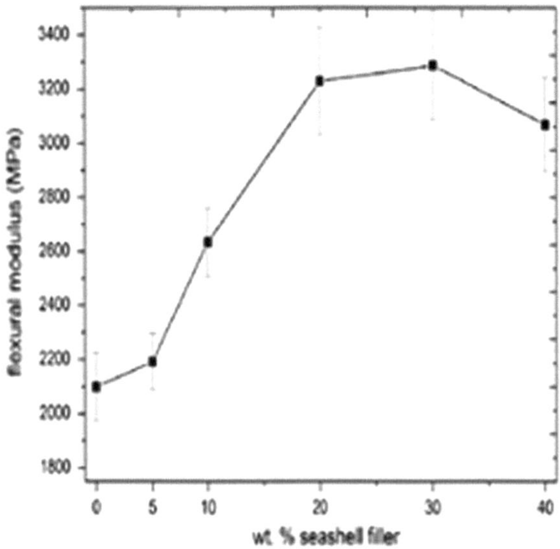

Fig. 12 Variation of the flexural modulus of the epoxy seashell powder (Fombuena et al. 2014)

From all the graphs, we can observe that tensile, impact and the flexural strength of the polyamide and nylon black composite increase in relation to the eggshell powder added to the composites. The change varies from 20 to 25%.

3.1.2 Seashells

Figure 12 illustrates the bending elastic modulus obtained from bending epoxy composites by varying the loads of biomass on marine cancers from 5 to 40% by weight.

The flexural modulus increases until a maximum filler content of 30% by weight is reached. Modulus of elasticity (in bending) for the unfilled epoxy resin is nearly 2.2 GPa, and the value of epoxy resin will continue to rise to a value nearly 3.3 GPa, for 30% by weight of waste sinks, that a percentage increment of 56% for blank material. It is distinguished that the relatively low filler content (10% by weight) can cause a significant increase of (25%) in bending.

3.1.3 Crab Shells

Young's modulus and tensile strength of the chitin film based on the nanometre scalar fibre unit (100 wt%) are 7.2 GPa and 137, 8 MPa, respectively. The chitin nanofibres incorporated in PMMA effectively improve the mechanical properties of PMMA. Increasing the fibre content from 40 wt% to 70 wt% will cause Young's modulus and

Fig. 13 Tensile stress–strain from cocoon (Zhao et al. 2005)

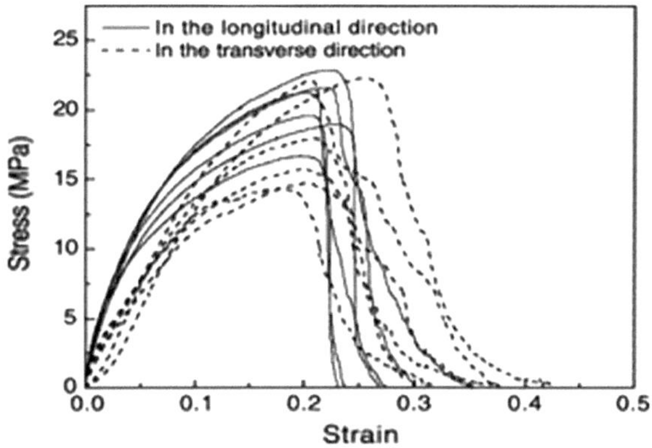

tensile strength of the corresponding nanocomposites to increase from 3.14 GPa–4.46 GPa and 77.5 MPa–95.8 MPa, while PMMA is at 1.5 GPa and 41.8 MPa, respectively. The addition of chitin nanofibres will improve the mechanical properties of the composite at a significant extent. This characterization will enable composites to further utilize high-strength nanopaper, like plastic currency, comprising of improved tear resistance and printability (Felix and Gatenholm 1991).

3.1.4 Silkworms

First, a quasi-static tensile test is carried out on a rectangular sample which is removed from a normal compact ridge along the transverse and longitudinal directions. Figure 13 the graph showing the tensile stress–strain curves from the same cocoon.

Elastic deformation exhibits significant nonlinearity, mostly at a region where high strains take place.

The tensile strength and Young's modulus of a cocoon are higher in longitudinal direction than the tensile strength and Young's modulus in transverse direction. For a cocoon composed of steric-bonded monofilament filaments, the anisotropic elastic properties and strength are mainly determined by the microstructure and silk orientation (Zhao et al. 2005; Zhao et al. 2005).

3.2 *Plant Source*

3.2.1 Lignocellulose

An increase in the filler loading results in an increasing in interphase region, which causes the deterioration of the interphase bonding between hydrophilic filler and hydrophobic matrix (Yang et al. 2006).

(a)

(b)

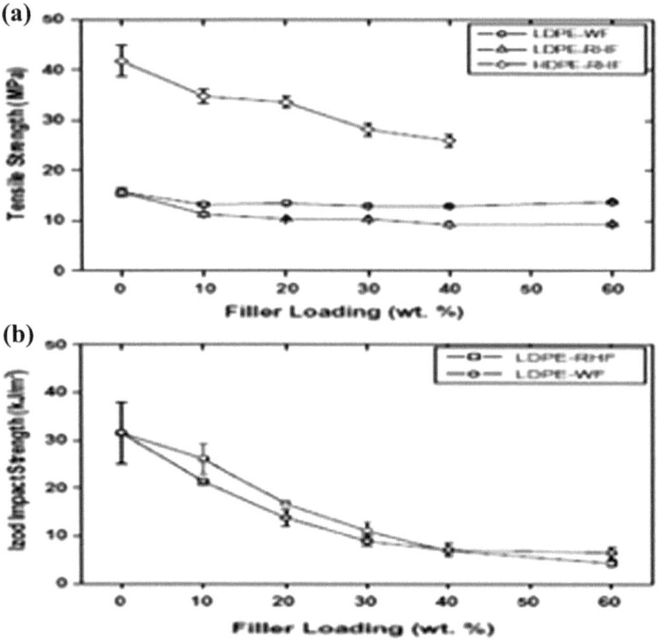

Fig. 14 **a** Tensile strengths of the lignocellulose filler–polyethylene biocomposites at crosshead speed of 100 mm/min at room temperature, **b** notched Izod impact strengths of LDPE-WF and LDPE-RHF biocomposites at different filler loadings (Yang et al. 2006)

The filler compositions exhibit lower tensile strength compared to PP-WF composites. Tensile strength of composites materials made up of lignocellulosic filler (RHF and WF), low-density and high-density polyethylene (LDPE and HDPE) at different filler loads are displayed in Fig. 14a. LDPE-RHF compositions also showed gradually decreased tensile strength than LDPE-WF composites.

Figure 14b shows the Izod impact strength of LDPE-WF and LDPE-RHF composites at different filler loads. Reduced shock forces decreased with an increase in filler load, and this is the results of increasing the size of the poorly bonded region between hydrophobic matrix polymer and hydrophilic filler (Yang et al. 2006).

3.2.2 Lignin

Mechanical Properties of Silt Treated with Lignin

To check for the strength and stiffness of soil treated, an unconfined compression test was done, and a resilient modulus test was conducted as per (ASTM (208b)) D4219 (Zhang et al. 2015) and the standard soil test method (GB/T 50123-1999), which is illustrated in Fig. 15. Cylindrical specimens were compacted in the sp2 model to measure the resilient modulus.

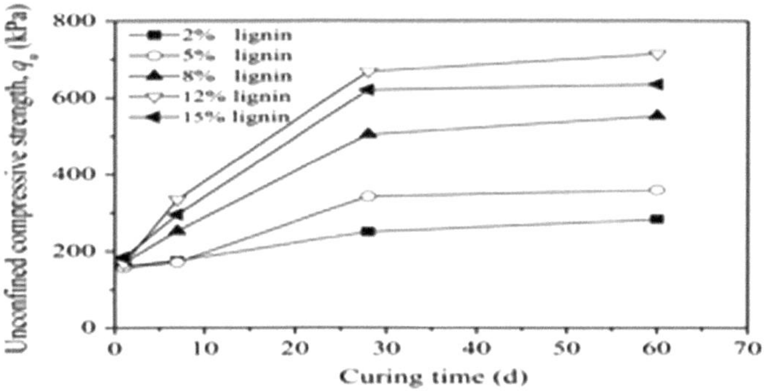

Fig. 15 Compressive strength against the curing time of the silt (Zhang et al. 2015)

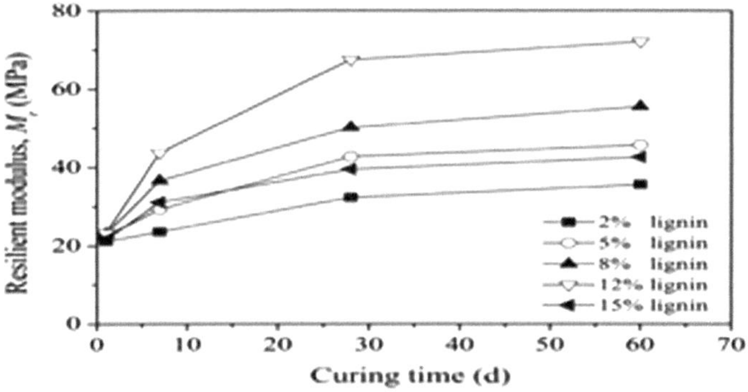

Fig. 16 Resilient modulus against the curing time of the silt (Zhang et al. 2015)

From this (Fig. 15), we can observe that the strength increases with the days that the slit took to cure. The strength also increases with the lignin content in the silt. The modulus increases as the number of days that the silt took to cure (Fig. 16). The modulus also increases with the lignin content in the slit.

3.2.3 Bagasse

The mechanical properties of chemically modified bagasse/polypropylene composites were compared in the case of using pre-treatment of bagasse with choline acetate (ChOAc) and untreated bagasse. The tensile strength was found to be 20 MPa only in the case of polypropylene, and the tensile strength of untreated bagasse was found to be 35 Mpa. In contrast, when ChOAc pre-treated bagasse was used, the tensile strength is increased up to 40 Mpa (Mustafa et al. 2013).

3.2.4 Wood

The surface above of the laminate is compressed, the centre part of it undergoes a shift and the lower face experiences stress (Mustafa et al. 2013). The rupture module obtained from the WPC samples in bending load at different levels of DPWF and different filler type have been mapped to each other and to the controls and the result is shown in Fig. 17(1); in the flexural output, the stress was calculated as the stress was matching to 3% strain.

Flexural Modulus (MOE)

When the filler content was increased to 60%, the modulus of elasticity started to bend (Figs. 17(2–4) shows that there are significant differences between plates from leaf and rickets flour, but both differ with a significance level of 1% (Sanchez-Jimenez et al. 2012).

Tensile Strength

The highest tensile strength among the samples was produced using a leaflet, a rachis and sheet flour, associated with samples which were made using leaf flour (Fig. 18).

This result agrees with the previous conclusions. Hanafi et al. showed that an increase in the content of OPWF4 in the ENR5 compound led to decreasing the tensile strength (Zhang et al. 2015).

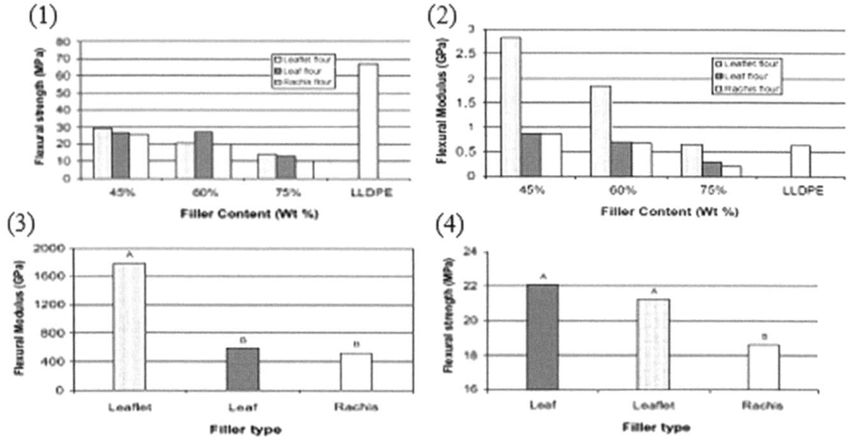

Fig. 17 Flexural modulus and strength of wood filler content in polymer (Polyethylene-based biocomposites and bionanocomposites 2016)

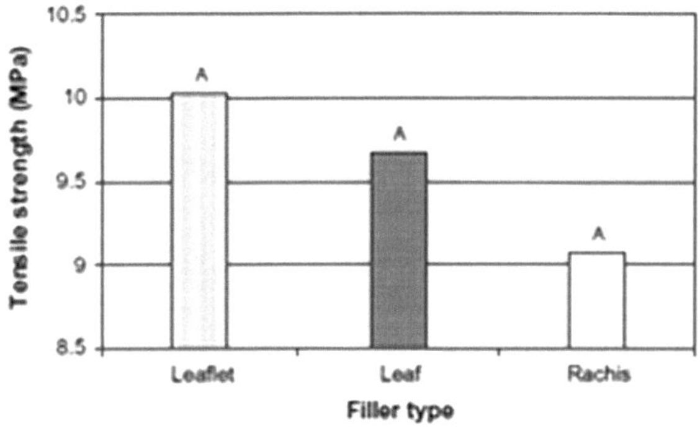

Fig. 18 Tensile strength of wood at different filler types (Polyethylene-based biocomposites and bionanocomposites 2016)

3.2.5 Nanocellulose

Nanocellulose fibres can be less than 100 nm in diameter with a few micrometres in length. Nano cellulose is a biodegradable fibre which is lightweight and has a very low density of about 1.6 g/cm^3 (Kamigaito 1991). It has a high elastic modulus up to 220 GPa which is found to be better than Kevlar fibre.

The crystalline structure of nanocellulose has strongly bonded hydrogen networks which make it tough with high strength and are insoluble in water.

3.2.6 Palm Oil Plant Ash

Figure 19 shows the compressive and flexural strength developed in self-compacting concretes containing POFA for the period of 28 and 180 days. Samples containing high contents of POFA showed lower compressive strength in early days of curing as compared to self-compacting concrete without POFA. The compressive strength of samples older than 28 days higher or comparable than corresponding plain concrete varies due to the replacement percentages. The incorporation of POFA in any amount lowered the flexural strength (Theivasanthi et al. 2018).

3.2.7 Starch

An experiment was done to test the mechanical properties of rubber reinforced EVM (ethylene-co-vinyl acetate) with starch at different particle sizes.

Micro-sized starch (MSt) and nano-sized starch (NSt) with dissimilar loads were tested as illustrated in Fig. 20 to examine the effect of the sizes of starch particles on the mechanical properties of the EVM rubber (Lee et al. 2008).

As shown above, the tensile strength of MSt is much greater than that of NSt as the starch loading is increased with time (Fig. 20a).

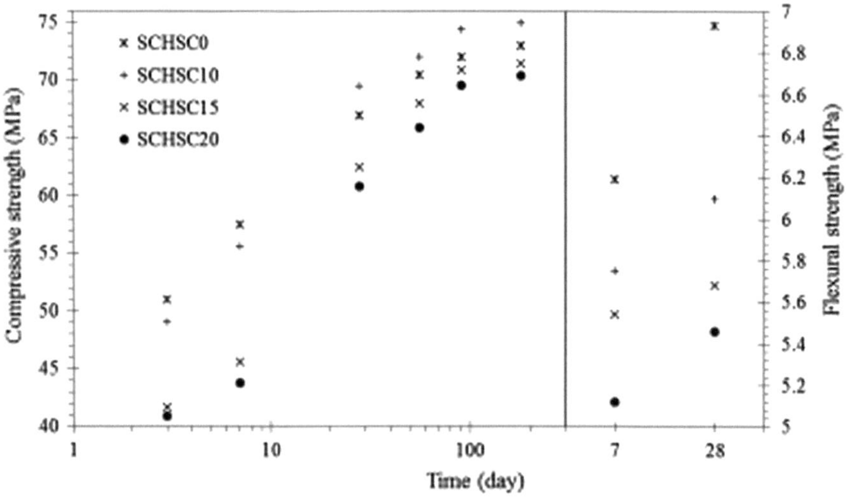

Fig. 19 Compressive and flexural strength development of POFA-based self-compacting concrete (Theivasanthi et al. 2018)

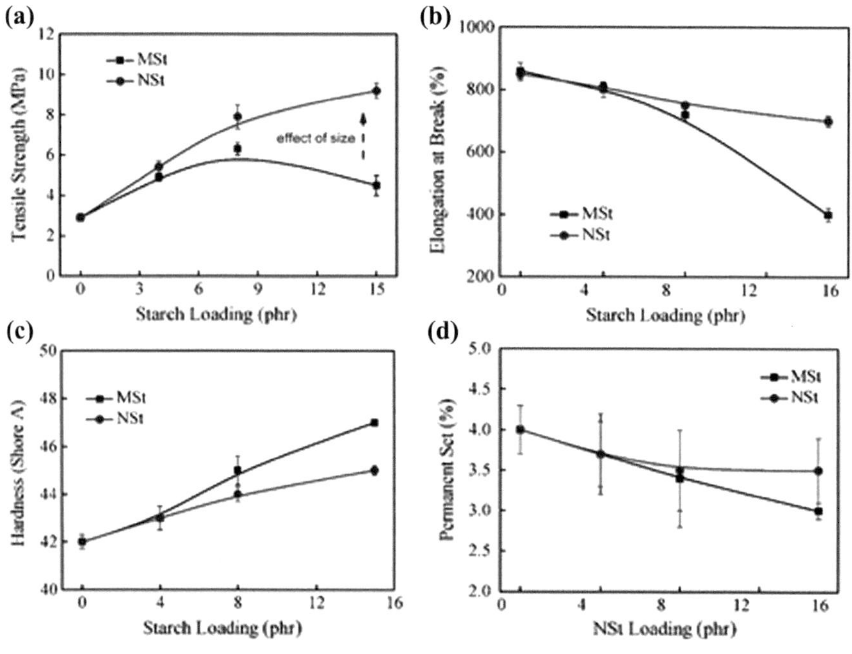

Fig. 20 **a** Tensile strength, **b** elongation at break, **c** hardness and **d** permanent set of EVM/MSt and EVM/NSt composites with various starch loadings (Lee et al. 2008)

Figure 20b shows the elongation graph of NSt significantly lower than that of MSt as the starch loading increases with time.

The hardness of MSt is greater than its counterpart NSt as displayed in Fig. 20c.

Figure 20d shows a close comparison between the MSt and NSt sets respectively as the NSt loading increases with time.

3.3 *Natural Source*

3.3.1 Carbon/Graphite

The location within the stability of the sp2 bonds forming a hexagonal lattice makes the mechanical properties of graphite superior to others.

A study showed that in a freestanding monolayer, graphene's mechanical properties were measured using nano-indentation that determined an intrinsic strength of 42 Nm in the monolayer membrane, amounting to strength of 130 GPa (Bergaya et al. 2013). A force–displacement response obtained from the graphene membranes attained both the elastic properties and breaking stress of graphene.

Due to a high surface activity and oxygen in nanographite, a strong interaction with the polymer matrix forms a good interphase. A layer of polymer, having a higher stiffness than the bulk polymer near the dispersed phase surface, is created from molecular mobility being restricted due to the interactions between phases (Saji et al. 2015).

Figure 21 shows the presence of aggregated and debonded nanographite particulates confirmed the theory that failure which occurs during the deformation of composites is primarily because of coalescence of the voids that forms critical cracks. A rough surface crack indicated that crack propagation is difficult. Filler concentrations give an excellent distribution of nanographite. While observing the polymer matrix, Fig. 22 shows that the nanographite became homogeneous as it dispersed but as the filler loading increased, aggregation took place.

Using atomic force microscopy, the force was measured to push the unanchored ends of a freestanding nanotube out of its equilibrium position. Currently, Young's modulus of a single-walled nanotube is recorded to be approximately 1 Tpa, but a value as high as 1.8 Tpa has been reported. Theoretically, the size and chirality of single-walled nanotubes have influenced its Young's Modulus, ranging from 1.22 to 1.26 Tpa. A value of 1.09 Tpa was found for a generic nanotube. Alternatively, when working with different multi-walled nanotubes, studies show that modulus measurements of multi-walled nanotubes using AFM techniques do not strongly depend on the diameter. Instead, there is an argument that the modulus of the multi-walled nanotubes correlated with the amount of disorder in the nanotube walls. Predictably, the outermost layers tend to break first when multi-walled nanotubes break (Zhang et al. 2015).

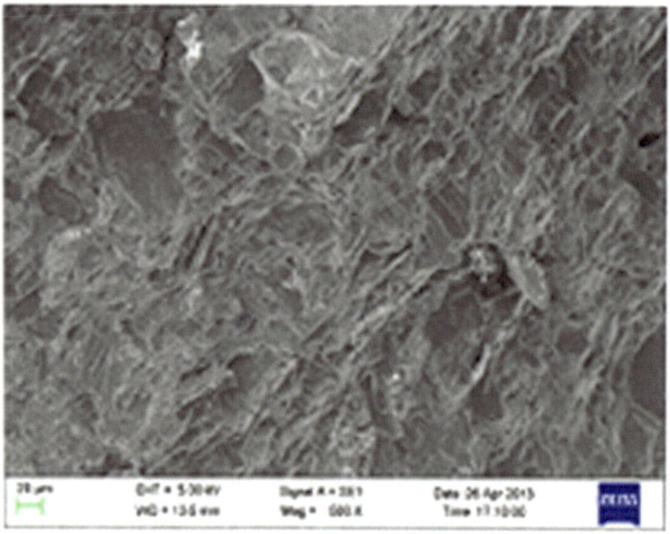

Fig. 21 Fractograph of 8 phr/nanographite loaded filler agglomerates initiate crack during tensile formation (Bergaya et al. 2013)

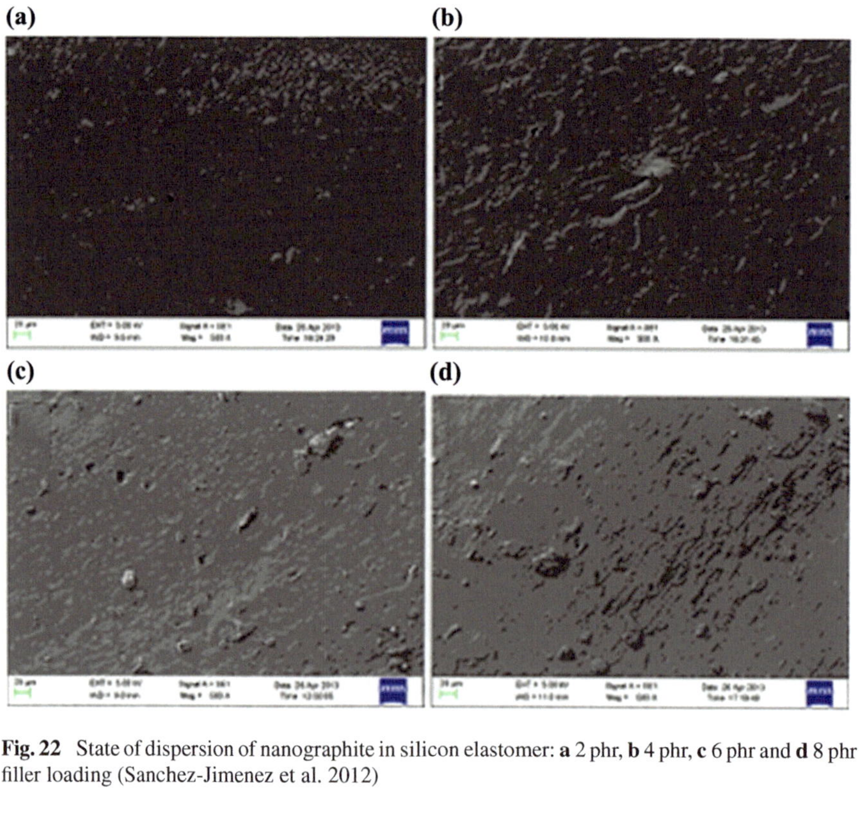

Fig. 22 State of dispersion of nanographite in silicon elastomer: **a** 2 phr, **b** 4 phr, **c** 6 phr and **d** 8 phr filler loading (Sanchez-Jimenez et al. 2012)

3.3.2 Natural Clay

A study was done to establish properties of a nanocomposite-based epoxy with clay loading. Improvements were seen in fracture toughness and a possible reduction in the maximum water uptake of the nanocomposite-based epoxy. Although, if the intercalated and fully exfoliated structures were partly obtained, the size of clay used in the polymer may have improved or reduced some of the properties in the nanocomposite. If nanoclay was introduced into the epoxy matrix, the short beam shear strength of the composites decreased, while the microhardness of the composite increased.

Figure 23 shows the effect of nanocomposite epoxy made with the direct mixing method (DMM) increased toughness and modulus than that of the filler composites, while the nanocomposites showed increase in fracture toughness and a minor decrease in temperature at low clay loading if made with the high-pressure mixing method (HPMM).

Tensile

See Fig. 24.

Fracture

Improvements were seen in fracture toughness and a possible reduction in the maximum water uptake of the nanocomposite-based epoxy shown in Fig. 24. Although, if the intercalated and fully exfoliated structures were partly obtained, the size of clay used in the polymer may have improved or reduced some of the properties in the nanocomposite. If nanoclay was introduced into the epoxy matrix, the short beam shear strength of the composites decreased, while the microhardness of the composite increased.

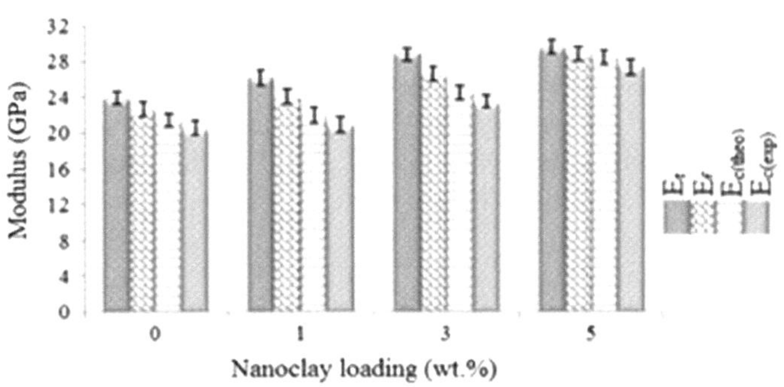

Fig. 23 Modulus versus nanoclay loading (Domun et al. 2015)

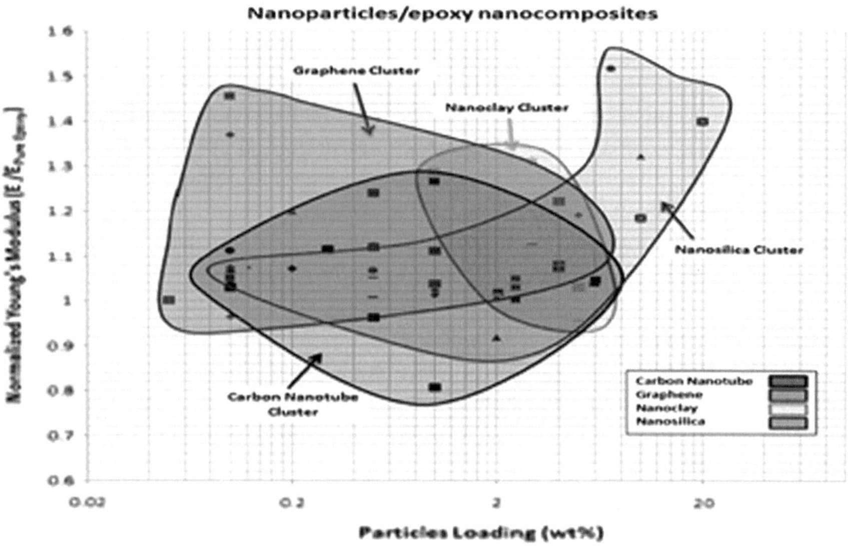

Fig. 24 Nanoparticle of different fillers (Domun et al. 2015)

The nanocomposite epoxy made with the direct mixing method (DMM) increased toughness and modulus than that of the filler composites, while the nanocomposites showed increase in fracture toughness and a minor decrease in temperature at low clay loading if made with the high-pressure mixing method (HPMM).

4 Thermal Properties

4.1 Animal Source

4.1.1 Seashells

Adding calcium carbonate from seashells results in an increase in the glass transition temperature (Tg). All compositions of epoxy seashells displayed as values of Tg above room temperature are shown in Fig. 25. Clearly defined peaks are observed at a temperature of 80.6 °C for unfilled material.

4.1.2 Silkworms

To characterize the thermal properties of cocoons, a dynamic mechanical thermal analysis was conducted. We can see in Fig. 26a that the temperature increases from 35 to 320 °C. In contrast, the loss modulus E'' changes faster with temperature, as

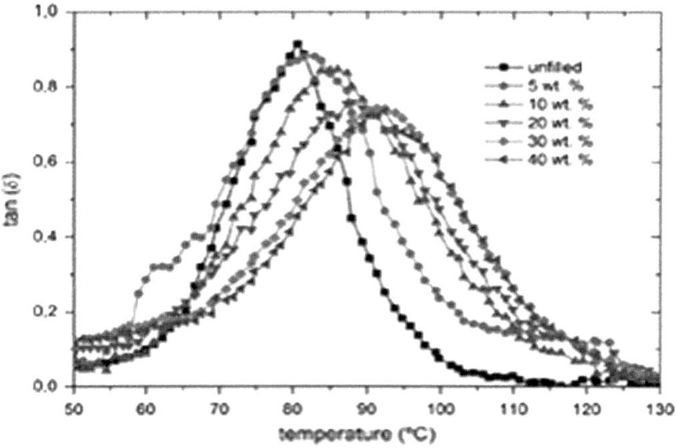

Fig. 25 Evolution of tan(σ) of epoxy–seashell powder composites in terms of the filler content obtained by DMA in torsion mode (Fombuena et al. 2014)

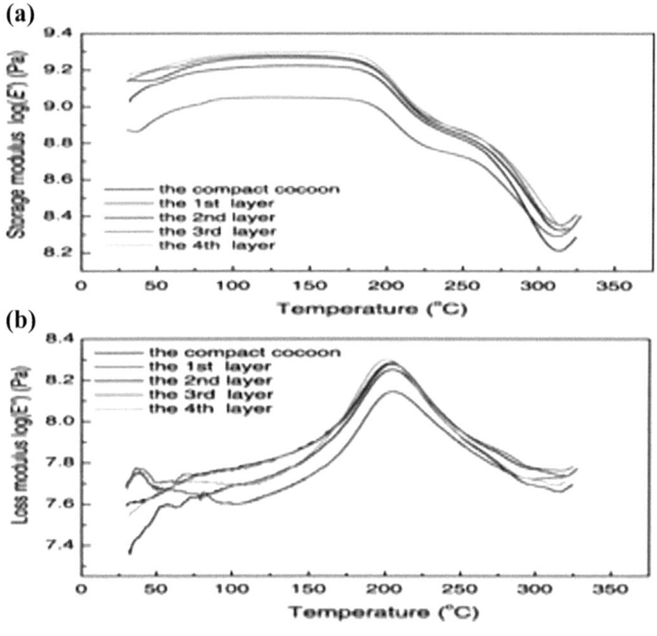

Fig. 26 **a** Storage moduli E and **b** loss modulus E (Zhao et al. 2007)

shown in Fig. 26b. E'' value first increases, peaks at an initial temperature of about 200 °C, then rapidly drops to 270 °C as temperature increases and finally shows another relatively slow growth phase. The endothermic peak of the loss modulus links to the glass transition temperature (T_g) of the cocoon in which is known as the onset temperature (Nordstrom et al. 2013; Cote 2005; Zhao et al. 2007).

4.2 Plant Source

4.2.1 Lignocellulose

Thermo-Gravimetric Analysis

The TG curves of the lignocellulosic fillers (RHF and WF) show two mass losses, the mass loss in the initial is under 100 °C, which is caused by a slow evaporation of the moisture particles and a second mass loss from around 150 and 500 °C. Lignocellulosic materials become chemically active and decompose thermo-chemically between 150 and 500 °C; hemicellulose is generally in the range of 150 and 300 °C, pulp ranges within 275 and 350 °C and lignin is between 250 and 500 °C (Yang et al. 2005).

4.2.2 Lignin

Thermal Resistivity of Silt Treated with Lignin Against Curing Time

An experiment was conducted to see the thermal resistivity of lignin treated silt against curing time. 2, 5, 8, 12 and 15% of lignin and water was mixed with the soil. In Fig. 27, the thermal resistivity of the mixture was measured with the use of a multi-purpose thermal needle system (MTN01) within 60-day curing time. A series of standard proctor tests (Kestur et al. 2013) were undertaken to determine the optimum moisture content (OMC) and maximum dry density (MDD) of all the mixtures (Zhang et al. 2015).

The lignin content greatly influences the initial thermal resistivity. The samples with more lignin had higher initial thermal resistance. Since thermal resistivity of natural lignin is usually greater than minerals found within soil (Kestur et al. 2013).

4.2.3 Bagasse

The glass transition temperature (Tg) was found by means of differential scanning calorimetry (DSC), and the thermal decomposition of the material studied was quantified by means of TGA. Figure 28 shows the thermo-grams obtained by TGA. The

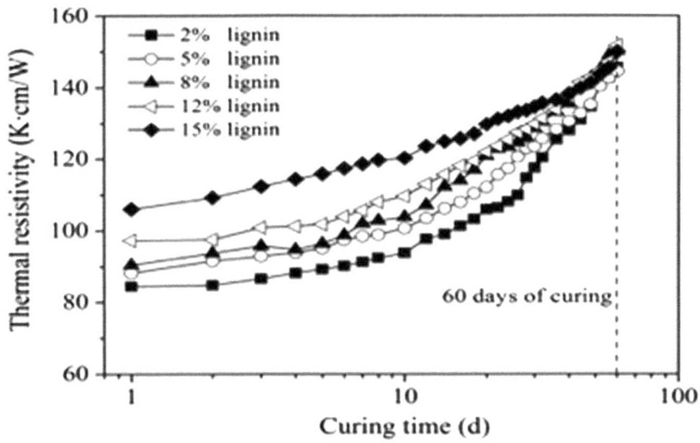

Fig. 27 Thermal resistivity of lignin treated silt against curing time (Zhang et al. 2015)

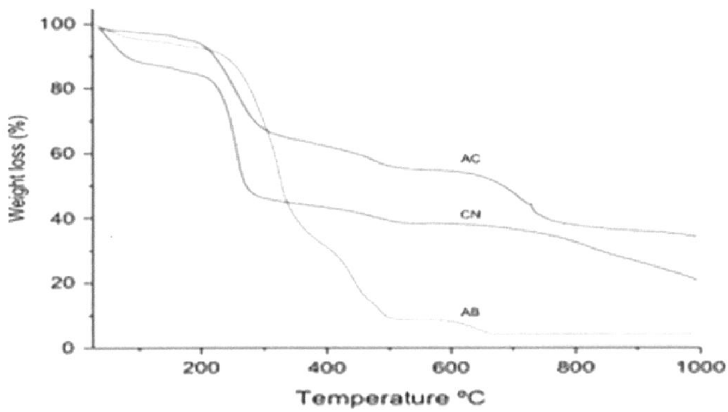

Fig. 28 Thermo-grams obtained by TGA (Cho and Park 2011)

thermal image shows a first weight loss at approximately 100 °C. This decrease in weight is linked to the reduction in the amount of moisture in the material.

The remaining 6.3% was consistent with the ash determination and was found to be 6.14% when analysed at 1000 °C. These results are caused by the presence of minerals whose degradation happens above these given temperatures (Cho and Park 2011; Abdullah and Ansari 2015).

4.2.4 Nanocellulose

The thermal stability of PVA nanocomposites reinforced using nanocelluose examined (Domun et al. 2015) with the use of a thermo-gravimetric analysis (TGA) (SDT

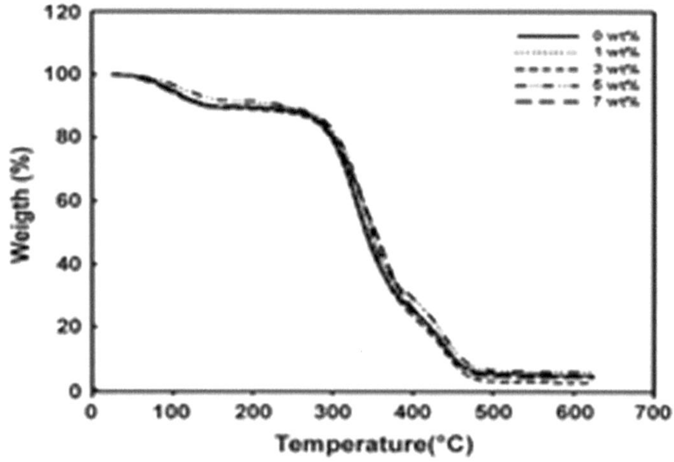

Fig. 29 TGA curves of PVA nanocomposites at varying nanocellulose contents (Abdullah and Ansari 2015)

Q600, TA instruments, USA). The temperature of samples of 10 mg was brought up from 25 to 600 °C with an increasing heating rate of 10C/min.

The reinforcing effect of the nanocellulose appeared at low temperature as shown in Fig. 29. This results in the improved stiffness of the nanocomposite due to the addition/introduction of stiff and strong nanocellulose (Abdullah and Ansari 2015). In Fig. 30, the 3 wt% has a higher E' modulus.

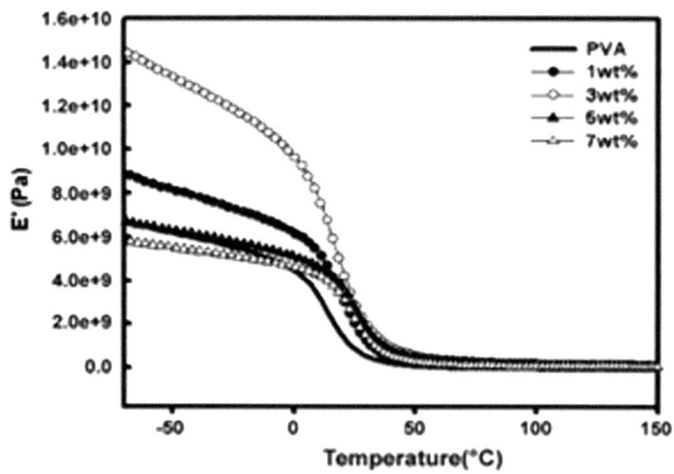

Fig. 30 Storage tensile modulus *E* versus the temperature for a nanocellulose reinforced PVA composite (Abdullah and Ansari 2015)

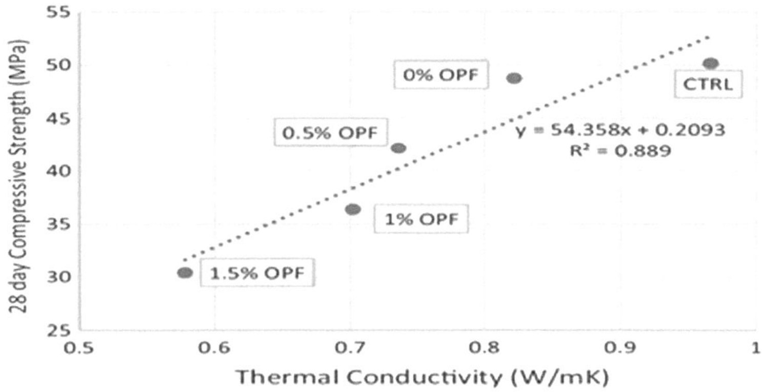

Fig. 31 Thermal conductivity versus compressive strength of fibre reinforced mortar (Xu et al. 2018)

4.2.5 Palm Oil Plant Ash

According to the results in Fig. 31, an inclusion of oil palm fibres into the mortar ensures the thermal insulation capacity of the concrete which becomes enhanced. The formation/presence of air voids throughout the porous fibres ensures the heat transfer which is reduced.

Studies from the past have shown that the compressive strength and thermal conductivity of the materials are directly proportional to each other. Therefore, we can conclude that fibre addition improves both the compressive strength and thermal conductivity (Xu et al. 2018).

4.2.6 Starch

An experiment was done to find the thermal stability of starch nanocrystals (SNC) under wet and dry environments. Starch nanocrystals are square-like, crystalline platelets approximately 10 nm in thickness and have equivalent diameters of 50–100 nm (Saba et al. 2016).

The SNC and the corresponding native starches' thermal performance were both examined in an excess of water (70% water content), to make it available in the use in aqueous suspension such as water-based composite treating or coating, and freeze-dried (0% water content) to evaluate process ability in "dry processes" like extrusion (Saba et al. 2016).

Classic responses are represented in Fig. 32a in an excess of water and Fig. 32b in a dry state. The starch and SNC granules are separated and displayed differing thermo-grams with different values of intensity, temperature and number of thermal transitions (Saba et al. 2016).

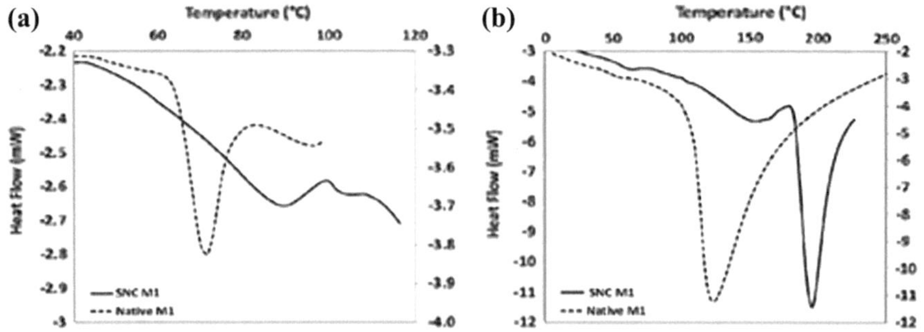

Fig. 32 DSC thermo-grams/for native and SNC from **a** in excess water and **b** in the dry state (Saba et al. 2016)

In conclusion, it was found that analysis of the thermal properties of SNC sourced from different DSCs, discovered that there are two different thermal transitions, whereas native starches show only one transition. In an excess amount of water, the first peak was credited to the first stage of crystallites melting and the second transition to the second stage of crystallites melting (Saba et al. 2016).

4.3 Natural Source

4.3.1 Carbon/Graphite

When the graphene layers are completely separated, a successful expansion process is obtained and is indicated by four aspects: (1) by an increasing surface area; (2) after a volume expansion of graphite oxide after thermal treatment; (3) by a formation of a porous, worm-shaped, nanostructure existing of nanosheets; and (4) by an eradication of all the diffraction peaks of the nanographite. Figure 33 shows the effect of nanographite loadings in silicon elastomer nanocomposites, as the frequency increases the impedance decreases.

CNTs have shown to have superconductivity below 20 K (approx. −253 °C). They have ability to copy the electrical properties of either semiconductors or perfect metals. They are exceptionally strong and stiff against axial strains. They are strong in-plane coupling and high flexibility against non-axial strains.

CNTs were ideal specimens for use in air, gas and water filtration. In certain ultra-high-purity applications, research is being performed in replacing activated charcoal with CNTs (Sui et al. 2009).

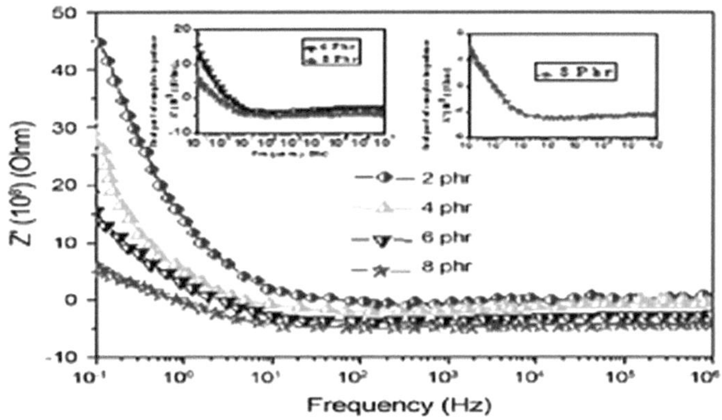

Fig. 33 Effect of nanographite loading on real part of complex impedance (Z') of silicon elastomer nanocomposites measured as a fraction of frequency (Saji et al. 2015)

4.3.2 Natural Clay

The TGA for the nanocomposites in nitrogen generally displayed that the loading of HNTs and HNT-silane combinations had very noticeable effects on the thermal stability. Temperatures at 5–85% of the loss in weight of the silica-filled composites were lesser than that of the HNT-filled NR nanocomposites. The thermal stability can be caused by multiple factors such as a strong interaction between HNTs, silica and natural rubber, the entrapment of degradation products of NR which was inside the lumen structure of the tubules, the interfacial and intratubular interactions between NR and HNTs. Silane coupling agent can increase the migration of polymer chains into the lumen of the HNTs. HNTs can be utilized by the silane which forms strong covalent and H-bonds with the layer of aluminium hydroxide and silicon hydroxide available in the HNT lumen (Singh and Swami 2015).

4.3.3 Organic/Inorganic Waste

The XRD graphs shown in Fig. 34 allow analysis of structures and the morphology of nano-OPEFB/epoxy nanocomposites and the pure epoxy composites. The XRD pattern of the pure epoxy composites shows sharp, very intense peaks at a 2θ value of 21° C, whereas all the nano-OPEFB/epoxy nanocomposites display relatively less peaks in intensity at the lower 2θ values. Thus, we can see that the introduction of the amorphous nano-OPEFB filler to the pure epoxy matrix reduced its crystallinity. Introduction of nanofiller in epoxy polymer reduced the magnitude of the diffraction crests and crystallinity, thus increasing elongational properties (Wijesena et al. 2012).

Thermo-mechanical analyses of the pure epoxy and nanocomposites at various nano-OPEFB filler loading were performed to determine the dimensional changes or the CTE, in both the rubber and glassy regions (Ranjbar et al. 2015).

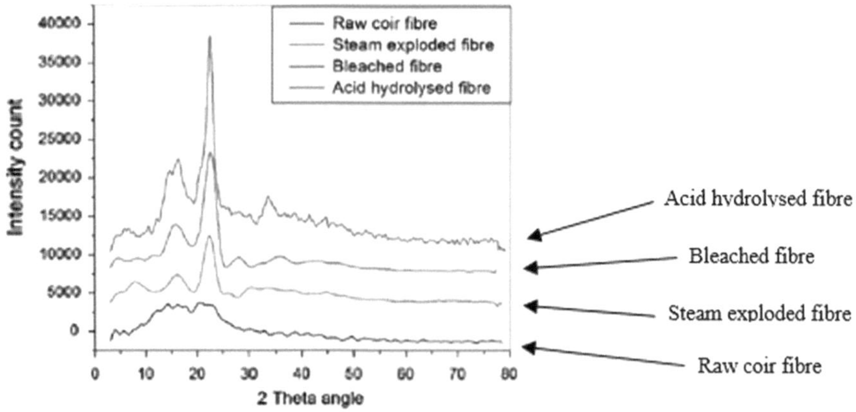

Fig. 34 XRD graph of inorganic waste under different phases (Ranjbar et al. 2015)

Inorganic Waste: Incineration Ash

Incineration ash, which is the fly ash and the bottom ash, is the ash remains after municipal solid waste (MSW) is burnt (Nwanonweyi and Chike-Onyegbula 2013). Reportedly, MSW incineration ash is abundant in the composition of oxide and carbonate ceramics that can provide improvements to the physical and mechanical properties of the composites.

5 Physical Properties

5.1 Animal Source

5.1.1 Eggshell

Water Absorption

From Fig. 35, two observation can be seen, first observation is that the water absorption of (HDPE/eggshell) composites with different filler content grain size (75 μm) is higher than for grain size (125 μm) for all filler contents, because of the surface area (Nwanonweyi and Chike-Onyegbula 2013). Second observation is that, as the filler content (for grain size (125 and 75 μm)) increases the water absorbed by the composites increases, this is because at high filler content it is difficult to achieve a homogeneous dispersion, since the increase in filler content increases the formation of agglomerations which increases the water absorption of the composites.

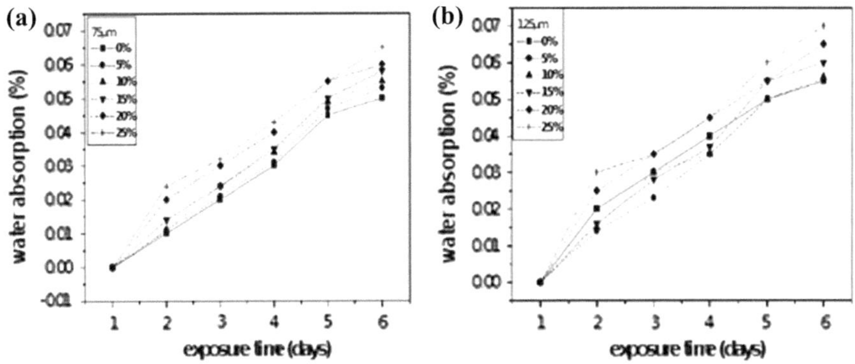

Fig. 35 Effect of exposure time on the water absorption of (HDPE/eggshell) composites with different filler content **a** 75 μm and **b** 125 μm (Nwanonweyi and Chike-Onyegbula 2013)

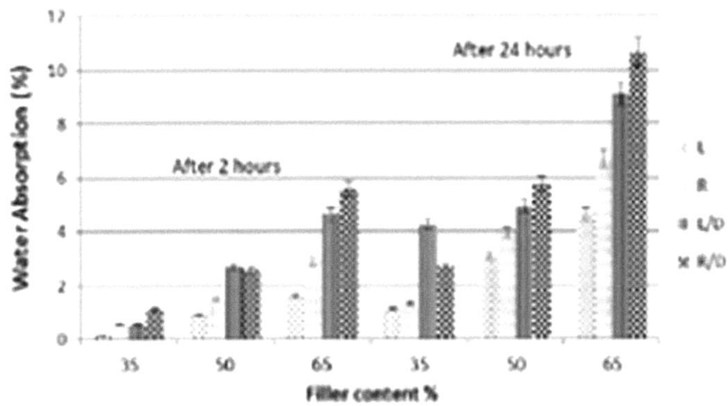

Fig. 36 Water absorption of composites with different DPLF contents and different filler types (Mirmehdi et al. 2015)

5.2 *Plant Source*

5.2.1 Lignocellulose

Water Absorption and Thickness Swelling

In Figs. 36 and 37, it is observed that the water absorption and thickness swelling increase with an increase in the filler content and composites with a high content of lignocellulose filler has high water absorption and thickness swelling when exposed to water for more than 2 h.

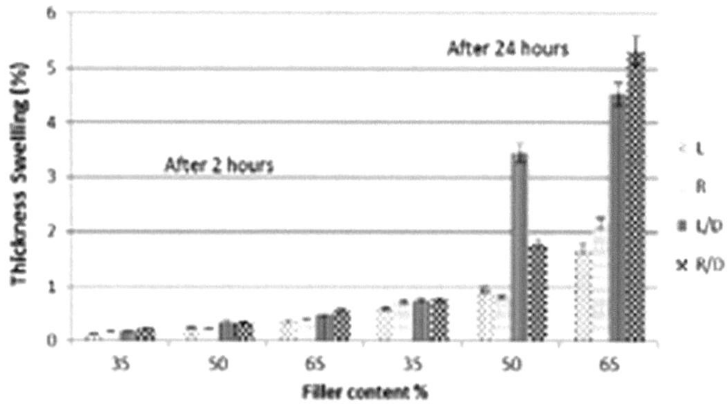

Fig. 37 Thickness swelling of composites with different DPLF contents and different filler types (Mirmehdi et al. 2015)

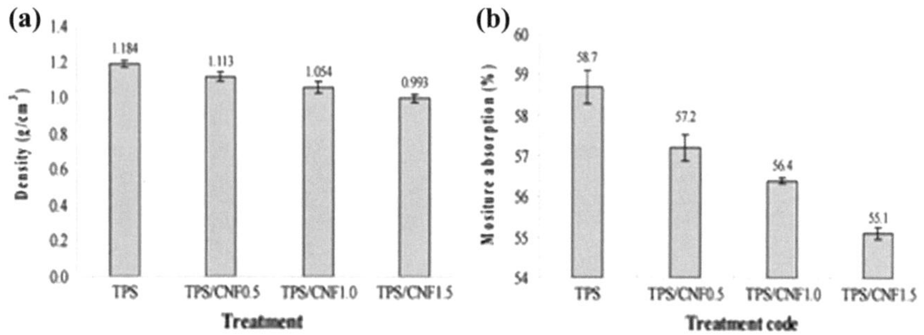

Fig. 38 Density (**a**) and water absorption (**b**) of TPS and TPS/CNF foamed composites (Ghanbaria et al. 2018)

5.2.2 Nanocellulose

In Fig. 38, the water absorption and density of thermo-plastic starch foamed (TPS) reinforced by cellulose nanofibres (CNF) decrease with an increase in the CNF content, and the unfilled TPS foamed composite had the highest water uptake and density as compared to TPS foamed composites filled with CNF. The decrease in water absorption for the TPS composites filled with CNF is due to the strong interaction between the hydroxyl groups of CNC with TPS.

5.2.3 Oil Palm Plant Ash

In Fig. 39, it is seen the density of pure epoxy composite increased with nano-OPEFB filler content, and this is due to the fact that nano-OPEFB filler is denser and contains a harder phase which increases cross-link density of the epoxy chains and improves

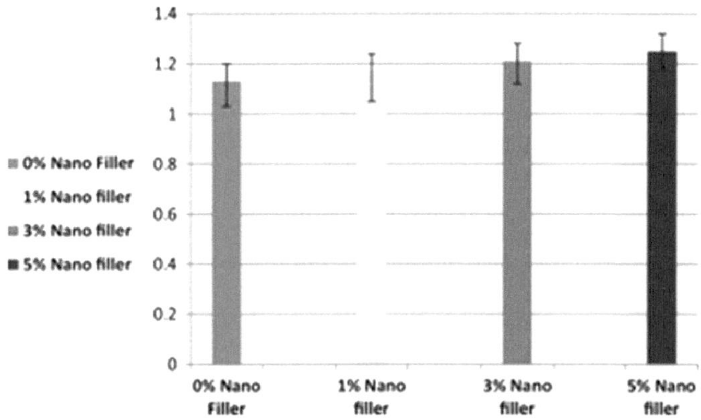

Fig. 39 Effect of nano-OPEFB filler loading on the density of epoxy composites (Saba et al. 2016)

hardness of resulting epoxy nanocomposites as compared to the epoxy polymer (Zhang et al. 2015). Excessive content of OPEFB nanofiller (loading of OPEFB nanofiller exceeding 3%) results in the aggregation of added nanofiller particles and porosity.

5.2.4 Bagasse

Bulk Density, Porosity and Water Absorption

Two observations can be seen, the first is that density of the composites filled with 3, 8 and 12% bagasse filler slight increases as the composite age increases from 28 days to 10 months. The second is that as the bagasse fibre content increases, the density of the composites decreases and the composites with 3% bagasse fibre content showed greatest density as compared to 8 and 12% bagasse fibre (Mustafa et al. 2013).

The apparent porosity for composites containing 3% bagasse fibre gradually decreased with time, while for composites containing 8 and 12% bagasse fibre rises slightly around 37%. This significant increase is caused by the increase in filler content which causes the composites with 8 and 12% bagasse fibre to be subjected to more air bubbles such that the void in the resulting composite cannot be filled by the created hydration product as composites with 3% bagasse fibre and the voids are filled with created hydration product which reduces porosity of the composite slowly as age increases (Mustafa et al. 2013).

The water absorbed by the composites increased steadily as time passed. 3% bagasse fibre had the lowest water absorption, while for high bagasse fibre content 8 and 12%, the water absorption of the composites increases, such at 10 months, the water absorption is 26.27% and 31.3%, which is 125.7% and 149.8%, respectively, of the composite with 3% bagasse fibre (Mustafa et al. 2013).

Table 1 Porosity and water absorption for pure geopolymer and geopolymer nanocomposites

Sample	Density in gm/cm^3	Porosity %	Water absorption %
Geopolymer	1.84	22.2	12.1
Geopolymer nanoclay-1	1.92	21.3	11.1
Geopolymer nanoclay-2	2.05	20.6	10.0
Geopolymer nanoclay-3	1.98	21.0	10.6

5.3 Natural Source

5.3.1 Graphite

Water Absorption

The water absorption decreases with an increase in the content of NG. WPCs filled with 1 wt% NG shows lowered water absorption when compared with 0.1 wt% NG. By filling the composite voids and lumens of BF with NG, the water is prevented from penetrating deep parts of composite, and thus, water absorption will be reduced (Ranjbar et al. 2014). The nature of NG surface tends to repel water which causes formation of the moisture.

Thickness Swelling

The content of NG has no significant effect on the decreasing of the thickness swelling, and however, the thickness swelling of composites can be increased by increasing the water exposure time. Fibre swelling creates stress on the surrounding matrix which leads to microcracking and eventually causes the composite to fail catastrophically the result, the fibre–matrix adhesion is venerable, and the dimensional stability of composites particularly for outdoor applications will be greatly affected (Ranjbar et al. 2014).

5.3.2 Nanoclay

Density, Porosity and Water Absorption

In Table 1, addition of nanoclay on geopolymer paste caused an increased in the density and reduction in the porosity and the water absorption of geopolymer nanocomposites compared to unfilled geopolymer paste. Adding 2.0 wt% of nanoclay reduces the porosity by 7.1% and the water absorption by 17% when compared to the control paste. It should be noted that excessive amounts of nanoclay cause an increase in porosity, water absorption and decreases the density of all samples.

6 Applications

Fillers are combined with a polymer to increase the physical and mechanical properties/qualities of a polymer.

There are certain factors to put into consideration when manufacturing polymers namely: the strength to weight, flexibility or rigidity, the behaviour of the polymer when subjected to very low or high temperatures, cost of the material and the environmental impact.

To select a filler, you must make deductions on the attributes that you deem are desirable in the filler, some of which could include particle shape and/or size; hardness; specific gravity; absorption; how reactive or inert it is; and temperature.

6.1 Green Fillers in Biodegradable Plastics

Bioplastic are plastics made from biomass (that is renewable) such as corn/potato starch, vegetable oils and microbiota. Nanocellulose is used to improve the modulus and tensile strength of the plastic, while reducing the moisture uptake of the film.

6.2 Green Fillers in Biodiesel

Diesel is any kind of liquid fuel which ignites without any spark due to high pressure and temperature and is compatible with diesel engines only. Diesel burns slower than gas and therefore is much more economical. There are technical studies that state lignin can be mixed with diesel through chemical process such as emulsion. Lignin is a good substitute for because it is a renewable and it needs no cultivation.

6.3 Green Fillers in Concrete

Concrete is a building material that is made by mixing sand and stones bonded together by means of fluid cement that hardens over a period. Gravel, stone and sand is added to concrete to increase the volume of the mixture which results in a reduction in the cost of the material.

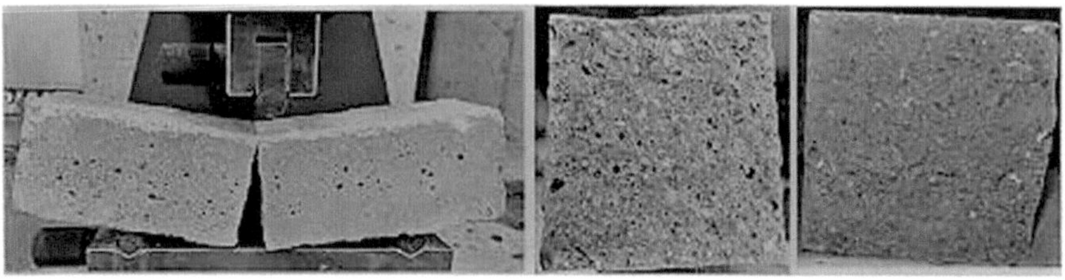

Fig. 40 Concrete block made from palm oil cement (Nayaka et al. 2018)

6.4 Green Fillers in Epoxy

Cotton flock also known as milled cotton fibre is used as a filler to thicken resin or to promote adhesion between laminate layers. Cotton flock is made from cotton by reducing the length of the fibre to microns. These fibres also help to prevent starvation resin starvation when bonding porous surfaces.

6.5 Green Fillers in Flame Retarding Polymers

Flame retarding polymers are polymers that resisting degrading and catching fire at very high temperatures. Nanocomposites are the most promising approach for making materials with flame retarding properties and eco-friendly.

6.6 Green Fillers in Medicine

Medicine is the study of diagnosing, treating and preventing diseases. The extractions of phytonanoparticles from *M. oleifera* can be used as anti-proliferative agents to combat cancer. Anti-proliferative agents are used to in inhibit the growth of cells such as tumours. *Moringa oleifera* is widely known for its health benefits and other uses like water purification. Figures 40 and 41 show few of the application of polymers filled with green fillers.

7 Conclusion

The aim of our study was to investigate the effects of "green" nanofillers on polymeric systems. Due to growing concerns with the environment, renewable polymeric

Fig. 41 Use of green fillers in the automotive industry (Clarke 2018)

materials have been modified with fillers to reduce environmental impact. Green composites are preferred as they are environmental-friendly, biodegradable and fully sustainable. In nanocomposites, the dimensions of the reinforcing material are in order of the nanometre range. Due to this characteristic, nanocomposites are superior over conventional composites considering its maximizing effect in interfacial adhesion.

Nanocomposites have substantially improved mechanical properties (e.g. strength, elastic modulus and dimensional stability), thermal properties, permeability (e.g. gases, water and hydrocarbons) and electrical properties. The addition of nanoparticles causes significant reduction in flammability and preserves the transparency of the polymer matrix. Two of the main parameters for nanocomposite properties are the dispersion of the filler and its interaction with the matrix, which may be influenced by chemical treatment at the surface. Chemical treatment prevents separation in the phases and creates strong interactions between the polymer and nanocomposite that leads to dispersion at a nanometre level.

References

Abdullah SI, Ansari MNM (2015) Mechanical properties of graphene oxide (GO)/epoxy composites. HBRC J 11:151–156

Asha A, Chandra Sekhar V (2014), Investigation on the mechanical properties of egg shell powder reinforced polymeric composites. Int J Eng Res Technol (IJERT) 3(12). ISSN: 2278-0181

Bedanokov AJ et al. (2011) Polymer nanocomposites. N Engl J Med p 31

Bergaya F, Detellier C, LambertJF LG (2013) Introduction to clay—polymer nanocomposites (CPN). In: Bergaya F, Lagaly G (eds) Handbook of clay science, vol 5, 2nd edn. UK Elsevier Ltd., Oxford, pp 655–677

Bootklad M, Kaewtatip K (2013) Biodegradation of thermoplastic starch/eggshell powder composites. Carbohyd Polym 97(2):315–320

Brody AL (2003) Nano, nano. Food packaging technology. Food Technol J 57(1):52–54

Cho M, Park B (2011) Tensile and thermal properties of nanocellulose-reinforced poly (vinyl alcohol) nanocomposites. J Ind Eng Chem 17:36–40
Clarke A (2018) United States Biodiesel Market (2018) Infinita renovables, neste oil. Diester Industries, Biopetrol and ADM, United States
Cote J (2005) Konrad thermal conductivity of base-course materials Can. Geotech J 42(1):61–78
Dana DA (2012) The nanotechnology challenge, vol 6. Cambridge University Press, New York, p 177
Domun N, Hadavinia H, Zhang T, Sainsbury T, Liaghat GH, Vahid S (2015) Improving the fracture toughness and the strength of epoxy using nanomaterials—a review of the current status. Nanoscale J 23:10294–10329
Felix JM, Gatenholm P (1991) The nature of adhesion in composites of modified cellulose fibers and polypropylene. J Appl Polym Sci 42(3):609–620
Fombuena V, Bernardi L, Fenollar O, Boronat T, Balart R (2014) Characterization of green composites from biobased epoxy matrices and bio-fillers derived from seashell wastes. Mater Des 57:168–174
Ghanbaria A, Tabarsaa T, Ashorib A, Shakeric A, Mashkour M (2018) Thermoplastic starch foamed composites reinforced with cellulose nanofibers. Therm Mech Prop
Giorgi R, Chelazzi D, Baglioni P (2005) Nanoparticles of calcium hydroxide for wood conservation, vol 21, issue no 23. Department of Chemistry and CSGI, University of Florence, pp 10743–10748
Guozhu M, Hongyang Z, Guanyi C, Huibin D, Jian Z (2015) Past, current and future of biomass energy research. Abibliometric analysis 52:1823–1833
Gusev AI (2007) nanomaterials, nanostructures, and nanotechnologies. N Engl J Med p 416
Hardy J, Römer L, Scheibel T (2008) Polymeric materials based on silk proteins. Polymer 49(20):4309–4327
Henry W (1991) The effect of starch on nano-materials. Ger Flatts 139–183
Hiremath P, Shettar M, Shankar M, Mohan N (2018) Investigation on effect of egg shell powder on mechanical properties of GFRP composites. Mater Today Proc 5(1):3014–3018
Kaisangsri N, Kerdchoechuen O, Laohakunjit N (2012) Biodegradable foam tray from cassava starch blended with natural fiber and chitosan. Ind Crops Prod 37:542–546
Kamigaito O (1991) What can be improved by nanometer composites? J Jpn Soc Powder Powder Metall 38(3):315–321
Kestur GS, Flores-Sahagún TH, Dos Santos LP, Mazzaro I, Mikowski A (2013) Characterization of blue agave bagasse fibers of Mexico. Composites Part A Appl Sci Manuf 45:153–161
Khanam PN, Al-Maadeed MA, Khanam PN (2015) Advances in silk science and technology, silk as a reinforcement in polymer matrix composites. Woodhead Publishing Series in Textiles, pp 143–170
Leceta I, Etxabide A, Cabezudo S, De La Caba K, Guerrero P (2014) Bio-based films prepared with by-products and wastes: environmental assessment. J Clean Prod 64:218–227
Lee C, Wei XD, Kysar JW, Hone J (2008) Measurement of the elastic properties and intrinsic strength of monolayer graphene. Sci Diction 321(5887):385–388
Li Z, Liu Z, Li B, Li D, Li Q, Wang H (2014), MnO_2 nanosilks self-assembled micropowders: facile one-step hydrothermal synthesis and their application as supercapacitor electrodes. J Taiwan Inst Chem Eng 45(6):2995–2999
Mirmehdi SM, Tonoli GHD, Dabbagh F (2015) Lignocellulose-polyethylene composite: influence of delignification, filler content and filler type
Mustafa AM, Liyana J, Noor MN, Abdullah MMAB (2013) Mechanical properties of polymer composites with sugarcane baggasse filler. Adv Mater Res 739–744
Nayaka R, Alengaram U, Jumaat M, Yusoff S, Alnahhal M (2018) High volume cement replacement by environmental friendly industrial by-product palm oil clinker powder in cement—lime masonry mortar. J Clean Prod 190:272–284
Ng P, Chia C-h, Zakaria S, Gan S (2015) Preparation of cellulose hydrogel from oil palm empty fruit bunch fibers cellulose. J Polym Res 9(4):449–459

Nordstrom Y, Norberg I, Sijoholm E, Drougge R (2013) A new softening agent for melt spinning of softwood kraft lignin. J Appl Polym Sci 129:1274–1279

Nwanonweyi SC, Chike-Onyegbula CH (2013) Water absorption, flammability and mechanical properties of linear low density polyethylene/egg shell composite, vol 4, issue no 1. Academic Research International. ISSN: 2223-9944

Ranjbar N, Mehrali M, Behnia A, JohnsonAlengaram U, Jumaat MZ (2014) Compressive strength and microstructural analysis of fly ash/palm oil fuel ash based geo-polymer mortar. Sci Moving Forward 59:532–539

Ranjbar N, Behnia A, Alsubari B, Birgani PM, Jumaat MZ (2015) Durability and mechanical properties of self-compacting concrete incorporating palm oil fuel ash, vol 12, issue no. 1. Department of Civil Engineering, Engineering Faculty, University of Malaya, 50603 Kuala Lumpur, Malaysia, pp 723–730

Saba N, Tahir PM, Abdan K, Ibrahim NA (2016) Fabrication of epoxy nanocomposites from oil palm nano filler: mechanical and morphological properties. Bioresources 11(3):7721–7736. https://doi.org/10.15376/biores.11.3.7721-7736

Saji J, Khare A, Choudhary RNP, Mahapatra SP (2015) Impedance and dielectric spectroscopy of nano-graphite reinforced silicon elastomer nanocomposites. Fibers Polym J 16(4):883–893

Sanchez-Jimenez PE, Perez-Maqueda LA, Perejon A, Criado JM (2012) Nanoclay nucleation effect in the thermal stabilization of a polymer nanocomposite: a kinetic mechanism change. J Phys Chem C 116:11797–11807

Singh K, Swami M (2015) Ecovision for a green growth: towards a renewable energy future, vol 2, issue no 8. M.S. College for Women, University of Delhi, Delhi, pp 828–833

Starch Europe (2019) What is starch? What is it used for? Why do we need it? [online] Available at https://starch.eu/starch/. Accessed 18 July 2019

Sue HJ, Wang S, Jane J (1997) Morphology and mechanical behaviour of engineering soy plastics. J Polym 38:5035

Sui G, Zhong W, Liu M, Wu P (2009) Enhancing mechanical properties of an epoxy resin using liquid nano-reinforcements. Mater Sci Eng 512:139

Šupová M, Martynková GS, Barabaszová K (2011) Effect of nanofillers dispersion in polymer matrices. Sci Adv Mat 3:1–25

Theivasanthi T, Anne Christma F, Toyin A, Gopinath S, Ravichandran R (2018) Synthesis and characterization of cotton fiber-based nanocellulose. Int J Biol Macromol 109:832–836

Vincent JFV, Wegst UGK (2004) Design and mechanical properties of insect cuticle. Arthropod Struct Dev 33(3):187–199

Visakh PM, María J, Martínez M (2016) Nanomaterials and nano-composites. Mahatma Ghandhi University, India. 10.1002/9783527683772

Wi K, Lee H, Lim S, Song H, Hussin M, Ismail M (2018) Use of an agricultural by-product, nano sized palm oil fuel ash as a supplementary cementitious material. Constr Build Mater 183:139–149

Wijesena R, Tissera N, Karunanayake L (2012), Preparation and characterization of α-Chitin nanofibers from crab shells of *Portunus pelagicus* (blue swimmer crab). In: Proceedings of international polymer science and technology symposium, vol 1

Xu P, Zhao X, Niu D, Hoch M, Ma P, Dong W, Chen M, Deshmukh Y (2018) Superior reinforcement of ethyl-co-vinyl acetate rubber composites by using nano-sized starch filler: the role of particle size and reactive compatibilization. Eur Polym J 105:107–114

Yang H-S, Wolcott MP, Kim H-S, Kim H-J (2005) Thermal properties of lignocellulosic filler-thermoplastic polymer bio-composites. J Therm Anal Calorim 82(1):157–160

Yang H-S, Kim H-J, Park H-J, Lee B-J, Hwang T-S (2006) Water absorption behavior and mechanical properties of lignocellulosic filler polyolefin bio-composites. Compos Struct Sci Direct 72(4):429–437

Zhang T, Liu S, Cai G, Puppala AJ (2015) Experimental investigation of thermal and mechanical properties of lignin treated silt. Eng Geol 196:1–11

Zhao H, Feng X, Yu S, Cui W, Zou F (2005) Mechanical properties of silkworm cocoons. Polymer 46(21):9192–9201

Zhao H-P, Feng X-Q, Cui W-Z, Zou F-Z (2007) Mechanical properties of silkworm cocoon pelades. Eng Fract Mech 74(12):1953–1962

Biosynthesis and Applications of Metal Nanomaterials

Shweta Kishen, Akshita Mehta and Reena Gupta

Abstract Biosynthesis of nanoparticles is an emerging technology with conventional methods. Green synthesis of metal nanoparticles (MNPs) has been done from various biological materials like bacteria, fungi and plants extracts. Leaves extracts contain phytochemicals which play a major role in formulating and enhancing bioactivity of nanoparticles. Two main approaches used to synthesize nanoparticles are "top-down" and "bottom-up" approaches. Studies on green biosynthesis of MNPs particularly focus on copper, cobalt, silver, gold, platinum, zirconium, palladium, iron, cadmium and metal oxides such as titanium oxides, zinc oxides. These various MNPs have diverse applications in food industries, drug delivery system, nanofiltration, UV-protectant paints, biomedical field, agricultural, environmental and physiochemical areas. They also have applications as artificial enzymes, nanocatalysts, lubricant additives and as therapeutic agents. This chapter mainly pivots around the biosynthesis and applications of green and advanced nanomaterials.

Keywords Nanoparticles · Nanocatalyst · Bioactive · Biomedical · Ultraviolet

1 Introduction

Nanotechnology sometimes also referred as "engineering at molecular level," involves the design, characterization, synthesis and application of nanomaterials. Nanotechnology provides an understanding at the atomic, molecular and supramolecular levels on fundamental properties of objects (Keat et al. 2015). Nanoparticles (NPs) are solid colloidal particles consisting of macromolecular substances that vary in size from 10 to 1000 nm (Kreuter 1994). New material engineered on a nanometric scale has great advantages to facilitate integration and miniaturization of materials, thus creating a direct connection among inorganic, organic and biological fields. Nanobiotechnology has in general had a great impact on material sciences.

S. Kishen · A. Mehta · R. Gupta (✉)
Department of Biotechnology, Himachal Pradesh University, Summerhill, Shimla, H.P. 171005, India
e-mail: reenagupta_2001@yahoo.com

S. Ahmed and W. Ali (eds.), *Green Nanomaterials*,
Advanced Structured Materials 126,
https://doi.org/10.1007/978-981-15-3560-4_6

Great attention has been gathered by biosynthesis of nanoparticles using biological agents (Malik et al. 2014). Biosynthesis of metal nanoparticles (MNPs) by plant extract is more appropriate for wide-ranging production under non-aseptic conditions (Makarov et al. 2014). The approach was applied to this study in the synthesis of nano-size gold (Au), platinum (Pt), palladium (Pd), silver (Ag) and cupric oxide (CuO) materials in simple liquid media. Plant-mediated synthesis of nanomaterials connects nanotechnology with plants (Khadeeja Parveen et al. 2016).

2 Nanoparticles and Nanomaterials

Nanomaterials are one of the main products of nanotechnology. Nanomaterials usually have dimensions that have internal structures measuring 100 nm or less (European Commission) and may be in the form of particles, rods, tubes or fibers. The nanomaterials may have different physiochemical properties at nanoscale in comparison with the same material in bulk form, due to smaller surface area in comparison with their larger size counterparts and may behave differently if enters in the body. Aggregated nanomaterials may exhibit same properties that those of the single NP.

Optical properties of the particles which according to the absorption in visible regions show different colors are determined by their size. Color dependence of AuNPs on the bases of size and shape is shown in Fig. 1. The reactivity and other properties of NPs are governed by their distinctive structure, shape and size. Their properties make them an appropriate candidate for wide range of applications.

3 Nanofabrication

Nanofabrication is the design and the synthesis of the nanomaterials. Metallic nanomaterials can be fabricated by chemical and physical techniques. However, these methods have many advantages such as reduced use of toxic solvents, no production of harmful by-products, less energy utilization. There is a need to develop environment-friendly process for the formation of MNPs. For the development of these procedures, there is a need to utilize biological resources in nature. In the past several years, for the production of most energy efficient, non-toxic and low-cost MNPs, many biological sources have been used (Thakkar et al. 2010a, b). Plants are being used for isolating various metals from land which are economically unreasonable to mine (Makarov et al. 2014). Different methods used for the synthesis of NPs are shown in Fig. 2.

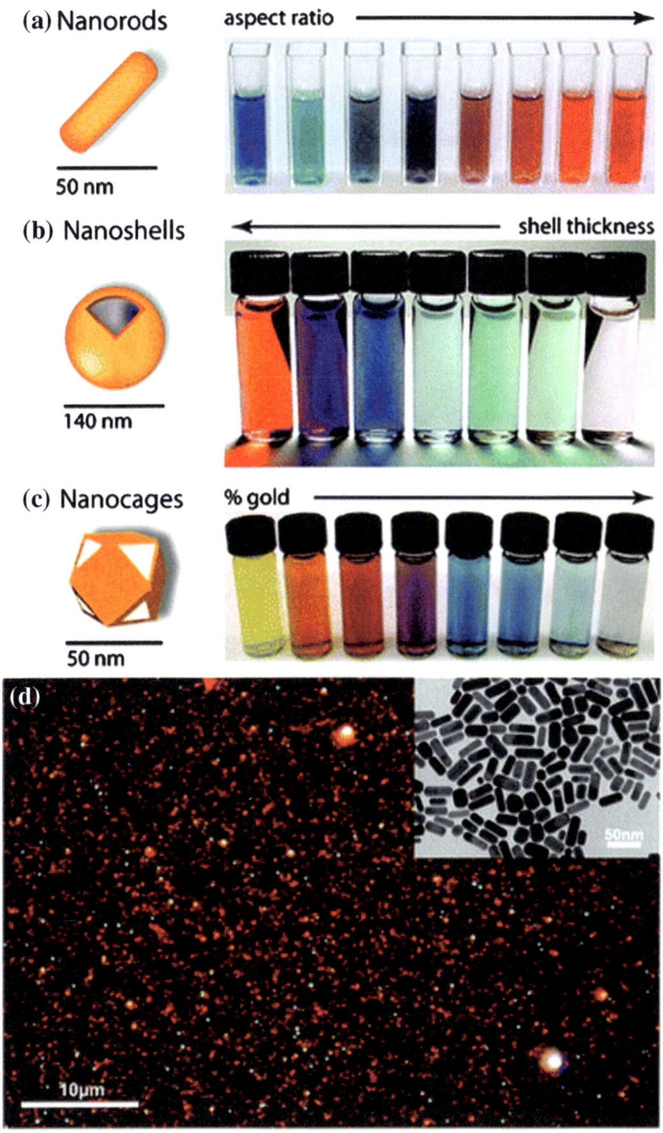

Fig. 1 Au NPs showing different colors on the bases of size and shape (Dreaden et al. 2012)

3.1 Different Methods Used for Synthesis of Nanomaterials

Nanomaterials can be synthesized by biological methods using biomolecules, biomembranes, microorganisms, plants. Nanomaterials are synthesized usually through two main approaches, i.e., "top-down and bottom-up" (Fig. 3) (Forough and Farhadi 2010).

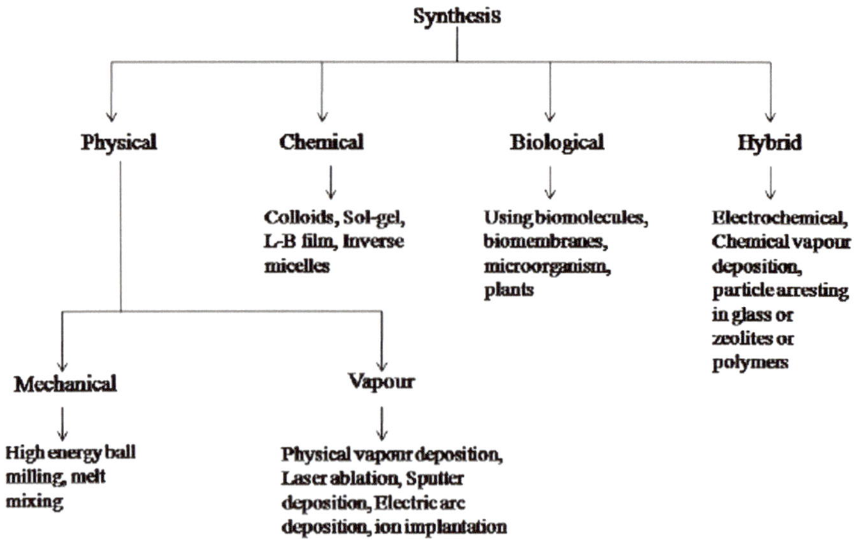

Fig. 2 Different methods for synthesis of nanoparticles (Aher 2013; with permission)

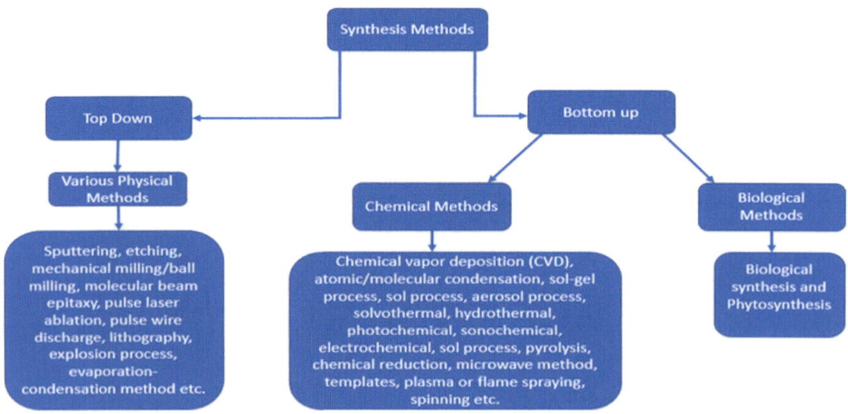

Fig. 3 General types of synthesis of nanomaterials (Nabi et al. 2018; with permission)

3.1.1 Top-Down Approaches

Destructive approach is applied in this method. It begins with a pattern generated on a large scale and then reduced to nanoscale. In this approach, nanostructures are synthesized by etching out crystal planes which are already present on the substrate. Bottom-up approach is more efficient and antagonistic to top-down approach, as it shows more possibilities of producing nanostructures with low defects and more homogenous chemical. Various top-down methods used for the synthesis of nanomaterials are mentioned below:

1. Optical lithography
2. E-beam lithography
3. Soft and nano-imprint lithography
4. Block co-polymer lithography
5. Scanning probe lithography.

3.1.2 Bottom-Up Approaches

Bottom-up approach is also known as building-up approach as the atoms and molecules are brought together to form nano-sized particles. It includes green synthesis, biochemical process, spinning and sol–gel process (method for producing solid materials from small molecules) (Iravani 2011). Mogilevsky et al. (2014) used graphite domains to synthesize TiO_2 anatase NPs through this technique. They synthesized the photosensitive complex for photoreaction degeneration of methylene blue using alizarin and titanium isopropoxide precursors. TiO_2 has high binding capacity with alizarin due to their axial hydroxyl terminal groups. X-ray diffraction (XRD) pattern confirmed the anatase form (Mogilevsky et al. 2014). Various bottom-up methods used for the synthesis of nanomaterials are mentioned below:

1. Atomic layer deposition
2. Sol–gel nanofabrication
3. Molecular self-assembly
4. Physical and chemical vapor-phase deposition
5. DNA scaffolding.

Due to some demerits of the above-mentioned techniques, there is always a search for reliable and environmentally friendly processes for the synthesis of nanoscale metal and metal oxide particles to minimize or even terminate the use of harmful chemicals. The only way to fabricate "green" processes is to redesign such production techniques that use mild reaction conditions and non-toxic reaction precursors.

3.2 *Green Synthesis of Nanomaterials*

MNPs are being widely used in various economic sectors, and there is increased awareness in their production, biological and environmental safety. Green synthesis of nanomaterials in bottom-up approach is achieved by using various biological agents such as microbes, plant extract, diatom, fungi, micro-algae including cyanobacteria, seaweed (macro-algae) and enzymes (Iravani 2011).

3.2.1 Synthesis of Metal Nanomaterials Using Plant Extracts

Plants act as bioreactors for the synthesis of metal nanoparticles (Makarov et al. 2014). Studies revealed that bioaccumulation process in plants usually accumulates metals as nano-sized particles. For example, *Brassica juncea* and *Medicago sativa* gather 50 nm silver NPs in excess (13.6% of their own weight) when grown on silver nitrate as a substrate (Harris and Bali 2008). Images of MNPs like iron, silver and gold produced in *Nicotiana benthamiana* extracts are shown in Fig. 4. The metals deposited by the plants can be isolated by harvesting via sintering and smelting methods.

MNPs of various shapes and sizes can be synthesized using different parts of plant by using biological approaches (Makarov et al. 2014). Novel secondary metabolites present in plant crude extract help in the reduction of ionic into bulk metallic nanoparticles formation (Aromal and Philip 2012). These secondary metabolites are continuously involved in the redox reaction to synthesize nature-friendly nanoparticles. Biomolecules present in plant extracts are used to reduce metal ions into nanoparticles in a single-step green synthesis process (Mittal et al. 2013). Individual monosaccharide components of disaccharides and polysaccharides form an open chain within an oligomer and therefore providing entry (of a metal ion) to an aldehyde group deciding the reducing ability of disaccharides and polysaccharides. The complete process of MNPs synthesized from plants and herbal extracts includes three main phases:

(1) The activation phase (reduction of metal ions and nucleation of metals)
(2) The growth phase (fusion of nanoparticles into larger ones through Ostwald ripening)
(3) The process termination phase (the final shape of nanoparticles is decided) (Glusker et al. 1999; Si and Mandal 2007).

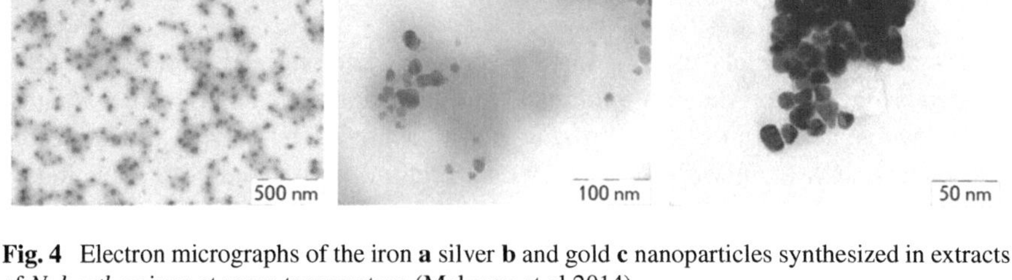

Fig. 4 Electron micrographs of the iron **a** silver **b** and gold **c** nanoparticles synthesized in extracts of *N. benthamiana* at room temperature (Makarov et al 2014)

Bio-reduction Mechanism

Gold: The plant extract contains different biomolecules such as proteins, enzymes, sugars, amino acids and other traces of metals. Bio-reduction process has strong involvement of these metabolites. The presence of various metabolites and redox enzymes reduces Au^{+} ions into metallic Au^{0} nanoparticles (Thakkar et al. 2010a, b).

$$HAu^{+}Cl_4 \cdot 4H_2O + \text{Plant extract} \rightarrow Au^{0}NPs + \text{by - products}$$

Silver: The reaction of silver nitrate with plant broth results in the formation of AgNPs (Tripathy et al. 2010), and the system of nanoparticle fabrication through plant extract is explained in the reaction below. Some other examples of synthesis of MNPs through reduction method using plant extracts are given in Table 1.

$$\text{Plant extracts} + \text{Biomolecules, } 2^{\circ}\text{metabolites} \rightarrow \text{NPs} \\ + \text{Non-hazardous by-products (at 30 to 60 °C, pH 4 to 7)}$$

Zinc oxide: A typical procedure was used in ZnO nanoparticles production. In aloe plant extract, zinc nitrate was dissolved to synthesize the nanoparticles. The reaction is as shown below (Sangeetha et al. 2011):

$$\text{Zinc oxide} + \text{plant extract} \rightarrow ZnO + \text{by-products}$$

Table 1 Metallic nanoparticles formed by various plant extracts

Plant and extracts	NPs produced	Synthesis location	Method	References
Coriolus versicolor	Ag	Intra- and Extracellular	Reduction	
Aspergillus flavus	Ag	Extracellular	Reduction	Saha et al. (2010)
*Jatropha curcas*latex	Ag	Extracellular	Latex mediated reduction	Thapa et al. (2004)
Acalypha indica leaf extract	Ag	Extracellular	Reduction	Krishnaraj et al. (2010)
Medicago sativa seed exudate	Ag	Extracellular	Reduction	Spadaro and Gullino (2005)
Cymbopogon flexuosus extract	Au	Extracellular	Reduction	Iravani (2011)
Live Alfalfa plants	Au	Intracellular	–	Mallikarjuna et al. (2011)

The table illustrates the location of the nanoparticles in relation to the cells and the method of synthesis (Pantidos and Horsfall 2014)

The *Mirabilis jalapa* flowers' aqueous extract acts as a reducing agent for synthesizing gold nanoparticles (Vankar and Bajpai 2014). Table 2 shows the plant metabolites which undergo the bio-reduction reaction for synthesis of MNPs and their pharmacological applications (Kuppusamy et al. 2016).

3.2.2 Synthesis of Metal Nanomaterials Using Bacteria

Biosynthesis of nanoparticles by microorganisms (both prokaryotes and eukaryotes) is a green and eco-friendly method. Microorganisms often produce metal or metal oxides of nano-size either extracellularly or intracellularly. Their system is capable of detoxifying heavy metals by reducing the metal ions or precipitate the soluble toxic ions into insoluble non-toxic metal nanoparticles. Bacteria can form nanoparticles both intracellularly via bioaccumulation and extracellularly on the cell wall using its enzymes. Intracellular NPs are with less monodispersity than extracellular particles. Hence, extracellular production of NPs has more commercial applications like in the field of optoelectronics, bio-imaging and sensor technology. Some examples of NPs synthesized by bacteria are given in Table 3.

NADH reductase is a suggested mechanism for the synthesis of AuNPs by the reduction of gold metals. The reduction of NADH is initiated by the electron transfer by NADH-dependant reductase as an electron carrier to form NAD+. The electrons result into the reduction of gold ions to elemental gold NPs, as these electrons are received by gold ions. Figure 5 shows the suggested reduction mechanism for the synthesis of gold NPs.

In addition to gold and silver NPs, sulfide nanoparticles like quantum dots (semiconductors) have also attracted the attention, for example, ZnS, CdS, PbS. These chemiluminescence dots are used for bio-imaging, biological detection and cell imaging. *Klebsiella pneumoniae* exposed to Cd^{2+} ions in the growth medium were found to form 20 to 200 nm CdS on the cell surface (Holmes et al. 1997).

Oxide nanoparticles are generally categorized into magnetic oxide and non-magnetic oxide nanoparticles. Magnetic oxide nanoparticles are produced by most of the species of magnetotactic bacteria and possess properties like super paramagnetic and have prospect for wide range of applications in biological separation and biomedicine. Magnetic nanoparticles are known to be biocompatible like Fe_3O_4 (magnetite) and Fe_2O_3 (maghemite). They have been extensively studied for targeted cancer treatment, stem cell sorting and manipulation, DNA analysis, guided gene therapy, drug delivery and magnetic resonance imaging (MRI) (Fan et al. 2009).

3.2.3 Synthesis of Metal Nanomaterials Using Fungi and Yeast

Synthesizing metal NPs have advantages over the processes used in the synthesis from bacteria. They are easy to scale-up and down-processing. For the large-scale production of certain NPs, yeast is preferred. The fungi produce napthoquinones and anthraquinones (Medentsev and Akimenko 1998; Bell et al. 2003) which act as

Table 2 Plant-derived metallic nanoparticles and their biomedical applications (Kuppusamy et al. 2016)

Plants	Nanoparticles	Plant parts	Size (nm)	Shapes	Plant metabolites involved in bio-reduction	Pharmacological applications	References
Acalypha indica	Ag, Au	Leaves	20–30	Spherical	Quercetin, plant pigment	Antibacterial	Krishnaraj et al. (2010)
Aloe vera	In_2O_3	Leaf	5–50	Spherical	Biomolecules	Optical properties	Maensiri et al. (2008)
Andrographispaniculata	Ag	Leaves	67–88	Spherical	Alkaloids, flavonoids	Hepatocurativ-e activity	Suriyakalaa et al. (2013)
Boswelliaserrata	Ag	Gum	7–10	Spherical	Protein, enzyme	Antibacterial	Kora et al. (2012)
Caria papaya	Ag	Fruit	15	Spherical	Hydroxyl flavones, catechins	Antimicrobial	Jain et al. (2009)
Cassia fistula	Au	Stem	55–98	Spherical	Hydroxyl group	Antihypoglycemic	
Cinnamon zeylanicum	Ag	Leaves	45	Spherical	Water-soluble organics	Antibacterial	Sathishkumar et al. (2009)
Citrulluscolocynthis	Ag	Calli	5–70	Triangle	Polyphenols	Antioxidant, anticancer	Satyavani et al. (2011)
Citrus sinensis	Ag	Peel	35	Spherical	Water-soluble compounds	Antibacterial	Kaviya et al. (2011)
Dioscoreabulbifera	Ag	Tuber	8–20	Rod, triangular	Diosgenin, ascorbic acid	Antimicrobial	Ghosh et al. (2012)
Gelsemiumsempervirens	Ag	whole	112	Spherical	Protein, amide, amine group	Cytotoxicity	

(continued)

Table 2 (continued)

Plants	Nanoparticles	Plant parts	Size (nm)	Shapes	Plant metabolites involved in bio-reduction	Pharmacological applications	References
Lippiacitriodora	Ag	Leaves	15–30	Spherical,	Isoverbascoside compound	Antimicrobial	Cruz et al. (2010)
Menthapiperita	Au, Ag	Leaves	90–150	Spherical	Menthol	Antibacterial	Mubarak Ali et al. (2011)
Mirabilis jalapa	Au	Flowers	~100	Spherical	Polysaccharides	Antimicrobial	
Iresineherbstii	Ag	Leaves	44–64	Cubic	Biomolecules phenolic compound	Biological activities	
Trigonella-foenumgraecum	Au	Seed	15–25	Spherical	Flavonoids	Catalytic	Aromal and Philip (2012)
Withaniasomnifera	Ag	Leaves	5–40	Irregular, spherical	Methyl 7-oxooctadecanoate	Antimicrobial	Nagati et al. (2012)

Table 3 Nanoparticles synthesised by bacteria (Pantidos and Horsfall 2014)

Name of organism	NPs produced	Synthesis location	Method	References
Thermomonospora sp.	Au	Extracellular	Reduction	Kasthuri et al. (2008)
Rhodococcus sp.	Au	Intracellular	Reduction	
E.coli	Pd, Pt	Extracellular	Reduction	Deplanche et al. (2010)
	Ag	Extracellular	Reduction	Kalimuthu et al. (2008)
Bacillus sphaerius JG-A12	U, Cu, Pb, Cd	Extracellular	Biosorption and reduction	
Lactobacillus sp.	Ag	Extracellular	Biosorption and reduction	Shahverdi et al. (2007)
Bacillus lincheniformis	Ag	Intracellular	Reduction	Schluter et al. (2014)

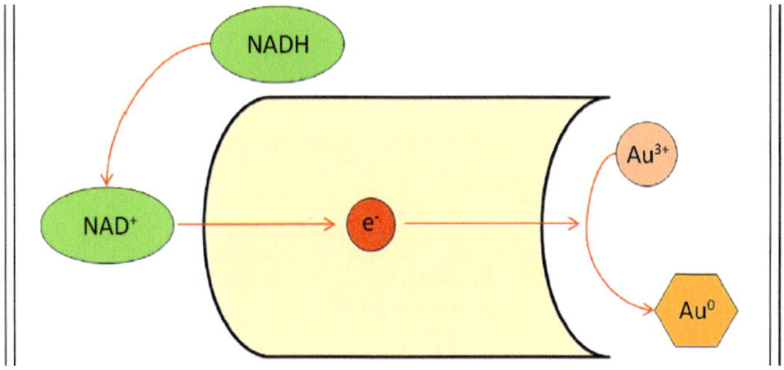

Fig. 5 Suggested mechanism for the NADH-dependent reduction of gold ions into AuNPs (Pantidos and Horsfall 2014)

reducing agents. When fungi are exposed to metal salts, in order to protect itself, enzymes (reducing agents) are released which reduce metal ions to their nanoscale form. Figure 6 shows some of the fungi and yeast used in the synthesis of NPs. Fungi can synthesize NPs either extracellularly or intracellularly. Table 4 shows examples of some fungi and yeast and the location of the NPs they synthesize.

Production of Fe_3O_4 and Fe nanoparticles employing *Aspergillus niger* in combination with the physical process is shown in Fig. 7.

Aspergillus niger is first cultured in the preferred media. Then, a homogenized *Aspergillus* mycelium solution is prepared with the cultured fungi. It is then mixed with the metallic solution (solution containing metal ions), and metallic precursor solution is added and kept for incubation for six days. After this, mycelium solution is

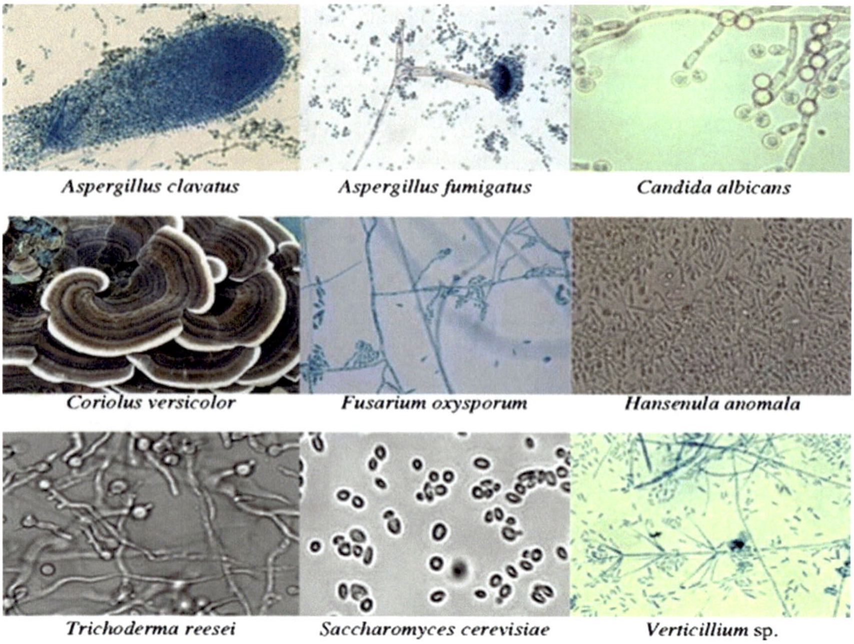

Fig. 6 Frequently used fungi and yeasts for metal nanoparticle synthesis (Siddiqi and Husen 2016)

Table 4 Location of synthesis and shape of the NPs produced by the fungi and yeast

Fungi/yeast	NPs produced	Location of synthesis	Shape	References
Candida albicans	Au	Cell-free extract	Monodispersed spherical	Ahmad et al. (2013)
C. glabrata	CdS	Intra- and Extracellular	Hexamer	Dameron et al. (1989)
A. niger	Au	Extracellular	Polydispersed	Xie et al. (2007)
Fusarium oxysporum	Pt	–	Spherical, triangular and aggregates	
A. fumigatus	ZnO	Extracellular	Spherical, hexagonal	Raliya and Tarafdar (2013)

centrifuged and dried, and in result, a dried power is obtained. Finally, heat treatment is given via supercritical pressure reactor at 300 °C temperature and 850 psi pressure, and as a result, in the end, magnetic nanoparticle powder is obtained.

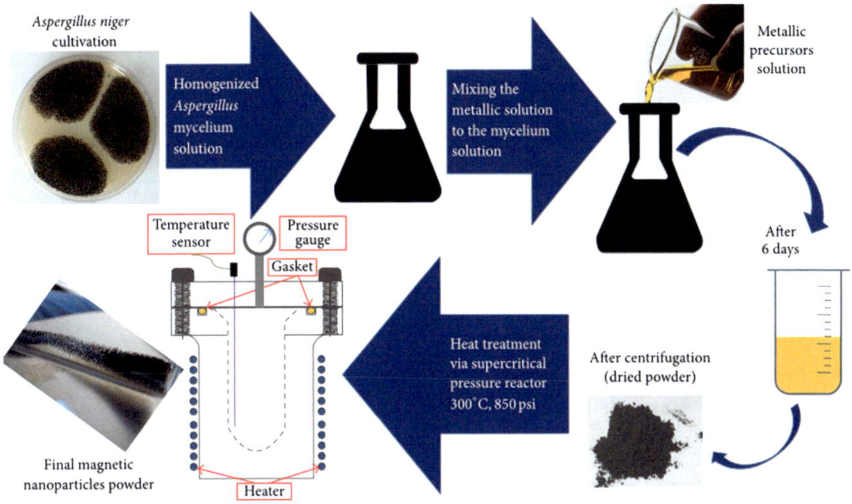

Fig. 7 Schematic diagram of the biological–physical method for synthesis of magnetic nanoparticles (Abdeen et al. 2016)

3.3 *Applications*

Noble nanoparticles have emerged in the field of biology, medicine and electronics due to their incredible and wide applications. Nanocapsules are nanoscale shells. They have wide range of uses, like medical promising applications for drug delivery, food enhancement, pharmaceuticals and for the self-healing of materials. Nanocapsules are being developed to study and mimic photosynthesis.

Dendrimers are nanomaterial polymer which has several types of branches. These are used in drug delivery as they have a symmetric morphology and an internal cavity.

Quantum dots are the other useful example of nanomaterials. They are semiconductor nanostructures or nanocrystals having 2–10 nm diameter range.

These nanostructures glow when illuminated by light and hence can be used as biosensors and in cell imaging. Following are some more areas in which nanomaterials are used (Fig. 8).

3.3.1 Plasmonic Biosensing

Lately, much attention is paid toward the synthesis of plasmonic nanoparticles owing to their unique optical, electronic and catalytic properties (Shen 2015). Surface plasmons (SPs) result from the collective oscillation of surface electron with the excitation of the magnetic field. SP sensors are usually fabricated by depositing thin layer of metal on a dielectric surface. Plasmonic MNPs including Au, Ag and Pt are highly efficient at absorbing and scattering light and exhibiting strong

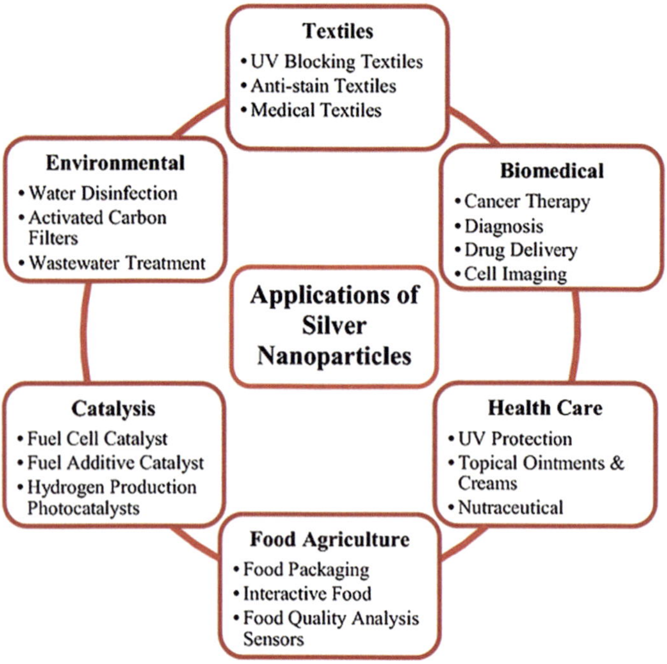

Fig. 8 Applications of silver nanoparticles (Keat et al. 2015)

absorbance band called localized surface plasmon resonance (LSPR) in the UV–VIS spectrum (Kumar and Govil 2014). LSPR is widely used for biosensing (Stuart et al. 2005). Many biosensors are being fabricated like colorimetric-based sensors, surface-enhanced Raman scattering (SERS)-based biosensors and two-photon photo luminescence-based biosensors (TPPL).

3.3.2 Mesoporous Nanoparticles

In 1990, silica and aluminosilicate materials with large and regularly ordered pore system were discovered, and numerous work has been done till date to control pore size of nanomaterial. Mesoporous silica NPs (MSN) are potential biomaterials as they have large surface area and pore volume size (Kumar and Govil 2014). MSN-based biomaterial exhibits excellent biocompatibility for biomedical applications such as drug and gene delivery carriers, biosensors and biomarkers for studying cell function (Haung et al. 2010). MSN has also been used as catalyst in many biochemical reactions and hence can also be used for biomedical applications.

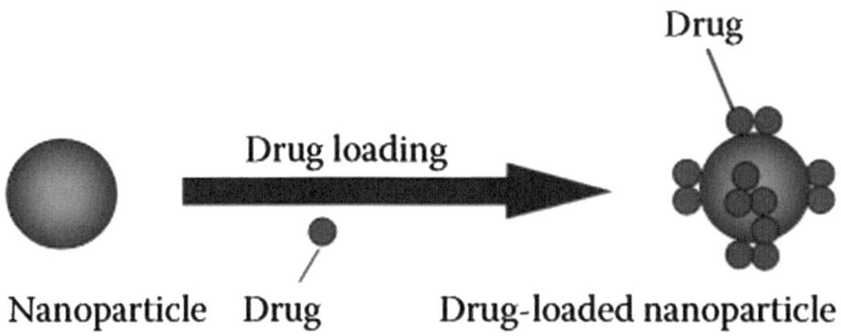

Fig. 9 Drug delivery through nanoparticles

3.3.3 Drug Delivery

Drug delivery system by nanoparticles has various advantages over traditional drug. Using these systems, a drug can directly be sent to the target area, and side effects can also be avoided. Several nanomaterials are used in the drug delivery system like dendrimers, nanocapsules, noble metal nanopartilces, etc. In drug delivery system, drug which is to be delivered is coated on the nanoparticle and is delivered on to the targeted area. Figure 9 shows the representation of drug delivery through NPs. It avoids the damage to the vital organs.

The other method of drug delivery is through nanocapsules. Nanocapsules have the internal cavity which is used to transport drug to the targeted area where drug needs to be delivered. They can also be used in strategic delivery of the drug in tumors, nanocapsule bandages to fight infection (Pavan et al. 2012).

3.3.4 Nanomedicine

Nanomedicine is the application of nanotechnology to the field of medicine by the use of materials at nanoscale. The common use of nanomedicine is for the enhancement of the action of drug (Xue et al. 2018). Nanomedicines have potential solutions for treatment of cancer, cardiovascular and neurodegenerative diseases. Many nanomedicines are now FDA approved and are available for clinical use including treatment for high cholesterol, autoimmune disease, fungal infections and hepatitis. Some FDA-approved nanomedicines are abelcet, abraxane, cimzia, depocyte and eligard (Ventola 2012).

4 Conclusion

Nanomaterials are chemical substances or materials having different physiochemical properties in comparison with the same material in the bulk form. They can be synthesized using bottom-up approach using biological methods. Plants, bacteria, fungi and other biological agents are used in the synthesis of nanoparticles. Metal nanomaterials have many applications like in health care, cosmetics, electronics and other areas. Nanomaterials have the potential to revolutionize the biomedical field. Nanomedicine promises an improvised and safe method of treatment against diseases like cancer, AIDS and other viral diseases. It may lead to the advancements in the treatment of cancer by synthesizing more effective and less harsh chemotherapy drugs. Nanobots do not exist yet but their practical use can help to boost the immune system of the patients suffering from cancer and AIDS like diseases where immune system becomes compromised. All the nanoparticles synthesized may not have the positive or desired properties and might result in unpredictable safety problems.

Acknowledgements The support from Department of Biotechnology, Ministry of Science and Technology, Government of India, to Department of Biotechnology, Himachal Pradesh University, Shimla (India), is thankfully acknowledged.

Conflict of Interests

The author(s) declare that there is no conflict of interest regarding the publication of this article.

References

Abdeen M, Sabry S, Ghozlan H et al (2016) Microbial-physical synthesis of Fe and Fe_3O_4 magnetic nanoparticles using *Aspergillus niger* YESM1 and supercritical condition of ethanol. J nanometer 7

Aher JA (2013) Characterization and application of silver nanoparticles biosynthesized under control and clinorotation conditions. Thesis https://doi.org/10.13140/RG.2.1.3076.5600

Ahmad T, Wani AI, Manzoor N et al (2013) Biosynthesis, structural characterization and antimicrobial activity of gold and silver nanoparticles. Colloids Surf B Biointerfaces 107:227–234

Aromal SA, Philip D (2012) Green synthesis of gold nanoparticles using *Trigonella foenum-graecum* and its size dependent catalytic activity. SpectroChim Acta A 97:1–5

Bell AA, Wheeler MH, Liu J et al (2003) Studies on polyketide toxins of *Fusarium oxysporum, F.vasinfectum*: potential targets for disease control. Science 59(6–7):736–47

Cruz D, Falé PL, Mourato A et al (2010) Preparation and physicochemical characterization of Ag nanoparticles biosynthesized by *Lippia citriodora* (Lemon Verbena). Colloids Surf B 81:67–73

Dameron CT, Reese RN, Mehra RK et al (1989) Biosynthesis of cadmium sulphide quantum semiconductor crystallites. Nature 338:596–597

Deplanche K, Caldelari I, Mikheenko IP et al (2010) Involvement of hydrogenases in the formation of highly catalytic Pd(0) nanoparticles by bioreduction of Pd(II) using *Escherichia coli* mutant strains. J Microbiol 156:2630–2640

Dreaden EC, Alkilany AM, Huang X et al (2012) The golden age: gold nanoparticles for biomedicine. Chem Soc Rev 41:2740–2779

Fan TX, Chow SK, Zhang D (2009) Biomorphic mineralization: from biology to materials. Prog Mater Sci 54:542–659

Forough M, Farhadi K (2010) Biological and green synthesis of silver nanoparticles. Turkish J Eng Env Sci 34:281–287

Ghosh S, Patil S, Ahire M et al (2012) Synthesis of silver nanoparticles using *Dioscorea bulbifera* tuber extract and evaluation of its synergistic potential in combination with antimicrobial agents. Int J Nanomed 7:483–496

Glusker JP, Katz AK, Bock CW (1999) Metal ions in biological systems. Rigaku J 16(2):8–16

Harris AT, Bali R (2008) On the formation and extent of uptake of silver nanoparticles by live plants. J Nanopart Res 10:691–695

Holmes JD, Richardson DJ, Saed S et al (1997) Cadmium-specific formation of metal sulfide "Q-particles" by *Klebsiella pneumoniae*. J Microbiol 143(8):2521–2530

Huang X, Teng Chen D, Tang F et al (2010) The effect of the shape of mesoporous silica nanoparticles on cellular uptake and cell function. Biomater Sci 31:438–448

Iravani S (2011) Green synthesis of metal nanoparticles using plants. Green Chem. https://doi.org/10.1039/C1GC15386B

Jain D, Daima HK, Kachhwaha S et al (2009) Synthesis of plant-mediated silver nanoparticles using papaya fruit extract and evaluation of their antimicrobial activities. Dig J Nanomater Biosyn 4:557–563

Kalimuthu K, Suresh Babu R, Venkataraman D et al (2008) Biosynthesis of silver nanocrystals by *Bacillus licheniformis*. Colloids Surf B: Biointerfaces 65:150–153

Kasthuri J, Kathiravan K, Rajendiran N (2008) Phyllanthin-assisted biosynthesis of silver and gold nanoparticles: a novel biological approach. J Nanopart Res 11:1075–1085

Kaviya S, Santhanalakshmi J, Viswanathan B et al (2011) Biosynthesis of silver nanoparticles using *Citrus sinensis* peel extract and its antibacterial activity. Spectrochim Acta A 79:594–598

Keat CL, Aziz A, Eid AM et al (2015) Biosynthesis of nanoparticles and silver nanoparticles. Bioresour Bioprocess. https://doi.org/10.1186/s40643-015-0076-2

Kora JA, Sashidhar BR, Arunachalam J (2012) Aqueous extract of gum olibanum (*Boswellia serrata*): A reductant and stabilizer for the biosynthesis of antibacterial silver nanoparticles. Proc Biochem 47:1516–1520

Kreuter J (1994) Encyclopedia of pharmaceutical technology in nanoparticles. Marcel Dekker Inc, New York, pp 165–190

Krishnaraj C, Jagan EG, Rajasekar S et al (2010) Synthesis of silver nanoparticles using *Acalyphaindica* leaf extracts and its antibacterial activity against water borne pathogens. Colloid Surf B: Biointerfaces 76:50–56

Kumar A, Govil NJ (eds) (2014) Biotechnology. Studium Press LLC, USA

Kuppusamy P, Yusoff MM, Maniam P et al (2016) Biosynthesis of metallic nanoparticles using plant derivatives and their new avenues in pharmacological applications—An updated report. Saudi Pharm J 24(4):473–484

Maensiri S, Laokul P, Klinkaewnarong J et al (2008) Indium oxide (In_2O_3) nanoparticles using *Aloe vera* plant extract: synthesis and optical properties. J Optoelectron Adv M 10:161–165

Makarov VV, Love JA, Sinitsyna VO et al (2014) "Green" Nanotechnologies: synthesis of metal nanoparticles using plants. Acta Nature, PubMed 1:35–44

Malik P, Shankar R, Malik V et al (2014) Green chemistry based benign routes for nanoparticle synthesis. J Nanopart. https://doi.org/10.1155/2014/302429

Mallikarjuna K, Narasimha G, Dillip GR et al (2011) Green synthesis of silver nanoparticles using *Ocumum* leaf extract and their characterization. Dig J Nanomater Biostruct 6:181–186

Medentsev AG, Akimenko VK (1998) Naphthoquinone metabolites of the fungi. J Phytochem 47(6):935–959

Mittal AK, Chisti Y, Banerjee UC (2013) Synthesis of metallic nanoparticles using plant extracts. Biotechnol Adv Nanomedicine 6:257–262

Mogilevsky G, Hartman O, Emmons ED et al (2014) Bottom-up synthesis of anatase nanoparticles with graphene domains, ACS Appl Mater Interfaces, pp 10638–10648

MubarakAli D, Thajuddin N, Jeganathan K et al (2011) Plant extract mediated synthesis of silver and gold nanoparticles and its antibacterial activity against clinically isolated pathogens. Colloids Surf B 85:360–365

Nabi G, Aain Q, Khalid NR et al (2018) A review on novel eco-friendly green approach to synthesis TiO_2 nanoparticles using different extracts, vol 28. J Inorg Organomet Polym Mater 4:1552–1564

Nagati BV, Alwala J, Koyyati R et al (2012) Green synthesis of plant-mediated silver nanoparticles using *Withania somnifera* leaf extract and evaluation of their anti-microbial activity. Asian Pac J Trop Biomed 2:1–5

Pantidos N, Horsfall LE (2014) Biological synthesis of metallic nanoparticles by bacteria, fungi and plants. J Nanomed Nanotechnol. https://doi.org/10.4172/2157-7439.1000233
Parveen K, Banse V, Ledwani L (2016) Green synthesis of nanoparticles: Their advantages and disadvantages. In: 5th National conference on thermophysical properties. APL Bioengineering. https://doi.org/10.1063/1.4945168
Pavan KK, Hemanth K, Niranjan R et al (2012) Nanocapsules: The weapons for novel drug delivery systems. Bioimpacts 2:71–81
Raliya R, Tarafdar JC (2013) ZnO nanoparticle biosynthesis and its effect on phosphorous-mobilizing enzyme secretion and gum contents in cluster bean (*Cyamopsis tetragonoloba* L.) Agirc Sci Res J 2:48–57
Saha S, Pal A, Kundu S et al (2010) Photochemical green synthesis of calcium-alginate-stabilized Ag and Au nanoparticles and their catalytic application to 4-nitrophenol reduction. Langmuir 26:2885–2893. https://doi.org/10.1021/la902950x
Sangeetha G, Rajeshwari S, Venckatesh R (2011) Green synthesis of zinc oxide nanoparticles by *aloe barbadensis* miller leaf extract: structure and optical properties. Mater Res Bull 46(12):2560–2566
Sathishkumar M, Sneha K, Won SW et al (2009) *Cinnamon zeylanicum* bark extract and powder mediated green synthesis of nano-crystalline silver particles and its bactericidal activity. Colloids Surf B 73(2):332–338
Satyavani K, Gurudeeban S, Ramanathan T et al (2011) Biomedical potential of silver nanoparticles synthesized from calli cells of *Citrulluscolocynthis L.* J Nanobiotechnology 9:43–51
Schlüter M, Hentzel T, Suarez C et al (2014) Synthesis of novel palladium nanocatalysts by microorganisms from heavy-metal-influenced high-alpine sites for dehalogenation of polychlorinated dioxins. Chemosphere 117C:462–470
Shahverdi AR, Minaeian S, Shahverdi HR et al (2007) Rapid synthesis of silver nanoparticles using culture supernatants of *Enterobacteria*: A novel biological approach. Process Biochem 42:919–923
Shen H (2015) Plasmonic nanoparticles: Towards the fabrication of biosensors. IOP Conf Ser Mater Sci Eng. https://doi.org/10.1088/1757-899X/87/1/012009
Si S, Mandal TK (2007) Tryptophan-based peptides to synthesize gold and silver nanoparticles: a mechanism and kinetic study. Chem Pubmed 13(11):3160–3168
Siddiqi KS, Husen A (2016) Fabrication of metal nanoparticles from fungi and metal salts: Scope and Application. Nanoscale Res Lett. https://doi.org/10.1186/s11671-016-1580-9
Spadaro D, Gullino ML (2005) Improving the efficacy of biocontrol agents against soil borne pathogens. Crop Protection 24:601–613
Stuart DA, Haes AJ, Yonzon CR et al (2005) Biological applications of localized surface plasmonic-phenomenae. IEE Proc-Nanobiotechnology 152:13–32
Suriyakalaa U, Antony JJ, Suganya S et al (2013) Hepatocurative activity of biosynthesized silver nanoparticles fabricated using *Andrographis paniculata*. Colloids Surf B 102:189–194
Thakkar KN, Mhatre SS, Parikh RY (2010a) Nanotechnology, biology and medicine. Nanomedicine 6(2):257–262
Thakkar KN, Mhatre SS, Rasesh YP (2010b) Biological synthesis of metallic nanoparticles. Nanomedicine. https://doi.org/10.1016/j.nano.2009.07.002
Thapa D, Palkar VR, Kurup MB et al (2004) Properties of magnetite nanoparticles synthesized through a novel chemical route. Mater Lett 58:2692–2694
Tripathy A, Raichur AM, Chandrasekaran N et al (2010) Process variables in biomimetic synthesis of silver nanoparticles by aqueous extract of *Azadirachta indica* (Neem) leaves. J Nanopart Res 12:237–246
Vankar PS, Bajpai D (2014) Preparation of gold nanoparticles from *Mirabilis jalapa* flowers. Indian J Biochem Biophys 47:157–160
Ventola CL (2012) The nanomedicine revolution. Pharm Ther 37(10):582–591

Xie J, Lee JY, Wang DIC et al (2007) High-yield synthesis of complex gold nanostructures in a fungal system. J Phys Chem C 111:16858–16865

Xue X, Huang Y, Bo R et al (2018) Trojan-horse nanotheranostics with dual transformability and multifunctionality for highly effective cancer treatment. Nat Commun 9(3653):1–15

Carbon Dots from Renewable Resources: A Review on Precursor Choices and Potential Applications

Perumalsamy Vishnukumar, Sivashunmugam Sankaranarayanan, Muruganandham Hariram, Singaravelu Vivekanandhan and Rodrigo Navia

Abstract In recent years, carbon dots have received immense attention not only due to their exciting structural, morphological as well as physicochemical properties but also because of the versatility in their fabrication and manipulation toward various applications. Traditionally, they have been synthesized using different carbon-rich precursors that are originated from petroleum resources. Emerging concerns in utilizing petroleum resources that include limited availability, poor cost stability, greenhouse gas effect and climate change that significantly impacted the synthesis of carbon dots from renewable feedstocks. The key advantages of utilizing renewable precursors as a carbon source in synthesizing carbon dots are their abundance and wider options for source selection. Recently, several renewable resource-based materials have been effectively utilized for the synthesis of carbon dots. Thus, the present chapter is ultimately aimed to summarize their recent developments, current potential applications, and emerging opportunities.

Keywords Nanomaterials · Carbon dots · Renewable resources · Sustainability

P. Vishnukumar · M. Hariram · S. Vivekanandhan (✉)
Sustainable Materials and Nanotechnology Lab (SMNL), Department of Physics, V.H.N.S.N. College (Autonomous), Virudhunagar, Tamil Nadu 626001, India
e-mail: svivekan@vhnsnc.edu.in

S. Sankaranarayanan · R. Navia
Scientific and Technological Bioresource Nucleus (BIOREN), Universidad de La Frontera, Av. Francisco Salazar 01145, Temuco 4780000, Chile

M. Hariram
Department of Physics, Bharathidasan University, Tiruchirappalli, Tamil Nadu 620024, India

R. Navia
Department of Chemical Engineering & Centre for Biotechnology and Bioengineering (CeBiB), Universidad de La Frontera, Av. Francisco Salazar 01145, Temuco 4780000, Chile

S. Ahmed and W. Ali (eds.), *Green Nanomaterials*,
Advanced Structured Materials 126,
https://doi.org/10.1007/978-981-15-3560-4_7

1 Introduction

Recent advancements in the science and technology of nanoscale materials create new opportunities for many materials; one of the best examples is carbon. Carbonaceous nanomaterials that include fullerenes (Kroto et al. 1985), carbon nanotubes (Iijima and Ichihashi 1993), graphenes (Geim and Novoselov 2007), carbon onions (Banhart and Ajayan 1996), nanodiamonds (Mochalin et al. 2012), carbon nanofibers (Zhang et al. 2014), and carbon dots (Lim et al. 2015) receive extensive attention due to their unique structural features as well as their size-dependent physicochemical properties. Among them, carbon dots (C-dots or CDs), which are also known as carbon quantum dots (CQ-dots), are unique material with a size of less than 10 nm and exhibit sp^2 structural features (Baker and Baker 2010; Wang and Hu 2014). The nanostructures are said to be quantum structures if it is one or more dimensions confined to the range of either the exciton Bohr radius (electron–hole pair separation) or the de Broglie wavelength of an electron (Pradeep 2007). In general, nanostructures having a size less than 10 nm are called quantum structures and if all the three dimensions of the materials belonging to this range, then they are said to be quantum dots (Q-dots). Q-dots receive extensive attention due to their unique photoelectric and luminescent properties. C-dots receive more interest due to their biocompatibility, less toxicity, chemical inertness, negligible cytotoxicity, abundant source, and tunable fluorescent emission while semiconductor Q-dots suffer from toxicity issues (Li et al. 2012). Their oxygen content is significantly high, which is estimated up to 40%, while they coupled with carbonyl, carboxyl, or epoxy groups (Baker and Baker 2010). Further, they can be modified with many other functional groups, passivation agents, which can enhance their optical property, chemical stability, biocompatibility, targeting ability, and solubility in water (Zuo et al. 2016a). Thus, C-dots exhibit many potential applications including sensors, biological labeling/imaging, drug delivery, sensors, catalysis, energy conversion, and in optoelectronic devices (da Silva and Gonçalves 2011; Wang and Hu 2014; Yang et al. 2014b).

1.1 Physics and Chemistry of Carbon Dots

C-dots have a quasi-spherical structure with sp^2 hybridized graphene and graphene oxide cores fused by the sp^3 hybridized diamond-like insertion (Lim et al. 2015). The tuneable fluorescent property is the major driving force behind C-dots research interest in which it enables their application potential in various fields. In general, this tuneable fluorescent property arises due to the quantum mechanical effects, since their dominance in lower dimensions (Bera et al. 2010). This is due to the fact that the electrons in the Q-dots are confined in all three dimensions and thus leads to the discreteness in its energy levels. As the particle size decreases, the band gap of the material slightly gets increased. When the electrons absorb energy, it jumps from valence band to conduction band and leaves a hole behind it. The fluorescent property arises due to the recombination of the electron–hole pair. In semiconductors, the increase in band gap due to quantum confinement leads to the change in its emission properties (Ilaiyaraja et al. 2018) but the mechanism is slightly different for C-dots. Here, the conjugated π domains and the surface or edge defects are the major parameters for their optical properties (Lim et al. 2015). The isolated π-conjugated domains of the C-dots have sp^2 hybridized islands rich in π-electrons. These π-electrons are the strongest light absorbers. A large number of this high-density π-electrons form excitonic states that result in the broad optical absorption. Commonly, C-dots have strong optical absorption in the ultraviolet region of 280–360 nm with the tail extending to visible region (Baker and Baker 2010). The sp^2 and sp^3 hybridized structures in C-dots have defective surfaces or edges. In general, the non-perfect sp^2 domains will lead to the surface energy trap (Demchenko and Dekaliuk 2013). While the excitation occurs, the electrons and holes get trapped at various surface sites, leading to their radioactive recombination which is responsible for the fluorescence emission of C-dots (Fernando et al. 2015) in UV/Vis to near-IR region due to their multiple surface defects. During passivating or functionalizing, C-dot surface defects become more stable permitting more effective radioactive recombination that results in brighter fluorescence emissions (Lim et al. 2015). The $-NH_2$ (amine), –CHO (aldehyde), –OH (hydroxyl), and –COOH (carboxyl) groups that present in the surface of the C-dots facilitate them to conjugate with various organic, inorganic, polymeric, and biological substances for effective surface passivation or functionalization (Mishra et al. 2018). C-dots along with the down-conversion fluorescence emission also possess up-conversion fluorescence emission after surface passivation process. The type of fluorescence in which the emission wavelength is shorter than that of the excitation wavelength is called as up-conversion fluorescence emission. It enables their applications in in vivo bio-imaging, drug delivery, anticancer therapy, and photoacoustic imaging since for improved photon tissue penetration and reduced auto-fluorescence, longer wavelength especially in near-IR region is required (Zuo et al. 2016a). Fluorescence properties are used to calculate the quantum yield (QY) of C-dots by comparing the integrated photoluminescence intensities at the same wavelength observed for C-dots as well as reference (e.g., quinine sulfate is used which has QY of 0.54) materials (Sahu et al. 2012). Photo-induced C-dots act as an excellent electron donor and acceptor, which leads to the excellent photo-induced electron transfer property (Wang et al. 2016e). This enables them for energy conversion, photocatalytic, bio-sensing, and photovoltaic device applications. Similar to semiconductor Q-dots, C-dots also possess excellent electrochemical luminescence which attributed to their sensing applications. Electroluminescence is mainly attributed to the surface state transitions, whereas photoluminescence is related to the transitions in its core state (Li et al. 2012). Figure 1 shows the schematic representation of the electrochemiluminescence (ECL) and photoluminescence (PL) mechanisms involved in C-dots (Zheng et al. 2009). All the above-discussed physicochemical and functional properties of the C-dots are mainly depended on the three important factors, which are the precursor material, synthesizing process and the

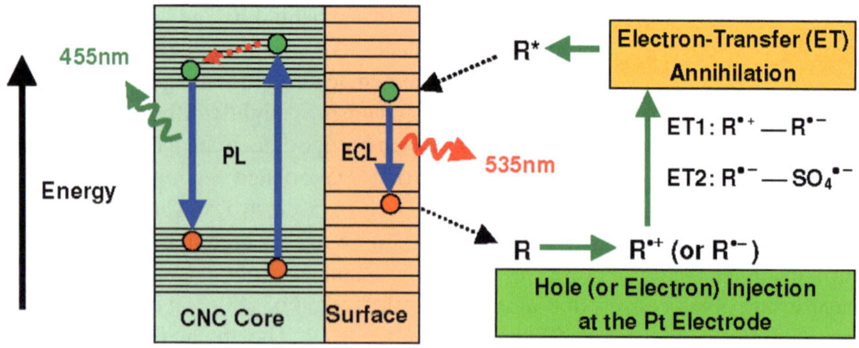

Fig. 1 Schematic representation of the ECL and PL mechanisms involved in C-dots (Zheng et al. 2009)

dopant elements. A wide range of precursor materials and synthesis processes have been explored for the carbon dots.

1.2 Processes Involved in Carbon Dot Synthesis

A small carbon nanoparticles (or) carbon dots exhibit the huge surface passivation and chemical modification with polymeric, inorganic, organic materials which are focusing for excellent features like high conductivity, improved chemical stability, broadband optical absorption, and other optical properties (Himaja et al. 2015; Lim et al. 2015). Thus, chemical, thermochemical, physicochemical tactics are involved to fabricate the efficient C-dots from the benign origins with the desired properties for diversified applications. C-dots can be synthesized by two different approaches such as "top-down" and "bottom-up" routes as shown in Fig. 2, and their details are discussed in the following sections.

1.2.1 Top-Down Approaches

The top-down approach generally deals with the breakdown of the biggest macro-sized structures into numerous nanostructures by various methods. In particular with the C-dots' synthesis, top-down route mentions the cutting of carbon materials into nanoparticle-sized carbon materials by using laser, arc, electrochemical, and plasma techniques. Proper selection of source molecule is the tricky point and nanostructures such as nanodiamonds, activated carbon, carbon nanotubes, carbon soot, graphite, and graphite oxide can be converted into C-dots by this top-down approaches (Lim et al. 2015; Shen and Liu 2016).

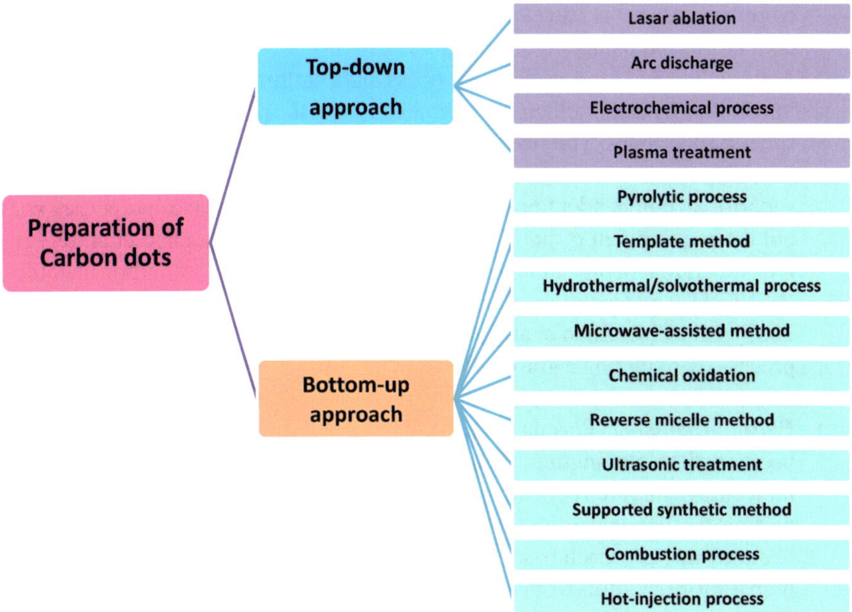

Fig. 2 "Top-down" and "bottom-up" routes for the synthesis of C-dots

(i) *Laser ablation*: Laser ablation is nothing but the removal of certain materials from a solid surface by using laser irradiation. In this technique, the laser flux takes main responsible, and in lower laser flux, the target material conversion happens by evaporation, whereas in higher laser flux, the material is converted into plasma. The removed materials often found to exhibit nanoscale dimension. This kind of approach was quite common for producing inorganic nanoparticles from solid substrates. C-dots were generated by laser ablation technique with argon gas flow in the presence of water vapor (Gonçalves et al. 2010; Sun et al. 2006). In general, the carbon target made from graphite powder and binder was used as the source for carbon dots. A wide range of laser sources like Q-switched Nd:YAG laser (1064 nm, 10 Hz) and UV pulsed laser irradiation (248 nm, KrF) were used for the ablation of carbon.

(ii) *Arc discharge*: Carbon dots were accidentally discovered while purifying the single-walled carbon nanotubes (SWCNT) produced by arc discharge process (Xu et al. 2004). In this process, the arc is produced using carbon-rich electrodes employing electrical discharge at the highest current density under the controlled atmosphere. By manipulating the composition of the carbon electrode source, it is possible to change the elemental dopant of the carbon dots in an effective manner. Dey et al. reported the synthesis of B- and N-doped graphene quantum dots by employing arc discharge process using doped graphene as the carbon source (Dey et al. 2014). In general, the yield of carbon dots produced through arc discharge process is significantly low, which consists of a variety of complex materials (Zuo et al. 2016b). Hence, it is difficult to purify the carbon dots fabricated using arc discharge.

(iii) *Electrochemical process*: In this process, bulk carbon material can be converted into C-dots by electrochemical fragmentation in the presence of electrolytes and specific electrodes. The size and the graphitization degrees of the C-dots can be precisely tuned by adjusting the applied potential. The resultant particles show superior size-dependent optical (excitation and fluorescence) properties without the requirement of purification and passivation steps. Zhou et al. reported the preparation of blue luminescent C-dots from multi-walled carbon nanotubes (MWCNTs) by electrochemical approach as a pioneer observation (Zhou et al. 2007). Likewise, Zhao et al. reported the synthesis of carbon dots through this process by oxidizing a graphitic column electrode in 0.1 M NaH_2PO_4 aqueous solution (Zhao et al. 2008).

(iv) *Plasma treatment*: Plasma is one of the states of matter, and generally, it has been used as etching/removing agent of certain molecules from the bigger molecules in the area of material science to get expected materials. Interestingly, plasma treatment can also be used for the synthesis of C-dots by obeying "bottom-up" approach based on the process involved. Wang et al. reported the preparation of C-dots from egg yolk and white by plasma treatment (Wang et al. 2012). Sung-Soo et al. reported the synthesis of Au nanoparticles as micellar core covered (or shield) by graphene nanodots by oxygen plasma etching (Sung-Soo et al. 2012). Among the various top-down methods, plasma-based synthesis is to be a simple and rapid process for the synthesis of carbon dots.

1.2.2 Bottom-Up Approaches

The bottom-up approach is nothing but build up a bigger molecule from small molecules by exploring various synthetic methods which are vice versa to the top-down approach. A wide range of bottom-up processes which incldues, pyrolysis, hydrothermal, microwave synthesis, chemical oxidation, and reverse micelle are extensively studied for the synthesis of carbon dots. In this approach, molecules such as citrates, carbohydrates, organic acids are generally used.

(i) *Pyrolytic process*: Pyrolytic (or) carbonization process refers to the pyrolysis/carbonization of carbon precursors into C-dots under high-temperature conditions in the presence of alkali or acid sources. In this method, the role of alkali or acid is to produce a colloidal state of C-dots by breakdown the carbon precursors. Wang et al. reported novel simple one-step carbonization process for the preparation of C-dots by heating octadecene and 1-hexadecylamine mixture at 300 °C under argon flow followed by sequential addition of citric acid (acid source) and heating at 300 °C for 3 h (Wang et al. 2010).

(ii) *Hydrothermal/solvothermal process*: Hydrothermal/solvothermal is the majorly explored method for the preparation of C-dots, which deals with the crystallization of various substances in aqueous system at desired temperature and pressure. This process mainly depends upon the solubility of the mineral in hot solvents which supports the growth of the particles with large quality. Even a simple heat treatment can result in the C-dots with nitrogen doping which has revealed the highly fluorescent behavior (Ren et al. 2017). Liu and the team achieved the $NiFe_2O_4$ integrated with ethylenediaminetetraacetic acid (EDTA) derived C-dots in nano-level scale via solvothermal method using high pressure and temperature of 200 °C for 10 h (Liu et al. 2017a).

(iii) *Microwave-assisted method*: Microwave-assisted synthesis of materials is known for their rapid heating (similar to hydrothermal process) that results in a tremendous increase in reaction rate. In this method, the carbon precursors can be quickly converted into C-dots without any difficulty in a short reaction time with improved QY. Combination of both hydrothermal and microwave techniques leads to the advanced process termed as the microwave hydrothermal process which further simplifies the synthesis process. In 2011, Wang et al. reported the synthesis of fluorescent C-dots from graphite oxide by microwave-assisted as well as microwave hydrothermal processes and found that the later mentioned process is better in terms of decrease in the reaction time along with improved fluorescence properties of C-dots (Wang et al. 2011).

(iv) *Chemical oxidation*: Chemical oxidation is the well-known process for the synthesis of C-dots from various carbon sources which are generally in the form of solids. The role of oxidant plays a vital role in this synthetic process. By using this method, C-dots can be prepared in large scale in an easy way. Sun et al. reported the synthesis of sulfur, nitrogen co-doped C-dots (S, N–C-dots) by carbonization and etching of hair fiber by using sulfuric acid as an oxidant (Sun et al. 2013).

(v) *Template method*: Template method is a well-known synthetic methodology in the area of materials science. This synthetic process follows two steps such as synthesizing C-dots by calcining the suitable mesoporous template and etching to remove the supports to get expected sized C-dots. Zong et al. reported the pioneer synthesis of C-dots/mesoporous silica composites by the impregnation method, and then the prepared composite was subjected to calcination to remove the template as well as to get the expected hydrophilic C-dots (Zong et al. 2011).

(vi) *Other methods*: A reverse micelle is a process for the preparation of nanometer-sized materials in the organic medium by the action of the surfactant molecules. This kind of synthetic approach helps to prepare the hydrophobic C-dots by a sequence of modification procedures. Linehan and Doyle developed a simple route for the synthesis of dodecane-capped CQ-Dots, by reverse micelles in the room temperature solution phase (Linehan and Doyle 2014). Ultrasonic method is a nondestructive experiment process based upon the distribution of ultrasonic waves in the target which is able to perform as highly automated operation. Oza et al. achieved quantum-sized C-dots from citrus limonene as a precursor by an ultrasonic pure wave with NaOH solution (Oza et al. 2015). The changeover on the material has indicated by using high pure wave applied in a range of 20 kHz for 1 h. Fabrication of green-emitted fluorescent of boron and nitrogen co-doped C-dots (B–N–C-dots) by the combustion method resulted in high resistance for both photobleaching and ionic strength (Rong et al. 2017). One-step chemical approach also reported for the preparation of N–ZnO-doped C-dots by using a modified Hummer's method, and the reaction mixture drives to the high thermal state of 240 °C for 2 h (Muthulingam et al. 2016).

1.3 Precursors for Carbon Dots

Selection of suitable precursor is the key point to obtain the desired physicochemical and functional properties in C-dots. In general, the sources which have acceptable amount of carbon content in it will be more suitable to convert into C-dots. Since the accidental discovery of C-dots in 2004, subsequent research works have been developed to prepare the C-dots from various precursors by various synthesizing techniques. In general, C-dots have been synthesized by using various carbon-rich precursors such as graphite (Sun et al. 2006), graphite oxide (Wang et al. 2011), carbon nanotube (Zhou et al. 2007), carbon black (Dong et al. 2012), activated carbon (Qiao et al. 2010), and carbon tetrachloride (Zhang et al. 2012). Organic solvents such as benzene and toluene (Jiang et al. 2010; Yu et al. 2015a), saccharides such as glucose and chitosan (Chowdhury et al. 2012; Ma et al. 2012), organic acids such as acetic acid and citric acid (Fang et al. 2012; Qu et al. 2014), vitamins such as folic acid and ascorbic acid (Bhunia et al. 2016; Wu et al. 2012), and amino acids such as serine, histidine, cystine, and glycine (Li et al. 2015; Pei et al. 2015) are the commonly used precursors. By taking the multiple sources, functionalized/modified C-dots can be prepared.

Surface modification/passivation of C-dots is the way to improve their properties which can be done by adding different functionalization sources along with precursors (Liang et al. 2014). Doping is one of the functionalization techniques used to introduce some components in the C-dot's matrix. The doping of heteroatom on carbon surface improves the fluorescent properties by controlling the pH values. In order to get nitrogen-doped C-dots (N–C-dots), the addition of nitrogen-containing compounds is needed during the synthesis process (Atchudan et al. 2016a). Mostly, organic nitrogen compounds like urea and amines are used for this purpose (Ding et al. 2015b). The organic precursors, which exhibit higher nitrogen content, can also be used for the preparation of N–C-dots without any additional nitrogen source. In the literature, maximum studies have been reported for the preparation of N–C-dots due to the easy handling of nitrogen source materials. Other nonmetal-doped C-dots, e.g., sulfur-doped C-dots (S–C-dots) using sulfonic acid derivatives (Lu et al. 2015) and phosphorous-doped C-dots (P–C-dots) using phosphorus pentoxide (Chen et al. 2018b), were also reported. Apart from nonmetal doping, few literatures are available for the preparation of metal (gadolinium (Gd) (Yu et al. 2016), terbium (Tb) (Xu et al. 2016a), iron (Fe) (Yang et al. 2017a)-doped C-dots using their corresponding metal salts as sources. Also, metalloid (boron (B))-doped C-dots are reported using boric acid as B source molecule (Bourlinos et al. 2015). Like single component-doped C-dots, dual components (nonmetal/metal/metalloid)-doped C-dots can also be prepared by taking separate/single sources (Gu et al. 2018; Nabid et al. 2016). Apart from the doping of nonmetal/metal/metal components, modification of the C-dots surface by numerous active molecules (Zhong et al. 2015) is also reported for specific applications as per the requirement.

Most of the precursor materials that have been used for the synthesis of C-dots are originated from non-renewable fossil resources. As the global concern arises in utilizing fossil resources as the feedstock for materials development due to various environmental issues including greenhouse gas emission, global warming, and climate change, there is a demand for eco-friendly alternative. Thus, a transformative research approach has been devoted in utilizing renewable precursors for the development of various carbon nanomaterials due to the presence of high carbon content in their compositions. As the result, a wide range of carbon-rich renewable precursors (agro/forestry derivatives/residues, industrial coproducts, clean municipal organic wastes, and marine products) have been explored for the synthesis of carbon nanomaterials that include carbon nanotubes, graphene, carbon nanofibers, and carbon dots (Ding et al. 2018; Sharma et al. 2017; Vivekanandhan et al. 2017; Zhang et al. 2018). In the case of C-dots preparation, renewable sources can be used effectively due to their plenty of availability as well as easy synthesis methods. Depending upon the components present in the renewable sources, modified/doped C-dots can be directly prepared without adding any chemical source. This chapter aimed to cover the reports on the preparation of C-dots from various renewable sources for diversified potential applications, and the overall outlay of the chapter is given in Fig. 3.

2 Carbon Dots from Renewable Sources

In recent years, the synthesis of C-dots from sustainable renewable bio-sources has received great attention because of their surplus availability and the diversity of their chemical compositions that are used for various applications. The whole process is considered as an eco-friendly, which does not require any noxious precursor to initiate the reaction. The carbon-rich materials are located in all kinds of renewable resources such as plants and its derivatives, animal-based products, marine products, and industrial waste streams with the distinct natures have been plenteously explored for the fabrication of C-dots.

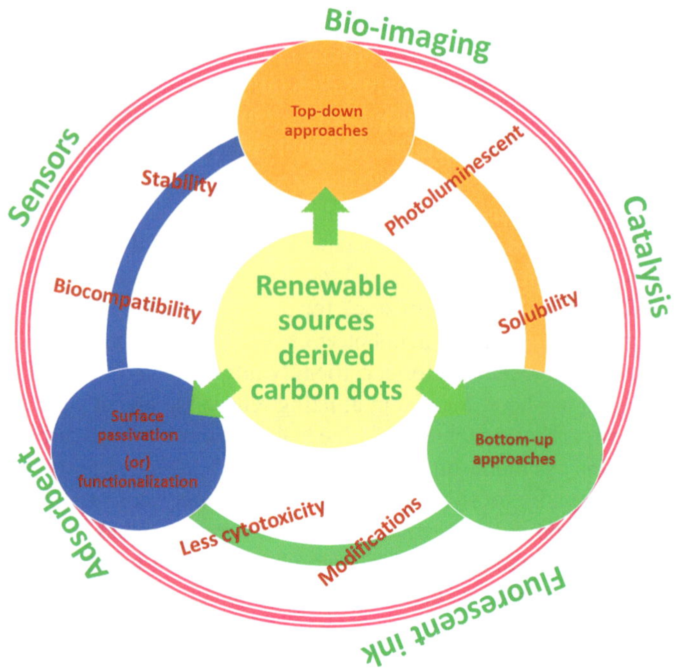

Fig. 3 Schematic diagram represents precursors, processes, modifications, and potential applications of bio-derived carbon dots

2.1 *Plant-Based Precursors*

Plant-based precursor materials that include biomasses, extracts and juices, plant derivatives, and plant wastes have been effectively explored for the synthesis of carbon dots with the diverse physicochemical and functional properties. Understanding the relationship between the phytochemical diversity of plant-based precursors and the physicochemical properties of the obtained carbon dots is very essential to achieve desired properties for extended applications. In this section, utilization of plant-based precursors toward the preparation of C-dots by using different synthetic methods is discussed.

2.1.1 Plant Biomasses and Their Extracts

Plant biomasses have been considered as suitable material for the fabrication of various carbonaceous materials including carbon dots due to their higher carbon content. Various biomasses including grass, leaf, stem, and their extracts were effectively utilized for the synthesis of carbon dots with varying physicochemical properties. Liu et al. investigated the utilization of hydrothermal process (180 °C for 3 h) for the fabrication of carbon-rich photoluminescent polymer nanodots (C-dots) by taking grass biomass as the carbon source (Liu et al. 2012). The obtained C-dots were found to have the size ranging between 3 and 5 nm. Jiang et al. reported the effective utilization of zoysia grass biomass for the synthesis of C-dots by employing hydrothermal process at 180 °C for 3 h (Jiang et al. 2014a). Further, they have adopted sonication process for 1 h at 313 K to get Ag/C-dots nanocomposites using different concentrations of $AgNO_3$ solution. Recently, Sabet and Mahdavi reported the preparation of N–C-dots from grass by hydrothermal treatment at 180 °C for 2 h (Sabet and Mahdavi 2019). Yuan et al. reported the synthesis of green-emitting C-dots from wheat straw as the feedstock by employing hydrothermal process at 250 °C for 10 h (Yuan et al. 2015). The obtained C-dots were identified as amorphous carbon, which found to have a size of ~2 nm. Very recently, wheat straw and bamboo residues (biorefinery process by-products) were used as carbon source along with urea (nitrogen source) for the preparation of N–C-dots by one-pot hydrothermal treatment at 180 °C for 4 h (Huang et al. 2019). In addition to that, Kumar et al. also demonstrated the hydrothermal process for the synthesis of carbon dots by using *Ocimum sanctum* leaf as the carbon source (Kumar et al. 2017). The obtained carbon dot found to have the size of ~3.0 nm. Along with the hydrothermal process, pyrolysis process also been explored for the synthesis of carbon dots. Zhu et al. explored the direct pyrolysis of various leaf biomass for the synthesis of fluorescent carbon dots with the size ranging 3.7 nm (Zhu et al. 2013). The obtained C-dots exhibited various potential applications including sensing, patterning, and coding. Assam tea leaf was explored as the carbon source for the synthesis of C-dots by Baruah et al. by adopting a simple calcination process at 200 °C for 5 h. C-dots were extracted by dispersing the calcined Assam tea leaf in acetic acid. The obtained fluorescent carbon dots were found to have the particle size of ~0.8 nm (Baruah et al. 2014). Recently, Zhai et al. reported the pyrolysis process for the synthesis of carbon dot from *Setcreasea purpurea* boom as a carbon source (Zhai et al. 2018). In addition to that, tulsi leaves also been used as the source for the synthesis of carbon dots by employing hydrothermal process (Bhatt et al. 2018). *Eichhornia crassipes*, a plant source, was used as powder form for the preparation of N–C-dots by hydrothermal process at different temperatures for 6 h in water medium (Yang et al. 2018). Chen et al. reported capsicum as a source for the synthesis of C-dots by hydrothermal treatment at 240 °C for 10 h in water medium (Chen et al. 2018a). Romero et al. used 18 different vegetables extract (such as broccoli, carrot, cauliflower, celery, chard, collard greens, courgette, endive, fennel, green beans, green Italian pepper, pumpkin, red bell pepper, romanesco, tomato, turnip greens, white cabbage, and yellow bell pepper) containing carbohydrates as a carbon source for the preparation of N–S-doped C-dots by following photochemical oxidation using NaOH and poly(ethylene glycol) (Romero et al. 2019). Likewise, garlic gloves were used as a source for the preparation of N,S,C-dots by hydrothermal method at 200 °C for 3 h in water medium (Zhao et al. 2015). Shasha et al. utilized celery stalk for the preparation of C-dots with blue photoluminescence by hydrothermal process at 190 °C for 6 h (Shasha et al. 2019).

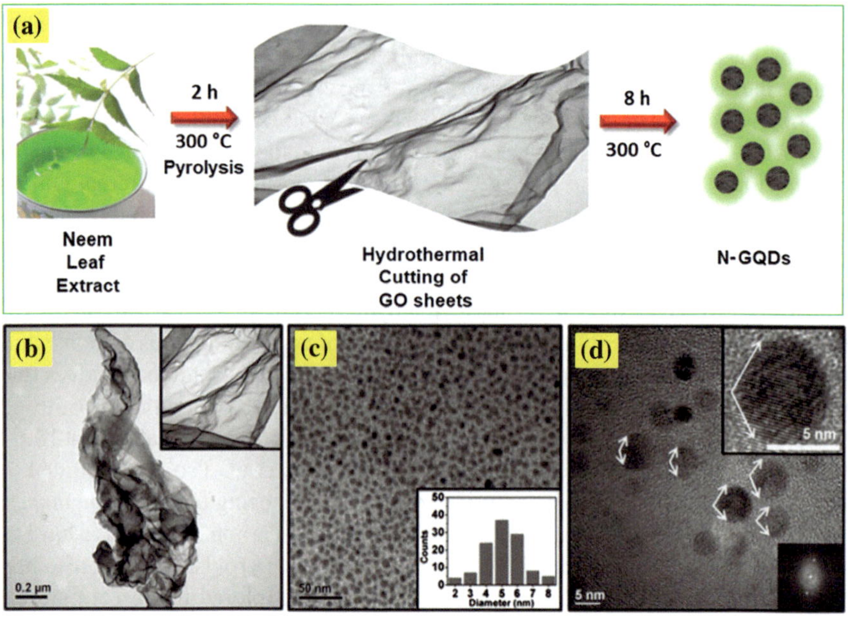

Fig. 4 **a** Schematic representation of the synthesis of GQDs from neem leaf extracts, **b** TEM image of the GO intermediate, and **c**, **d** synthesized GQDs (Roy et al. 2014)

Apart from the plant biomasses as the direct source for carbon dots, a wide range of plant extracts have also been explored for the synthesis of carbon dots by employing various techniques. Roy et al. reported the synthesis of graphene quantum dots (GQDs) using the neem and fenugreek leaf extract by employing hydrothermal process at 300 °C for 8 h without using any passivating and reducing agents (Roy et al. 2014). TEM analysis confirmed the formation of GQDs with the size ranging between 5 and 7 nm, respectively, synthesized from neem and fenugreek leaf extracts. Figure 4 shows the schematic of GQDs synthesis from neem leaf extract along with the TEM images of the synthesized GQDs. Sachdev and Gopinath explored the possible utilization of coriander leaves extract for the synthesis of C-dots employing the hydrothermal process for 4 h at 240 °C (Sachdev and Gopinath 2015). The synthesised carbon dots were found to have a mean diameter of 2.387 nm with size distribution ranging between 1.5 and 2.98 nm. It was found that the leaf extract consists of the hydrocarbons with low oxygen content which is suitable for the synthesis of graphene quantum dots. Further, Konwar and Chowdhury prepared carbon dots composite using tea-derived C-dots and alginate with different bivalent and trivalent cross-linker ions (Konwar and Chowdhury 2015). The fabricated alginate–carbon dot nanocomposite showed enhanced properties than the pristine alginate. In addition to that, Mehta et al. reported the utilization of carbon-rich *Saccharum officinarum* (sugar cane) juice for the preparation of C-dots by employing hydrothermal treatment at 120 °C for 180 min in ethanol medium (Mehta et al. 2014). They were able to obtain the C-dots with an average size of ~3 nm. Freshly chopped garlic (commonly identified as the modified underground stem) contains allicin, which serves as the efficient carbon, nitrogen, sulfur, and oxygen sources, and it has been converted into N–S-doped C-dots employing hydrothermal approach (Chen et al. 2016). Very recently, Pudza et al. reported the synthesis of fluorescent carbon dots from tapioca (starch extract from cassava plant roots) by hydrothermal process (Pudza et al. 2019).

2.1.2 Flower, Fruit, and Seed Extracts

The phytochemical composition of the plant components that include leaf, root flower, fruit, and seed is highly diversified, which can effectively alter the properties and functionalities of the carbon dots derived from them. Hence, research efforts have been paid to explore the various parts of the plant for carbon dots synthesis.

Flower: Feng et al. reported the synthesis of blue-emitting carbon dots from rose flower by employing microwave process (Feng et al. 2015b). The obtained carbon dots were found in the size range between 4 and 6 nm. Atchudan et al. investigated the utilization of *Magnolia liliiflora* flower as the carbon source for the synthesis of carbon dot by employing hydrothermal process (Atchudan et al. 2018). The hydrothermally synthesized C-dots were found to have the size ranging about 4 nm with nearly spherical in shape. Murugan and Sundramoorthy reported the synthesis of carbon dots from *Borassus flabellifer* flower through pyrolysis at 300 °C followed by ultrasonication (Murugan and Sundramoorthy 2018). TEM analysis found that the size of the C-dots ranged between 3 and 8 nm. Ensafi et al. reported the preparation of C-dots using saffron as a carbon source by hydrothermal treatment at 200 °C for 14 h (Ensafi et al. 2017).

Fruit: A wide range of fruits have been explored as the carbon source for the synthesis of C-dots. Apple juice was explored as the carbon source by Yue et al. and Mehta et al. for the synthesis of carbon dots by employing hydrothermal process, respectively (Yue et al. 2014; Mehta et al. 2015). From their reports, it was understood that lowering the reaction temperature results in the higher particle size of the carbon dots. Yue et al. employed 180 °C for 12 h and obtained the particle of 2.8 $\pm$ 0.4 nm, whereas Mehta et al. obtained the CD particles with the size of 4.5 $\pm$ 1.0 nm while applying the reaction temperature of 150 °C for 12 h. Jin et al. reported the synthesis of C-dots from carrot juice by hydrothermal process at 180 °C for 5 h (Jin et al. 2017). Further, they dispersed the polyethyleneimine (PEI) and Nile blue (NB) solution into the C-dots aqueous suspension in order to get C-dots/PEI/NB nanocomposites. Recently, Pakkath et al. reported the synthesis of C-dots doped with transition metal ions (Mn^{2+}, Fe^{2+}, Co^{2+}, and Ni^{2+}) through microwave-assisted pyrolysis employing 800 W for 6 min using fresh lemon juice and ethylenediamine (EDA) as sources (Pakkath et al. 2018). Oza et al. explored the ultrasonic process for the synthesis of blue-emitting carbon dots from the citrus limonene juice followed by the sucrose density gradient centrifugation-based purification process (Oza et al. 2015). Yu et al. explored the *Jinhua bergamot* fruit as the precursor for the preparation of C-dots by hydrothermal process at 200 °C for 5 h and attained the quantum yield of 50.78% (Yu et al. 2015b). Along with the fruit extracts, aqueous ammonia was used as the nitrogen source for the fabrication of nitrogen-doped C-dots. Edison et al. reported the preparation of N–C-dots using *Prunus avium* fruit extract as carbon source and ammonia as a nitrogen source by hydrothermal carbonization at 180 °C for 5 h and that attained the average size of 7 nm (Edison et al. 2016b). Similarly, Arul and team used *Hylocereus undatus* extract with aqueous ammonia for the preparation of N–C-dots by hydrothermal carbonization at 180 °C for 12 h (Arul et al. 2017). Also, the same approach was reported by Atchudan and researchers toward the synthesis of N–C-dots using the unripe fruit extract of *Prunus persica* (peach) (Atchudan et al. 2016a), *Prunus mume* (Atchudan et al. 2016b), *C. retusus* (Atchudan et al. 2017) fruit extracts and aqueous ammonia by hydrothermal treatment at 180 °C for 5 h. Recently, Lu et al. reported the watermelon juice-assisted hydrothermal synthesis of nitrogen-doped carbon dots at 180 °C for 3 h (Lu et al. 2018). Figure 5 shows the schematic illustration of carbon dot synthesis from watermelon juice for the Fe^{3+} ion deduction, TEM micrograph, Raman, FTIR, and XPS spectra. In addition to the above discussed, various other fruit pulps/extracts/juices were also been explored for the synthesis of C-dots, which include banana (Ghosh et al. 2016), guava (Wang et al. 2014b), papaya (Kasibabu et al. 2015a; Wang et al. 2016d), winter melon (Feng et al. 2015a), pomegranate (Kasibabu et al. 2015b), *Carica papaya* (Kasibabu et al. 2015a), tamarind (Jhonsi and Thulasi 2016), lime (Barati et al. 2015; Suvarnaphaet et al. 2016), orange (Amjadi et al. 2015), tomato (Liu et al. 2017c; Miao et al. 2016), tender coconut water (Purbia and Paria 2016), and quince fruit (Ramezani et al. 2018).

Seed Products: To some extend, few biomasses are having in-build nitrogen content, which makes them a suitable source for the preparation of nitrogen-doped C-dots by single-step approach. Soybean consists of 5.6% of nitrogen, which is higher than the other natural products. Hence, soybean can be used as the feedstock for the synthesis of nitrogen-doped carbon dots. Meng et al. reported the preparation of C-dots from soybean powder by hydrothermal treatment at 170 °C for 4–30 h (Meng et al. 2017). Soy milk was explored as the source for carbon dot by Zhu et al. for metal-free electro-catalytic applications (Zhu et al. 2012). Jiang et al. reported the existence of photoluminescent carbon dots in commercial Nescafe original instant coffee (Jiang et al. 2014b). Interestingly, Fang et al. explored the soy flour as the precursor for the synthesis of carbon dots by employing hydrothermal process at 200 °C for 3 h (Fang et al. 2016). The synthesized C-dots were made of graphitic carbons and found to have a mean diameter range of 2.5 nm. In addition to that Xu et al. also reported the fabrication of C-dots from the soybean powder employing pyrolysis process in a pipe (tubular) furnace at 200 °C for 3 h under the argon atmosphere (Xu et al. 2016b). Wang et al. reported the preparation of graphene Q-dots from coffee grounds employing hydrazine hydrate-assisted hydrothermal method at 150–200 °C for 6–10 h (Wang et al. 2016c). The resulted carbon dots were treated with poly(ethylene

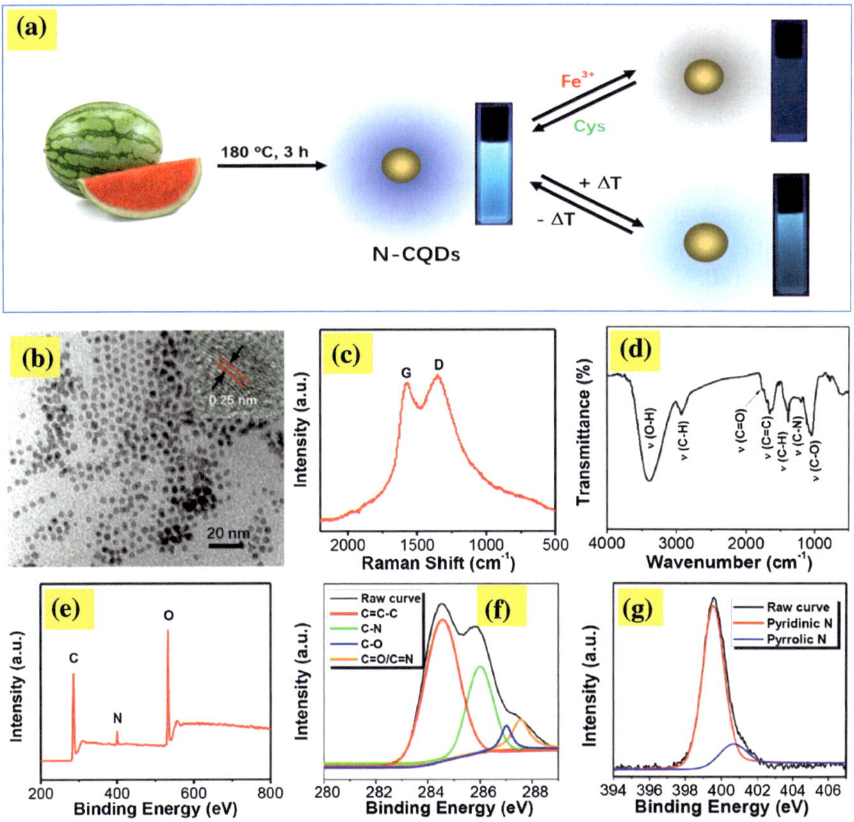

Fig. 5 **a** Schematic illustration of N-CQDs synthesis from watermelon juice by hydrothermal treatment for Fe^{3+} ion deduction, cysteine (Cys) and temperature, **b** TEM micrograph, **c** Raman, **d** FTIR, and **e**–**g** XPS spectra. (Lu et al. 2018)

imine) to enlarge their fluorescent yield for the efficient bio-imaging and sensing of heavy metals. Likewise, kidney beans, which contain extensive protein, also been explored for the synthesis of carbon dots by employing pyrolysis process at 450 °C for 2 h (Tripathi et al. 2017). The fluorescent performance of the synthesized carbon dots was enhanced through nitric acid treatment. In 2019, Li et al. explored different natural seeds such as kiwi, white sesame, and black sesame as an effective source for the preparation of N–C-dots by pyrolysis method (Li et al. 2019).

2.1.3 Root/Tuber-Mediated Carbon Dots

As the exploration for the biogenic source for carbon dots expands, various types of root/tuber feedstocks (flours and extracts) were investigated. Teng et al. investigated the effective utilization of konjac tuber flour (heteropolysaccharides) as the source for the synthesis of nitrogen-doped carbon dots employing mild pyrolysis process followed by a simple ethanol and water extraction (Teng et al. 2014). Li et al. reported the synthesis of carbon dot from ginger juice employing hydrothermal treatment at 300 °C for 2 h and the obtained carbon dots were found to have a size of 4.3 nm (Li et al. 2014). Likewise, lotus root has also been utilized for the preparation of N–C-dots by applying one-pot microwave treatment without using any other surface passivating agents (Gu et al. 2016). Further, the flour of salep tuber was explored by Jahanbakhshi and Habibi for the synthesis of carbon dots employing hydrothermal carbonization process at 180 °C for 12 h (Jahanbakhshi and Habibi 2016). The obtained C-dots were decorated with silver nanoparticles (NPs) on the surface in order to form nano-hybrids. Liu and the team reported the synthesis of N–C-dots from rose-heart radish which has rich amount raw protein, amino acid, carbohydrates, and vitamins, through autoclave carbonization at 180 °C for 3 h (Liu et al. 2017b). Further, sweet potato has also been explored for the synthesis of carbon dots in the range of 2.5–5.5 nm by Shen et al. for Fe^{3+} sensing and cell imaging applications (Shen et al. 2017).

2.1.4 Plant Derivatives and Plant-Based Products

Camphor, a plant-derived material was explored for the synthesis of water-soluble C-dots using three-step processes such as acid oxidation (using nitric and sulfuric acids), alkaline hydrolysis (by sodium hydroxide), and sodium borohydride reduction (Oza et al. 2016). Gholinejad et al. reported the synthesis of carbon quantum dots from vanillin (extract of vanilla bean) in the presence of ethanol as solvent by hydrothermal treatment at 180 °C for 12 h (Gholinejad et al. 2017). Further, they have also prepared the CQ-dots-coated Fe_3O_4 nanoparticles (CQ-dots@Fe_3O_4) as well as palladium loaded on CQ-dots@Fe_3O_4 (Pd@CQ-dots@Fe_3O_4) by simple chemical processes. Ma et al. reported the preparation of C-dots from sodium citrate and glutathione (antioxidant present in plants, animals, fungi, and some bacteria) under water medium by hydrothermal treatment at 200 °C for 4 h, and they achieved the QY of 21.03% (Ma et al. 2016). Glucose as carbon source along with PEG-200 was reported for the preparation of C-dots by microwave pyrolysis for 5 min at 500 W (Xu et al. 2016a), and the prepared C-dots were vigorously stirred (room temperature, 2 h) with aqueous terbium chloride ($TbCl_3$) solution to get Tb–C-dots. Amjadi et al. reported the preparation of C-dots from glucose by microwave heating at 720 W for 10 min without using any further chemicals (Amjadi et al. 2017). Further, they studied the chemical reduction of $AgNO_3$ with C-dots at room temperature in order to achieve the C-dots/Ag nanocomposites. Interestingly, photochemical treatment (UV irradiation) of sucrose/NaOH/poly(ethylene glycol) system was reported for the preparation of carbohydrate-based C-dots within 1 min, and this approach resulted in the monodisperse C-dots with an average size of 8 nm (Costas-Mora et al. 2015). Dry heating method was explored by Weng et al. for the synthesis of mannose-modified fluorescent CQ-dots by using ammonium citrate and mannose as precursors (Weng et al. 2015). They obtained high soluble C-dots having an average particle size diameter of 3.1 ± 1.2 nm which exhibits the QY of 9.8% with the wavelength of 365 nm (excitation) and 450 nm (emission). Starch was heated at high temperature with pressure to obtain black carbon source which was then treated with mixed oxidants (acetic acid and 30% hydrogen peroxide) and refluxed for 8 h at 120 °C to prepare green C-dots (Yan et al. 2016). Liang et al. reported the preparation of C-dots from xylan (rich in hemicellulose) which was subjected to various chemicals to get NH_4OH passivated C-dots by hydrothermal treatment at 200 °C for 12 h (Liang et al. 2014). Water dissolved neem (*Azadirachta indica*) gum was taken as carbon source along with NaOH/ethanol (1:1 ratio) and subjected to sequential incubation (at room temperature for 3 h) and dialysis to get C-dots (Phadke et al. 2015). In addition, cashew gum was also explored for the preparation of C-dots by microwave-assisted method at 800 W for 30–40 min in which the formation of C-dots follows two steps such as partial depolymerization (results 5-hydroxymethyl furfural) and polycondensation/polymerization processes (Pires et al. 2015). Gaddam et al. achieved C-dots by oxidative treatment (piranha solution; an oxidative mixture of H_2SO_4 and H_2O_2 in 7:3 ratio) method of carbonized (350–400 °C; till the material degraded) gum olibanum resins as a source by refluxing for 5 h (Gaddam et al. 2017).

2.1.5 Plant Residues and Industrial Coproducts

During the agricultural harvesting/post-harvesting/industrial processing, a huge amount of plant residues and industrial coproducts are produced, which can be considered as the feedstock for chemicals and materials. As they exhibit significant carbon content, those plant residues and industrial coproducts can be effectively explored for the production of various carbonaceous materials including carbon dots. Watermelon peel was utilised as a source for the synthesis of C-dots at 220 °C for 2 h under air atmosphere followed by sonication for 30 min (Zhou et al. 2012). Likewise, Prasannan and Imae also reported the synthesis of C-dots from orange waste peels by hydrothermal carbonization at 180 °C for 12 h (Prasannan and Imae 2013). Further, the obtained C-dots were loaded on the ZnO (C-dots–ZnO) by dispersing ZnO and C-dots in 2-propanol with the stirring for 4 h followed by oven heating at 100 °C for 3 h. The extract of *Trapa bispinosa* peel was used by Mewada et al. for the preparation of C-dots by refluxing the biomass at 90 °C for 2 h in distilled water without the addition of oxidizing agents (Mewada et al. 2013). The resulted C-dots were ranging from 5 to 10 nm and exhibited prominent green fluorescence under UV radiation. Vandarkuzhali et al. utilized pineapple peel as a source for the preparation of C-dots (Vandarkuzhali et al. 2018). In detail, pineapple peel water extract and ethanol mixture were subjected to hydrothermal treatment at 150 °C for 2 h. Onion waste extract along with ethylenediamine (EDA) was utilized by Bandi et al. for the preparation of C-dots by hydrothermal treatment at 120 °C and 15 lbs pressure (Bandi et al. 2016). Park et al. reported the synthesis of water-soluble green carbon nanodots (~4 nm) in large-scale from food waste-derived sources as a novel attempt (Park et al. 2014). Similarly, the peels of Indian yellow cucumber and the lemon fruit were used for the synthesis of C-dots, respectively, by employing hydrothermal process (Himaja et al. 2014; Tyagi et al. 2016).

Zhao et al. utilized the industrial citrus pomace residues as the source for citrus pectin and effectively converted into cyan fluorescence C-dots through hydrothermal carbonization (Zhao et al. 2014). The synthesized C-dots were found to have 2.7 nm and well dispersed in water with the advantages of high photostability and good biocompatibility, which is suitable for bio-imaging. Rodrigues et al. reported the effective utilization of the brewing industry waste of wort bagasse for the synthesis of C-dots (Rodrigues et al. 2015). They have converted the wort bagasse into C-dots by carbonization at 300 °C for 3 h followed by two different methods such as sequential oxidation and extraction process. These processes result in the C-dots with the sizes, respectively, ranging between 2.5 and 5.9 nm (sequential oxidation) and 2.5 and 5.3 nm (extraction). Thambiraj and Sankaran utilized the sugarcane bagasse pulp for the preparation of C-dots through their combustion at 60 °C in an open atmosphere followed by exfoliation with toluene under ultrasonication for 1 h (Thambiraj and Shankaran 2016). Figure 6 shows the schematic representation of the synthesis of carbon dots from sugarcane bagasse pulp along with the TEM images, particle size distribution and elemental composition of the obtained C-dots. The carbon dots were found in the size range of 4.1 ± 0.17 nm with excellent stable fluorescent properties. Following this, Eslami et al. also explored the sugarcane bagasse as the source for the synthesis of C-dots (Eslami et al. 2018). Tan et al. reported the synthesis of C-dots from sago industrial waste using thermal pyrolysis process with the temperature ranges from 250 to 450 °C, and then the heated sago waste was dispersed in water followed by homogenization under sonication for 2 h (Tan et al. 2014). Xue et al. used the peanut shells as a precursor for the synthesis of C-dots by pyrolysis/carbonization at 250 °C for 2 h, and then the obtained sample was ultrasonically dispersed in water to get homogeneous solution (Xue et al. 2016). Interestingly, waste frying oil (WFO) was also converted into S–C-dots in the presence of H_2SO_4 by one-step carbonization processes (Hu et al. 2014). Further, paper waste also been explored as the source for the synthesis of carbon dots (Lin et al. 2018).

2.1.6 Others

Tan et al. reported the synthesis of multicolor C-dots from plant soot by HNO_3 oxidation process by refluxation of the mixture at 110 °C for 20 h followed by neutralization by Na_2CO_3 aqueous solution (Tan et al. 2013). Further, they have synthesized the four functionalized C-dots by using various surface passivating agents such as 4,7,10-trioxa-1,13-tridecanediamine, poly-L-lysine, cysteine, and chitosan. Also, they have reported the preparation of reduced C-dots by stirring the prepared C-dots with the reducing agent $NaBH_4$ at room temperature for 24 h (Tan et al. 2013). Li et al. reported the preparation of nano-sized C-dots by refluxing the incomplete combustion soots of candles and corn stalk in nitric acid with stirring for 20 h followed by neutralization with NaOH (Li et al. 2016). Also, pectin in the form of

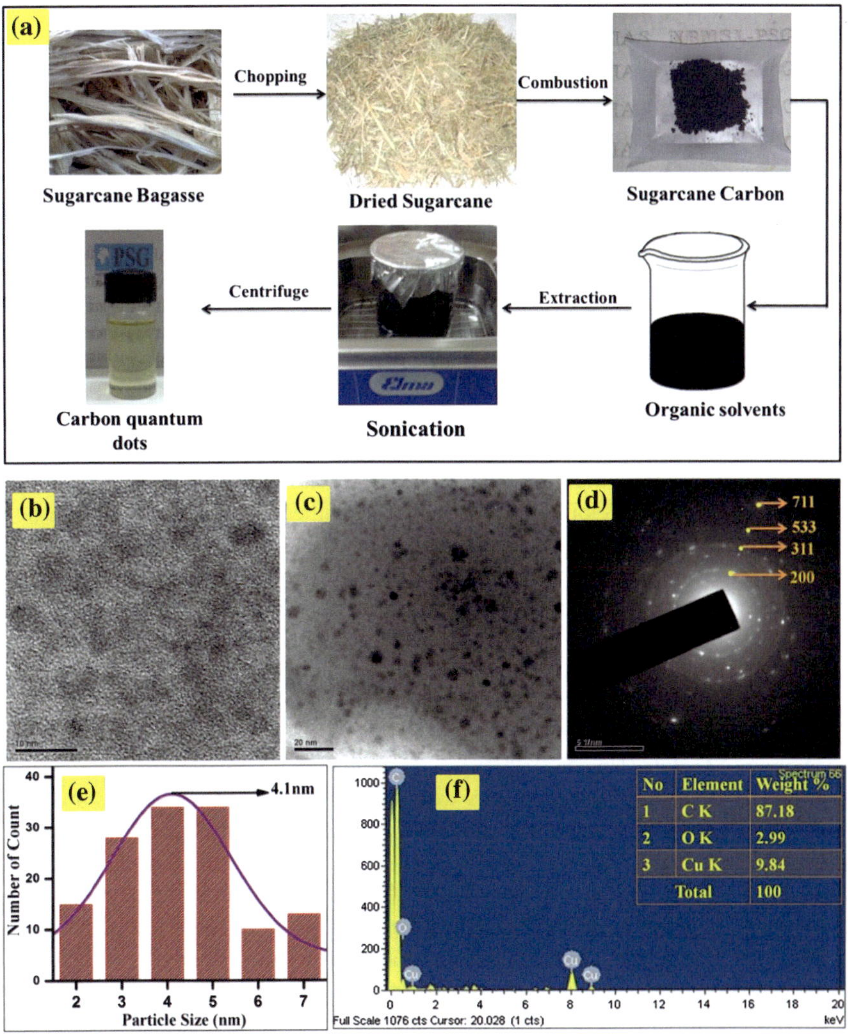

Fig. 6 **a** Schematic representation of the synthesis of carbon dots from sugarcane bagasse pulp, **b**, **c** TEM images, **d** SAED pattern, **e** particle size distribution, and **f** elemental composition of the obtained carbon dots (Thambiraj and Shankaran 2016)

alkaline solution (prepared by stirring pectin and NaOH) was converted into blue-emitting fluorescent C-dots by boiling for 60 min followed by dialysis to remove the unreacted NaOH (Ahmed et al. 2016).

2.2 *Animal-Based Precursors*

As the need for carbon dots with various physicochemical and functional properties increases for the various applications, researchers also been explored various animal products and waste materials as the source. Like plant biomass/derivatives, animal products (cow milk, bee pollen, silk protein, etc.) as well as animal wastes (such as dung and shells) are also rich in carbon contents and can be used for the preparation of C-dots due to their plenty availability as well as chemical compositions.

2.2.1 Animal Products

Dey et al. reported the preparation of C-dots via thermal carbonization (also known as caramelization) of clotted cream (processed cow milk product) followed by solubilization in water medium (Dey et al. 2013). Then they used the prepared C-dots as reducing as well as stabilizing agents for the synthesis of palladium nanoparticles (Pd NPs), in which the C-dots formed a thin layer around the nanoparticle surface. In 2014, Wang and Zhou reported the synthesis of C-dots with the size of 3 nm using milk as a carbon source by hydrothermal process at 180 °C for 2 h without any harsh synthetic conditions (Wang and Zhou 2014). In the same year, Wang et al. reported the preparation of C-dots, S–C-dots, and N–C-dots from milk by hydrothermal process at 180 °C for 8 h without any solvent medium (Wang et al. 2014a). Similarly, Han et al. reported the hydrothermal treatment of cow milk at 180 °C toward the formation of C-dots (Han et al. 2014).

Further, cow milk-derived C-dots were also synthesized by hydrothermal treatment at 180 °C for 12 h in water medium with the yield of 34.5% and the prepared C-dots were subjected to ethyl acetate extraction to get amphiphilic C-dots (yield 18%) (Han et al. 2015). By keeping amphiphilic C-dots as both reducing agent and template, supported silver nanoparticles (amphiphilic C-dots–Ag nanocomposites) were prepared by simple process. Also, further modifications were done to get amphiphilic C-dot–Ag/polymethylmethacrylate nanocomposite film with antibacterial property by solvent casting method. In 2015, Zhang et al. reported the preparation of fragrant N–C-dots from different (rapeseed flower, camellia, and lotus) bee pollens and neat C-dots from rapeseed flower bee pollen by hydrothermal treatment at 180 °C for 24 h (Zhang et al. 2015a, b). Another animal-based precursor, silk protein (source: *Bombyx mori*) was used as a precursor for the preparation of amphoteric N–C-dots by hydrothermal treatment at 190 °C for 3 h in the presence of aqueous NaOH (Wu et al. 2013). In 2016, Feng et al. utilized silkworm chrysalis as a source for the preparation of N–C-dots by microwave heating process at 210 °C for 45 min (Feng et al. 2016). Wang et al. reported the synthesis of C-dots from egg white or yolk by plasma treatment under ambient conditions for few minutes (Wang et al. 2012). Later, Zhang et al. reported the preparation of C-dots from egg white (albumens) separated from chicken eggs by hydrothermal treatment at 220 °C for 48 h (Zhang et al. 2015c). Further, the derived C-dots were used for the preparation of microgels (P(NIPAM-co-VIM)-C-dots) by using emulsion polymerization method.

Lin et al. reported the synthesis of novel shrimp egg-derived spherical-shaped C-dots by thermal heating at 180 °C for 25 min (Lin et al. 2014). Ye et al. utilized pigeon products for the preparation of N–S–C-dots by pyrolysis at 300 °C (Ye et al. 2017). The C-dots obtained by using feather, egg white, egg yolk, and manure resulted in high fluorescent QY of 24.87, 17.48, 16.34, and 33.50%, respectively. Figure 7 shows the schematic representation of the synthesis of N–S–co-doped C-dots by pyrolysis carbonization process using pigeon products along with their TEM images and UV–Vis absorption spectra. Yang et al. reported the preparation of high fluorescent C-dots from honey solution with 30% H_2O_2, and then the resulted solution was hydrothermally treated at 100 °C for 2 h (Yang et al. 2014a). Finally, the obtained solution was neutralized with NaOH, and then C-dots were collected by simple physical processes. Ding et al. reported the synthesis of C-dots using the genomic DNA isolated from *Escherichia coli* bacteria by incubation (hydrothermal)

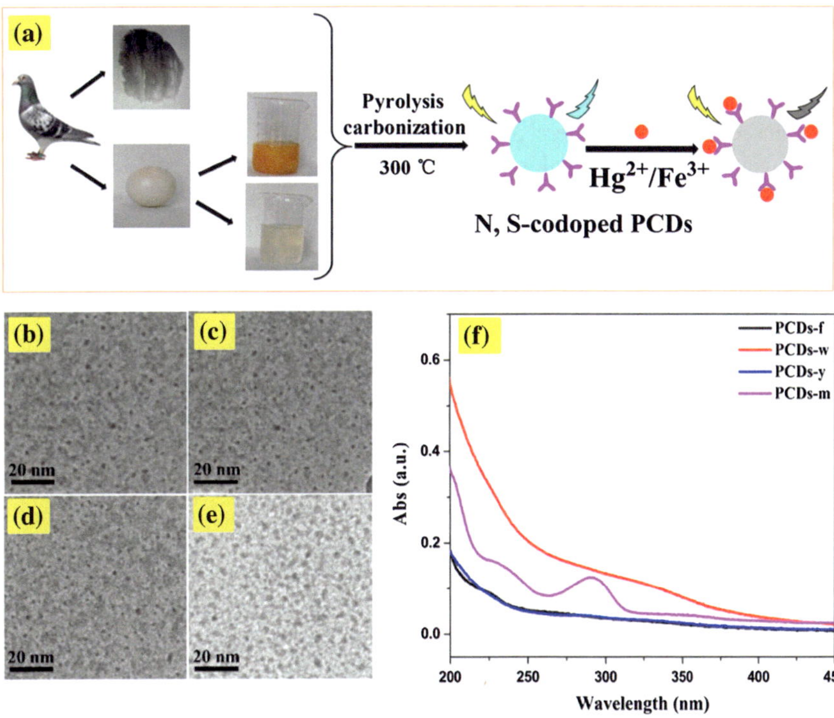

Fig. 7 **a** Schematic representation of the synthesis of N–S-co-doped C-dots by pyrolysis carbonization process using pigeon feather as well as egg as sources, TEM images of **b** PCDs—feather, **c** PCDs—egg white, **d** PCDs—egg yolk, **e** PCDs—manure and **f** UV–Vis absorption spectra of the PCDs (Ye et al. 2017)

treatment at 180 °C for 12 h (Ding et al. 2015a). It was found that in the synthesis of DNA–C-dots, ribose was collapsed along with nitrogen release, and several new bonds were formed, whereas most of the covalent bonds in DNA were retained. Liang et al. used the gelatin as a source for the preparation of C-dots by hydrothermal treatment in water medium at 200 °C for 3 h. They obtained 1.7 nm sized C-dots which exhibits the QY of 31.6% (Liang et al. 2013). In addition to that they were prepared the gelatin–C-dots composites at 45 °C under agitation method with the fluorescent properties. In 2018, Zhao et al. reported the synthesis of C-dots using pork as a carbon source by hydrothermal treatment process at 200 °C for 10 h (Zhao et al. 2018).

2.2.2 Animal Wastes

Apart from animal products, researchers also utilize animal wastes as a precursor for C-dots due to their carbon-rich content. In 2013, Sun et al. effectively utilized the human hair fiber for the synthesis of S–N–co-doped carbon dots by sonicating it with sulfuric acid for 30 min in different temperatures (40, 100, 140 °C) (Sun et al. 2013). They obtained the amorphous C-dots having a size in the range of 3–8 nm. In the next year, Liu et al. reported the synthesis of C-dots from hair by pyrolysis at 300 °C for 2 h under nitrogen atmosphere, and further, they prepared C-dots/polymer composites (Liu et al. 2014). Liu et al. reported the microwave hydrothermal carbonization of goose feathers for the synthesis of highly photoluminescent C-dots (Liu et al. 2015). By heating the mixture of goose feather and water in a microwave autoclave at 180 °C for 40 min and dialyzing at a cellulose ester membrane, they obtained the oxygen, nitrogen, and sulfur content C-dot having ~21.5 nm diameter and ~4.5 nm height. It exhibits high photoluminescence of ~17.1% efficiency. In 2016, Hou et al. reported the synthesis of C-dots from feathers by hydrothermal reaction in autoclave at 160 °C for 7 h in the presence of H_2O_2 (30%) and ammonia (25–28%) (Hou et al. 2016).

In 2016, Essner et al. used urine-based dehydrated residue as a precursor for the preparation of C-dots by carbonization process at 200 °C for 12 h (Essner et al. 2016). Barbosa et al. utilized cow manure as a precursor by calcination at 300 °C for 3 h followed by chemical oxidation process using nitric acid and then functionalized it with ethylenediamine (Barbosa et al. 2015). Wang et al. employed wool as a carbon source for the synthesis of C-dots by microwave-assisted pyrolysis process (Wang et al. 2016a). The synthesized C-dots having the average size of 2.8 nm. Shi et al. also utilized wool biomass for the synthesis of nitrogen-doped C-dots by pyrolysis and microwave treatment (Shi et al. 2016). They obtain spherical, well dispersed, crystalline C-dots having the average diameter of ~6 nm and exhibits high fluorescent quantum yield of 22.5%. Figure 8 shows the diagram of the synthesis of N-CDs from wool, along with their TEM images and spectral details. In 2018, Chatzimitakos et al. explored human fingernail as precursors for synthesizing N–S–C-dots by microwave irradiation at 400 W for 2 min. The obtained C-dot having narrow size distribution of around 2.2 nm exhibits good stability and high quantum yield (Chatzimitakos

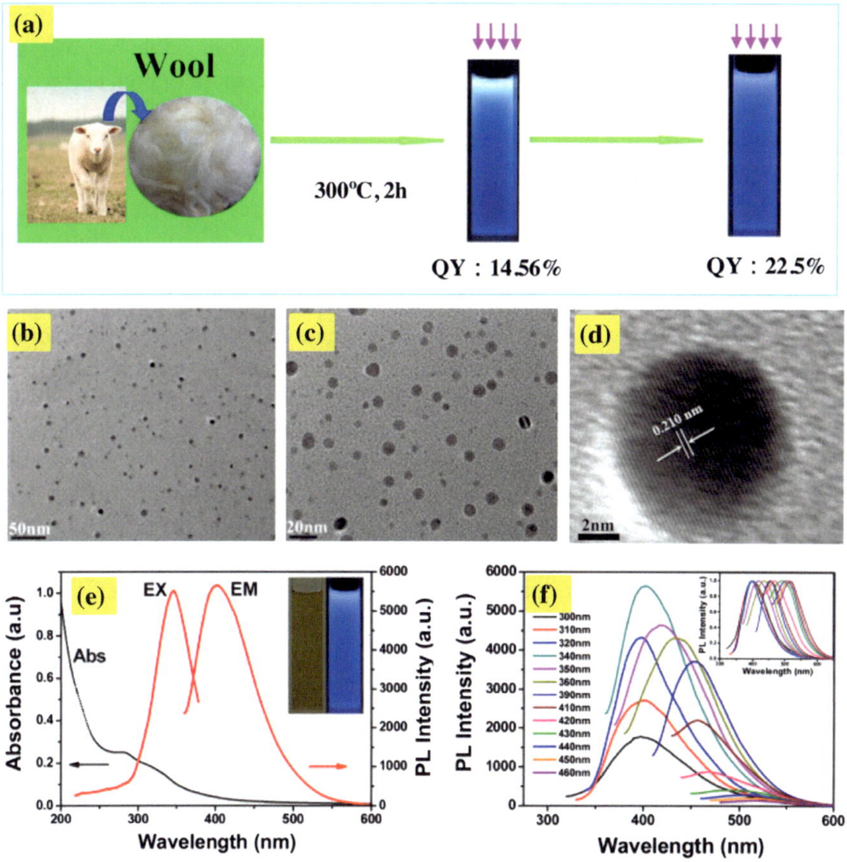

Fig. 8 **a** Synthesis of N-CDs by pyrolysis and microwave irradiation of wool and their photographs under UV light exposure, **b, c** TEM images and **d** HRTEM image of N-CDs, **e** absorption, excitation, and emission spectra and excitation dependent emission spectra of N-CDs and its corresponding normalized spectra (inset) (Shi et al. 2016)

et al. 2018a). The same group also used the human fingernails for the preparation of N–S–C-dots by simple carbonization process at 200 °C for 3 h (Chatzimitakos et al. 2018b).

2.3 *Aquatic and Marine Precursors*

Basically, the marine-based precursor illustrates as effective bio-resources for preparing of many active components and that can be synthesized even in quantum scale to enhance its capability. Though numerous marine precursors are present in the ocean, chitin as well as its deacetylated derivative, chitosan gained more interest toward the preparation of bio-based materials including carbon materials due to their structural nature. Basically, chitosan comes under the category of biocompatible polymer, and it is structured by unique amino groups that acts as the precursor of both carbon and nitrogen. Shchipunov et al. used chitin to synthesis C-dot by hydrothermal treatment in dilute nitric acid at 180 °C for 3 h (Shchipunov et al. 2015). They obtained the luminescent crystalline C-dots of 4–8 nm. In 2012, Chowdhury et al. synthesized chitosan gel by treating chitosan with acetic acid and glycerol mixture as solvent, and further, they reported the preparation of fluorescent C-dots from chitosan gel under microwave heating (Chowdhury et al. 2012). They prepared the C-dots at different pH conditions and found that C-dots became darker at lower pH level. Shen et al. reported the conversion of chitosan into fluorescent C-dots by hydrothermal treatment at 180 °C for 12 h in the presence of acetic acid medium (Shen et al. 2013). Further, the prepared C-dots were used for the reduction of Ag^+ to Ag^0 toward the formation of silver nanoparticles (Ag NPs) without the addition of any further reducing agents. In the same year, Xiao et al. reported that chitosan as a source for the preparation of microwave-assisted synthesis of fluorescent carbon nitride dots (CN-dots) functionalized with amino gropus in water medium at 700 W for 9.5 min (Xiao et al. 2013). Sachdev et al. reported a one-step synthesis of multicolor, surface passivated C-dots (PEG–C-dots) from chitosan and poly(ethylene) glycol (PEG) through rapid microwave-assisted pyrolysis process (Sachdev et al. 2013).

Gong et al. reported the preparation of N–C-dots by microwave-assisted pyrolysis at 700 W for 15 min using chitosan as a carbon source along with acetic acid and 1,2-ethylenediamine as condensation agent and N-dopant, respectively (Gong et al. 2015). Wang et al. used chitosan as a source for the preparation of N–C-dots by one-step hydrothermal process at 220 °C for 12 h (Wang et al. 2016b). Similarly, Liu et al. reported the simple process of carbonization of chitosan at 300 °C for 2 h and the resulted ground powders were dispersed in water medium to produce the amino-functionalized C-dots (Liu et al. 2016a, b). Bhaisare et al. reported the preparation of magnetic C-dots by mixing chitosan with Fe_3O_4 NPs (laboratory made) in acetic acid medium followed by hydrothermal treatment at 120 °C for 12 h (Bhaisare et al. 2016). Wang et al. used hydro-soluble chitosan as carbon source for the preparation of good crystalline structure of N–C-dots by one-step hydrothermal process at 220 °C for 12 h, and they achieved a high yield of 31.8% (Wang et al. 2016b). Recently, Janus et al. reported the preparation of N–C-dots using chitosan as carbon source and various amino acids or casein hydrolysate as nitrogen sources by microwave treatment process with 300 W radiation for 3–5 min (Janus et al. 2019).

The investigation by D'souza et al. reports the effective synthesis and characterization of fluorescent nitrogen-doped carbon dots using dried shrimps as the carbon source. Gedda et al. used prawn shells as a carbon source for the preparation of C-dots (4 nm in size) by sequential chemical treatment (base/acid treatment and refluxed at 80 °C for 8 h in base medium) followed by process at 200 °C for 8 h (D'souza et al. 2016). Yao et al. explored crab shell as a carbon source for the fabrication of multifluorescent Gd-, Mn-, and Eu-doped C-dots by microwave-assisted pyrolysis at 220 °C for 10 min (Yao et al. 2017). Figure 9 shows the schematic illustration of preparation of multifluorescent (MF) CQDs using a microwave-assisted

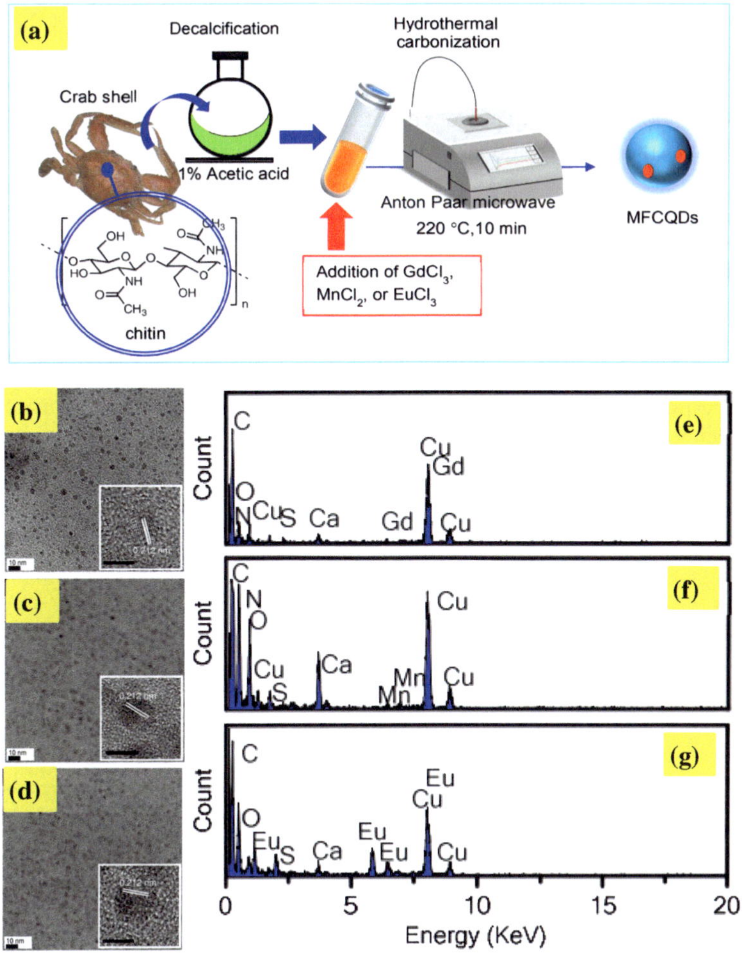

Fig. 9 **a** Preparation of multifluorescent CQDs using a microwave-assisted hydrothermal approach, TEM images and EDS spectra of **b**, **e** Gd@CQDs, **c**, **f** Mn@CQDs, and **d**, **g** Eu@CQDs. Inset (**b**–**d**): corresponding HRTEM images of MFCQDs (Yao et al. 2017)

hydrothermal approach, TEM images, and EDS spectra of Gd@CQDs, Mn@CQDs, and Eu@CQDs. Zhang et al. reported the preparation of N–S–C-dots from *Nannochloropsis* bio-crude oil (NBO) via Maillard reaction (Zhang et al. 2017). In detail, NBO was heated in the microwave at 750 W for 30 min and then the resulted polymer was stirred well with concentrated sulfuric acid for 30 min at 80 °C followed by dilution with water and pH neutralization with NaOH. S. Sankaranarayanan et al. explored brown macro-algal bio-oil (*Macrocystis pyrifera*) for the synthesis of graphene oxide-carbon dot composite by hydrothermal carbonization at 170 °C for 4 h (Sankaranarayanan et al. 2019). It has been found that low molecular weight and high molecular weight compounds present in the bio-oil are responsible for the formation of CD and graphene oxide respectively. Guo et al. prepared the C-dots from formaldehyde-treated microalgae powder by hydrothermal process at 180 °C for 10 h (Guo et al. 2017). Recently, Wang et al. utilized the dried cyanobacteria (known as blue-green algae) powder as a carbon source for the preparation of carbon dots by hydrothermal method at 180 °C for 12 h (Wang et al. 2019).

3 Applications

C-dots found many application potentials in various fields not only due to their unique physicochemical, functional, and tunable properties but also for their easy modifications as per the requirement. As the demand enhances for sustainable materials in many technological applications, carbon dots synthesized using renewable resources found increasingly important. In this section, few of the important applications studied using the renewable source-derived C-dots are discussed.

3.1 Sensors

Fluorescence properties of carbon dots significantly vary with their chemical environment, and hence, they have been widely used for various sensing applications. Sensing of distinct molecules is one of the significant areas of research in environmental as well as pharmaceutical. The carbon dot-based sensors can be classified into ion sensors, biosensors, chemical sensors, and gas sensors.

3.1.1 Ion Sensor

Detection of ions in any kind of systems is always a challenging task. Studies are clearly showing that renewable source-derived C-dots are effectively used as sensors for both cations (e.g. Hg^{2+}, Cu^{2+}, Fe^{3+}, Pb^{2+}, Ag^{+}) and anions (e.g. $C_2O_4{}^{2-}$, $PO_4{}^{3-}$, CN^{-}, F^{-}, S^{2-}, ClO^{-}, I^{-}). C-dots derived from coriander leaves were used for sensing Fe^{3+} ions with good sensitivity and selectivity. (Sachdev and Gopinath 2015). The developed C-dots/PEI/NB nanocomposites were used for the ratiometric TPF turn-on sensing of sulfide anion (S^{2-}) ions in various biological fluids (fetal bovine serum (FBS), human serum, pig urine, and human urine) as real samples (Jin et al. 2017). The N–S–C-dots derived from vegetables were used as a fluorescent nanoprobe (turn-on) for sensitive and selective detection of highly oxidant periodate anion ($IO_4{}^{-}$) in wastewater samples (Romero et al. 2019). Interestingly, they have demonstrated the integrated broccoli-based N–S-co-doped C-dots' synthesis and detection of periodate anion ($IO_4{}^{-}$) in one-step process. *Borassus flabellifer* flower-derived C-dots showed excellent fluorescent-sensing properties for Fe^{3+} ions (Murugan and Sundramoorthy 2018). Similarly, C-dots derived from wheat straw

(Yuan et al. 2015) and soy flour (Fang et al. 2016) were also used as a sensor for the detection of ferric ion (Fe^{3+}). Also, C-dots prepared from limeade were used for the selective detection of Fe^{3+} in the presence of various metal ions (Suvarnaphaet et al. 2016). Graphene Q-dots and poly(ethylene imine)/graphene Q-dots were used as a fluorescent probe for the sensitive determination of Fe^{3+} and Cu^{2+} ions (Wang et al. 2016c). The Jinhua bergamot plant-derived C-dots were used for detecting Hg^{2+} and Fe^{3+} ions respectively by the buffer solutions of HAC-NaAC and Tris-HCl (Yu et al. 2015b). Egg white-derived C-dots were used as a sensor for the selective detection of Fe^{3+} in the presence of various metal ions (Zhang et al. 2015c). Further, the honey-derived C-dots were explored for the detection of Fe^{3+} in water and blood samples because of their quenching effects (Yang et al. 2014a).

Alternatively, N–C-dots derived from rose-heart radish were used as a fluorescent probe for the detection of Fe^{3+} in water samples (Liu et al. 2017b). Also, N–S–C-dots prepared by using DL-malic acid as carbon source were effectively used for the selective detection of Fe^{3+} metal ion in the stock solution having sixteen metal ions (Lu et al. 2015). The N–C-dots prepared from *Prunus avium* fruit extract were used as the fluorescent detector for biologically important Fe^{3+} ions in water (Edison et al. 2016b). The wool-based N–C-dots were effectively utilized for the detection of Fe^{3+} in living cell systems due to their easy cellular uptake behavior (Shi et al. 2016). N–S–C-dots prepared by one-step hydrothermal process from garlic were used for the selective detection of Fe^{3+} in tap as well as lake water (Chen et al. 2016). Celery stalk-derived C-dots were used for the detection of Cu^{2+} ions along with various competitive ions (Shasha et al. 2019). Grass-derived N–C-dots were effectively used for the removal of Cd^{2+} and Pb^{2+} ions from water (Sabet and Mahdavi 2019).

Lotus root-derived C-dots were used as a sensor for the detection of Hg^{2+} in a water sample (Gu et al. 2016). C-dots derived from sodium citrate and glutathione were used for the selective and sensitive recognition of mercury ion (Hg^{2+}) in aqueous solution (Ma et al. 2016). Also, for the detection of Hg^{2+} in environmental water samples, apple juice-derived C-dots were used (Yue et al. 2014). Chitosan-derived C-dots were used for the highly selective and sensitive detection of Hg^{2+} (Wang et al. 2016b). N–C-dots were used for as a fluorescent probe for the sensitive and selective detection of Hg^{2+} ions, and further, the PL ratios of the N–C-dots aqueous solution were also studied in the presence and absence of various metal ions as well as anions in the same system (Barati et al. 2015). N–S–C-dots derived from pigeon products and wastes were utilized for the selective detection of mercury and iron ion (Ye et al. 2017), whereas their manure source-derived C-dots were not favorable for detection purposes. C-dots derived from *Ocimum sanctum* source were reported as a probe for label-free, sensitive detection of Pb^{2+} ions in normal as well as real samples (Kumar et al. 2017). Gum olibanum resins-derived C-dots were used for the detection of heavy metals like Pb^{2+}, Hg^{2+}, and Cd^{2+} in aqueous medium (Gaddam et al. 2017). Interestingly, sage industry waste-derived C-dots were used as fluorescent probe for the detection of various metal ions, whereas the better quenching effects were observed in both Cu (II) ions and Pb (II) ions (Tan et al. 2014).

Human fingernails-derived N–S–C-dots were used as a ultra-sensitive probe for the detection of Cr(VI) (Chatzimitakos et al. 2018b). Onion waste extract-derived

C-dots were used as a fluorescent probe for the selective and sensitive detection of Fe^{3+} ions (Bandi et al. 2016). CQ-dots obtained from lemon peel were used as a fluorescent probe for the selective detection of Cr^{6+} in water purification processes (Tyagi et al. 2016). Pectin-derived C-dots were used for the selective detection of silver ions (Ag^{+}) without any irradiation or heating or additional reducing agents and cyclic voltammetry confirmed the reduction of silver ions to silver nanoparticles in the process (Ahmed et al. 2016). In all the above-discussed cases, neat C-dots were used as a probe for the detection of different metal ions. The N–S–C-dots prepared using citric acid and cystamine dihydrochloride mixture were effectively used as a sensor for the detection of Cr (VI) (Chen et al. 2017). Amphiphilic fluorescent C-dots prepared using citric acid and methionine as a preliminary source were used for the detection as well as imaging of Al^{3+} in HUVEC cells (Kong et al. 2017). Ag NPs prepared using chitosan-derived C-dots as reducing agent were used for the detection of silver in water samples (Shen et al. 2013). N–C-dots prepared using chitosan and 1,2-ethylenediamine were used as a selective and sensitive probe for the detection of Fe^{3+} (Gong et al. 2015). Prawn shell-derived C-dots showed excellent sensitivity toward Cu^{2+} ions from various sources such as drinking water, river water, and seawater with the low detection limit of 5 nM (Gedda et al. 2016).

3.1.2 Chemical Sensor

Like ion sensors, detection of chemicals (including specific biological compounds) is also an important area of research in which renewable source-derived C-dots have been extensively reported. L-tyrosine methyl ester-derived C-dots were used as a sensor for the detection of methyl parathion in the real system analysis of cabbage, milk, and fruit juice samples (Hou et al. 2015). Bright-blue fluorescent C-dots (derived from citrate and acrylamide) were used as a fluorescent probe for the determination of pyrophosphate (PPi) (Liu et al. 2017d). Carbohydrate-derived C-dots (developed by photochemical synthesis pathway) were used as an optical sensor for the detection of H_2O_2 and antioxidants such as ascorbic acid and glutathione (Costas-Mora et al. 2015). Tender coconut water-based C-dots were used for the fluorometric detection of thiamine in aqueous solution and vitamin capsule (real sample) (Purbia and Paria 2016). Orange juice-derived C-dots were effectively used for the detection of cysteine (concentration range from 6.0 to 300 nM) in human plasma and urine samples (Amjadi et al. 2015). L-cysteine and glycerol mixture-derived C-dots were used as a fluorescent sensor to determine the fluazinam fungicide (Zou et al. 2017). Saffron-derived C-dots were used as a sensor for the detection of prilocaine (Ensafi et al. 2017). Natural seeds-derived N–C-dots were effectively used as an electrochemical sensor for the detection of nitrites in real samples (Li et al. 2019).

The glassy carbon electrode containing AgNPs/C-dots nanohybrids was used as a sensor for the electrochemical detection of H_2O_2 in the disinfected fetal bovine serum samples (Jahanbakhshi and Habibi). The Ag/C-dots nanocomposites modified glassy carbon electrodes developed using zoysia grass as a preliminary source were explored as a good non-enzymatic H_2O_2 sensor having high sensitivity and low detection limit (Jiang et al. 2014a). Polymer-coated C-dots were used for the detection of bisphenol A in the laboratory made as well as real water (river) samples (Liu et al. 2016a, b). Terbium containing glucose-derived C-dots (Tb-CDs) were used for quantitatively monitoring adenosine 5'-triphosphate (ATP) by the dispersed/agglomerated gold nanoparticle (AuNP) interactions. (Xu et al. 2016a). Figure 10 shows the schematic representation of fluorescent autosensing platform for quantitatively monitoring ATP by AuNP and Tb-CDs interaction. C-dots/Ag nanocomposites were used for the colorimetric detection of methimazole (an antithyroid drug) in solution as well as urine samples (Amjadi et al. 2017). C-dots obtained from starch as a source were used for the determination of the imatinib molecule which is used for cancer treatment (Yan et al. 2016). Feather-derived C-dots were used as a fluorescence sensor for the determination of organophosphorus pesticide (DDVP) in cabbage and fruit juice samples with high selectivity and sensitivity (Hou et al. 2016). Based on the inner filter effect sheep of wool-derived C-dots and quenching nature by silver nanoparticles (AgNPs), fluorescence system C-dots/AgNPs were developed for the detection of glyphosate (Wang et al. 2016a). Pork-derived C-dots were used as nanosensors for the detection of uric acid in human serum and urine samples (Zhao et al. 2018).

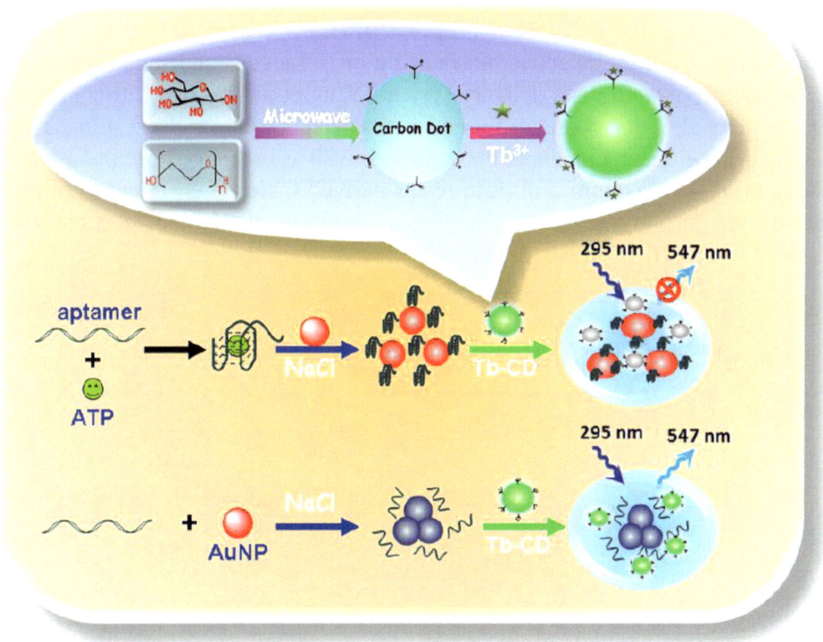

Fig. 10 Schematic illustration of fluorescent autosensing platform for quantitatively monitoring ATP by AuNPs and Tb-CDs interaction. (Xu et al. 2016a)

3.2 Biomedicine

3.2.1 Bio-labeling/Bio-imaging

C-dots chelated with alizarin red S (ARS) was used as an inner filter-mediated illuminant nano-fluorescent reporter for imaging of intracellular Ca^{2+} in the biological system (Chen et al. 2018a). Figure 11 shows the schematic representation of the preparation of C-dots from garlic, TEM image of the C-dots and bright field, fluorescent images of A549 cells incubated with CDs were obtained at the different excitation wavelengths along with particle size distribution histograms (Zhao et al. 2015). Coriander leaves-derived C-dots were used for the bio-imaging of lung normal (L-132) and cancer (A549) cell lines (Sachdev and Gopinath 2015). The developed transition metal ion-doped C-dots were used as a magneto-fluorescent probe for the bio-imaging of human colon cancer cells (SW480) (Pakkath et al. 2018). C-dots are used in the field of imaging in both in vivo and in vitro due to their non-toxic nature. Kidney beans-derived self-passivated water-soluble C-dots were used for the multicolor cellular imaging of HeLa cells (Tripathi et al. 2017). Wheat straw-derived C-dots were used for the bio-labeling of layer-by-layer capsules with $CaCO_3$ and also for the bio-imaging of *E. coli* bacteria (Yuan et al. 2015). *Ocimum sanctum*-derived C-dots were used for the fluorescence bio-imaging of triple negative breast cancer MDA-MB 468 cells (Kumar et al. 2017). Folic acid-derived C-dots were effectively used for the selective bio-imaging of HeLa cells and supported the further possibility toward gene delivery system (Yang et al. 2017b). Citric acid–urea mixture-derived C-dots were also used for the detection of *Staphylococcus aureus* bacteria by bio-imaging in laboratory culture as well as in orange juice (real sample) (Zhong et al. 2015). Gum olibanum resins-derived C-dots showed a bright green fluorescence property in the bio-imaging of B16F10 cells (Gaddam et al. 2017).

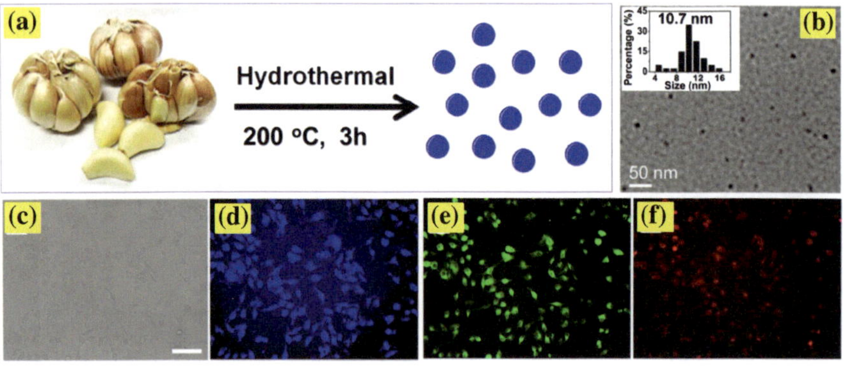

Fig. 11 **a** Schematic representation of the preparation of C-dots from garlic by hydrothermal treatment, **b** TEM image and particle size distribution (Inset) of the CDs, **c** bright field, fluorescent images of CDs incubated A549 cells at **d** 330–385 nm, **e** 450–480 nm and **f** 510–550 nm (Zhao et al. 2015)

Camphor-derived C-dots were used for the bio-imaging of SHSY5Y cells at different concentrations (Oza et al. 2016). C-dots derived from citrus limonene juice with two different sizes (two fractions) were used as a fluorescent probe for the bio-imaging of Hep-2 cells (Oza et al. 2015), and interestingly, pectin-derived C-dots (Zhao et al. 2014) were also used for the same purpose. The synthesized carbon dots of *Prunus avium* fruit extract exhibited less cytotoxicity and good biocompatibility against L-929 and MCF-7 cells (Edison et al. 2016b). Similarly, C-dots derived from *Punica granatum* (pomegranate) fruit juice were studied as probes for the imaging of bacteria (*Pseudomonas aeruginosa*) and fungus (*Fusarium avenaceum*) cells (Kasibabu et al. 2015b). *Carica papaya*-derived C-dots were used as fluorescent probes for the bio-imaging of bacterial (*Bacillus subtilis*) and fungal (*Aspergillus aculeatus*) cells (Kasibabu et al. 2015a). Graphene C-dots derived from humic acid were used as a fluorescent probe for the bio-imaging of chicken embryo cells (Shi et al. 2015). The C-dots prepared from apple juice were used as alternative biocompatible fluorescent probes for the bio-imaging of bacteria (*Mycobacterium tuberculosis* and *Pseudomonas aeruginosa*) and fungal (*Magnaporthe oryzae*) without any toxic effect on both of the cell lines (Mehta et al. 2015). The tender coconut-derived C-dots were also used for the bio-imaging of fungal (*Aspergillus niger*) cells (Purbia and Paria 2016). Due to their photostability and biocompatibility, C-dots derived from limeade were used for the bio-imaging of yeast cell morphology (Suvarnaphaet et al. 2016). Sugar cane juice-derived C-dots were utilized as fluorescent probes in cellular bio-imaging of bacteria (*E. coli*) and yeast (*Saccharomyces cerevisiae*) (Mehta et al. 2014). *Trapa bispinosa* extract-derived C-dots were used to study the cytotoxic effects on MDCK cells and found to be exceptionally biocompatible (Mewada et al. 2013). Peanut shell-derived C-dots were applied for the multicolor living cell imaging of HepG2 cells due to their low cytotoxicity and excellent biocompatibility (Xue et al. 2016).

Onion waste extract-derived C-dots were used for the bio-imaging of HeLa cells (Bandi et al. 2016). Due to their quenching nature with Fe^{3+} and recovered with L-lysine, N–C-dots prepared from konjac flour was used for as a fluorescence probe for the bio-imaging of HeLa cells (Teng et al. 2014). N–C-dots prepared by one-step process from winter melon were used for the bio-imaging of HepG2 (liver hepatocellular carcinoma) cells (Feng et al. 2015a). Rose-heart radish-derived N–C-dots were used for the bio-imaging of human uterine cervical squamous cell carcinoma (SiHa) cells (Liu et al. 2017b). Garlic-derived N–S–C-dots were used as a fluorescent probe for the bio-imaging of RAW264 7 cells (Chen et al. 2016). Lotus root-derived direct N–C-dots were used for cytotoxicity study to use them as a probe for living cell bio-imaging of A549 cells (Gu et al. 2016). Poly(ethylene imine)/graphene Q-dots (carbon source: coffee grounds) were used for the bio-imaging of HeLa cells because of their highly efficient fluorescent and low cytotoxicity properties (Wang et al. 2016c). N–C-dots derived using L-ascorbic acid and β-alanine were used as staining probe for the confocal bio-imaging of MDCK and HeLa cells (Edison et al. 2016a). The N–S–C-dots derived from DL-malic acid (as carbon source) were used for the cell imaging of human cervical carcinoma HeLa cells (Lu et al. 2015). N–S–C-dots derived using citric acid and cystamine dihydrochloride were used for cellular imaging of HeLa cells (Chen et al. 2017). Similarly, gadolinium (Gd)-doped C-dots (carbon source: citric acid) were used as a probe for the bio-imaging of HeLa cells (Yu et al. 2016). Mannose-modified fluorescent CQ-dots were used for the bio-labeling of *E. coli*, and it was found that CQ-dots could selectively bind to the FimH lectin unit in the flagella of the wild-type 1 *E. coli* K12 strain (Weng et al. 2015). NH_4OH passivated C-dots were demonstrated for the bio-imaging of murine embryonic stem cells (Liang et al. 2014). *Prunus avium* fruit extract-derived N–C-dots were also used for bio-imaging of MDA-MB231 cells (Edison et al. 2016b). N–C-dots prepared using *Prunus persica* and *Prunus mume* fruit extracts as carbon sources were used for the bio-imaging of MDA-MB-231 cells and showed strong fluorescence within the cells (Atchudan et al. 2016a, b). Cellular uptake and cytotoxicity of the N–C-dots (carbon source: lime juice; nitrogen source: NH_4HCO_3) were employed for the Rat pheochromocytoma-derived cell line PC12 cells (Barati et al. 2015). Surface passivated C-dots prepared from plant soot were used as fluorescent bio-probes for the imaging of Chinese hamster ovary (CHO) cells as well as ex vivo guppy fish (Tan et al. 2013). Saffron-derived C-dots were used as a probe for the imaging of bone marrow cells and olfactory mucosa cells from rats (Ensafi et al. 2017). Biorefinery process by-products (wheat straw and bamboo residues)-derived N–C-dots were used for the imaging of murine myeloma cells line SP2/0 and also for in vivo bio-imaging of tumor cells in mouse (Huang et al. 2019).

Milk-derived C-dots were used as a probe for the bio-imaging of U87 cells, a human brain glioma cancer cell line (Wang and Zhou 2014). Neat as well as S-(or) N-doped C-dots was used for the bio-imaging of live SMMC-7721 cells (Wang et al. 2014a). Bee pollen-derived C-dots were used as a fluorescent probe for the bio-imaging of LoVo human colon carcinoma cell (Zhang et al. 2015b). *Bombyx mori* silk protein-derived amphoteric N–C-dots and silkworm chrysalis-derived C-dots were used as a probe for the bio-imaging of A549 cells (Wu et al. 2013) and HeLa cells (Feng et al. 2016), respectively. Egg white-derived C-dots were used for the bio-imaging of HeLa and KB cells (Zhang et al. 2015c). Shrimp egg-derived carbon dots were used for the bio-imaging of Human hepatoma cell line SK-Hep-1 cells (Lin et al. 2014). The honey-derived C-dots were used for the bio-imaging of Hep-2 as well as HeLa cells (Yang et al. 2014a). DNA-derived C-dots were used as a novel source for the fluorescent bio-imaging for both prokaryotic and eukaryotic cells as well as they used for drug delivery studies too (Ding et al. 2015a). Gelatin-based highly luminescent C-dots were used for live cell imaging of A549 cells, in which the cultivation of A549 cells in the presence of C-dots for 6 h indicated the strong blue fluorescence property (Liang et al. 2013). Interestingly, membrane and cytoplasmic area of the A549 cells were showed luminescent spots, whereas its central region involved with nucleus was showed very weak fluorescence. The G-dots prepared from using food wastes were also used for the bio-imaging of HepG2 cells (Park et al. 2014). Wort bagasse source-derived C-dots were used as a probe for the bio-imaging of MCF-7 breast cancer cells (Rodrigues et al. 2015). The S–N–C-dots derived by using hair as carbon source were used for the bio-imaging of HeLa cells because of their good luminescence stability, solubility and biocompatibility (Sun et al. 2013).

The N–C-dots derived from wool were used for the bio-imaging of HeLa cells due to their strong fluorescence and low toxicity (Shi et al. 2016). The surface-modified C-dots prepared from cow manure were used as bio-probes for selectively staining the cell nuclei of the breast cancer cell lineage MCF-7 (Barbosa et al. 2015). Also, the surface-modified C-dots were extended to other cell lineages such as HUVEC, MDA-MB-231, Caco-2, and DU145 for live cell imaging (Barbosa et al. 2015). Waste frying oil-derived S–C-dots were used as a fluorescent probe for the bio-imaging of HeLa cells due to their high photostability and low cytotoxicity nature (Hu et al. 2014). Urine-derived C-dots were used as a nanosensors for detecting heavy metal ion pollutants including Hg^{2+} by fluorescence quenching (Essner et al. 2016). Human fingernails-derived N–S–C-dots were used as a fluorescent source for sunset yellow detection as well as live cell imaging of various cell lines (Chatzimitakos et al. 2018a, b). PEG–C-dots prepared using chitosan as carbon source were effectively used for the bio-labeling of gram-negative *E. coli* and gram-positive *S. aureus* bacterial model systems (Sachdev et al. 2013). N–C-dots prepared using chitosan and 1,2-ethylenediamine were used for imaging of the human renal carcinoma 786-0 cells in the bio-system (Gong et al. 2015). Amino-functionalized C-dots prepared using chitosan was used for the imaging of onion root tip cells (Liu et al. 2016a, b). N–S doped C-dots derived from *Nannochloropsis* bio-crude oil were used as bio-imaging agent for *Arabidopsis thaliana* guard cells and root tissues (Zhang et al. 2017). Microalgae powder-derived C-dots were successfully employed for the imaging of MCF-7 cells (Guo et al. 2017). Cyanobacteria-derived C-dots were used for the imaging of PC12 cells (Wang et al. 2019).

3.2.2 Bio-therapeutics

Gum olibanum resins-derived C-dots exhibited cytotoxicity to cancer cell lines during in vitro experiments (Gaddam et al. 2017). Ginger extracted C-dots were studied in the human hepatocellular carcinoma cell (HepG2) proliferation and thus reduced the weight of HepG2 cell-induced tumor in nude mice (Li et al. 2014). The cytotoxicity of N–C-dots derived from *Hylocereus undatus* source was evaluated on L-929 and MCF-7 cells (Arul et al. 2017). Tomato juice-derived C-dots were used as a label-free sensitive assay for detecting carcinoembryonic antigen (CEA) by adsorption and desorption of aptamers on the surfaces of C-dots through a competitive mechanism in solution as well as blood samples (Miao et al. 2016). Amphiphilic C-dots-Ag nanocomposites showing the good biocidal effect on both Gram-positive (*S. aureus*) and Gram-negative (*E. coli*) bacteria (Han et al. 2015). They also revealed that developed novel amphiphilic C-dots–Ag/polymethylmethacrylate nanocomposite antibacterial film can be used for many potential applications due to their antibacterial, light admitting, and flexible properties (Han et al. 2015). Magnetic C-dots prepared using chitosan as carbon source was effectively used for the detection and identification of pathogenic bacterial strains *S. aureus* and *E. coli* in PBS and urine matrix, respectively (Bhaisare et al. 2016). Chitosan-derived N–C-dots showed the lack of cytotoxicity on human dermal fibroblast cells (Janus et al. 2019).

3.3 *Catalysis*

Renewable source-derived C-dots were reported as a catalyst for many chemical transformations, and the following sections will give some idea about that.

3.3.1 Electro-catalysis

NH_4OH passivated C-dots derived by using hemicellulose-rich xylan source were used for the electrochemical studies (Liang et al. 2014). Ag/C-dots nanocomposites prepared from zoysia grass as carbon source were utilized as an electro-catalyst for the reduction of hydrogen peroxide (H_2O_2) (Jiang et al. 2014a). N–C-dots derived from *Prunus persica* were used as an effective catalyst toward the oxygen reduction reaction (ORR) of approximately 0.72 V against reversible hydrogen electrode (RHE). It reduced into hydrogen peroxide in alkaline medium by following 2-electron pathway (Atchudan et al. 2016a).

3.3.2 Photo-Catalysis

The catalytic reduction of methylene blue was performed by using N–C-dots derived from *Hylocereus undatus* as an active catalyst and $NaBH_4$ as a reducing agent (Arul et al. 2017). TiO_2/C-dots nanocomposites that derived by using spinach as carbon source were also used as the photocatalyst for the production of hydrogen (H_2) by water splitting reaction (Wang et al. 2014b). Grass-derived N–C-dots were effectively used for the photocatalytic degradation of organic dyes such as Acid Blue, Acid Red, Methylene Blue, Methyl Orange, Eosin Y, and Eriochrome Black T (Sabet and Mahdavi 2019). Prepared C-dots–ZnO composite using orange waste peel as a carbon source was used as photocatalyst for the degradation of naphthol blue-black azo dye under UV irradiation and showed superior activity (Prasannan and Imae 2013). TiO_2 nanofibers and CQ-dots composite (carbon source: lemon peel waste) were used as photocatalyst for the degradation of methylene blue dye (Tyagi et al. 2016).

3.3.3 Other Catalysis

Very few studies were carried out using bio-derived C-dots-based materials as heterogeneous catalysts for chemical transformations. Pd@CQ-dots@Fe_3O_4 were used as an effective catalyst for the Suzuki–Miyaura cross-coupling reaction of aryl bromides/chlorides at room temperature in aqueous ethanol at 120 °C in PEG200 under low catalyst loading in the air (Gholinejad et al. 2017). Palladium nanoparticles reduced using cow milk product-derived C-dots were used as a catalyst for the Heck and Suzuki–Miyaura coupling reaction of various aryl halides (Dey et al. 2013).

3.4 Fluorescent Ink

Egg-derived C-dots were used as piezoelectric material-based fluorescent ink in inkjet or silk-screen printing (Wang et al. 2012). N–C-dots were used as fluorescent ink (by loading in the rollerball pen) that exhibited bright-blue luminescence and readily flow without any leakage or coagulation within the pen while writing. (Lu et al. 2015). N–C-dots prepared from wool was also used as fluorescent inks for drawing multicolor patterning (Shi et al. 2016). Figure 12 shows the photographs of the patterns of the images used C-dots as fluorescent ink under UV as well as white light conditions. Gelatin–C-dots composites prepared by using gelatin were used as fluorescent ink on a microscope slide which showed emission under visible and UV light excitations (Liang et al. 2013).

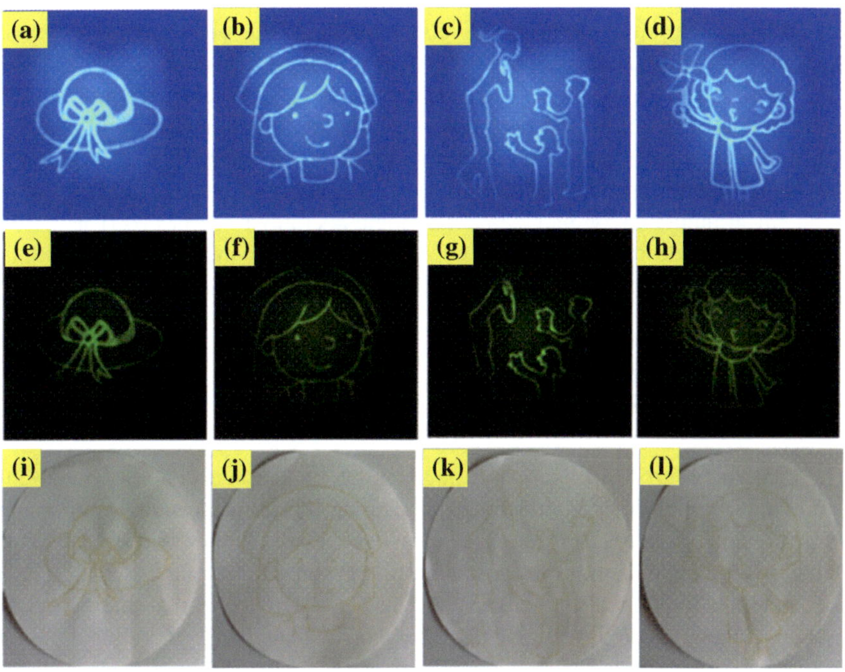

Fig. 12 **(a–d)** Photographs of various patterns created using C-dots under UV light excitation,**(e–h)** 470 nm excitation and **(i–l)** white light (Shi et al. 2016)

3.5 *Other Applications*

The alginate-C-dots nanocomposite (carbon source: tea powder) was reported as a very good UV blocker due to their higher absorption in the UV region (Konwar and Chowdhury 2015). (WPU)/C-dots nanocomposites showed great potential as an environmentally friendly surface coating material because of their low volatile organic compounds (VOCs) (Ghosh et al. 2016). The by-products obtained along with G-dots derived from food wastes significantly improved the growth rate of germination on 3DW-supplemented gauze (Park et al. 2014). Similarly, brown macroalgal bio-oil (*Macrocystis pyrifera*) derived GO/CD composite showed better seed germination property for mung bean seeds (Sankaranarayanan et al. 2019). C-dots/polymer hybrid materials prepared using hair as carbon source was used to construct the multidimensional materials such as one-dimensional (1D) microfibers, 2D patterned films, and 3D microbeads (Liu et al. 2014). In solar cell applications, *Eichhornia crassipes* source-derived N–C-dots sensitized TiO_2 photo-anodes improved the solar-to-electrical conversion efficiency of 0.17% with a good fill factor of 67% (Yang et al. 2018). Similarly, soybean powder-derived C-dots sensitized m-TiO_2/LPP photo-anodes maximized the dark power conversion efficiency as high as 7.97% along with persistent electricity output for several hours (Meng et al. 2017). Due to the spectral change characteristics, by using C-dots as probe and Hg^{2+} and L-cysteine as chemical inputs, individual elementary logic operations such as NOT as well as IMP gates were designed. Further, the C-dots were used in memory device with the idea of switching (Vandarkuzhali et al. 2018).

4 Future Perspectives

The continuous growth of research publications on the conversion of renewable sources into carbon dots for numerous applications indicating the notable importance of these unique materials in the society as well as research world. Even though numerous synthetic methods are possible, still now only few methods were explored for the utilization of renewable sources. This clearly indicates that finding out the proper synthetic methods can give more heights in the area of preparation of C-dots from renewable sources. Development of processes for the preparation of renewable source-derived C-dots by obeying green chemistry aspects can create a global market for them. In general, hydrothermal treatment is the preferred approach due to its easy, rapid, and scalable possibilities. Though microwave and ultrasonication are reported for the rapid preparation of C-dots, their own demerits such as energy-intensive as well as limitations in scalable processes make them to occupy the next place to hydrothermal processes.

Only selective renewable sources are till now explored in this area of research and open the wide scope for variety of renewable sources which differ from one source to another (especially plant resources due to climate issues and soil effects).

In these aspects, regional-based resources (mainly non-edibles) can be effectively used for the conversion to C-dots, and thus, more opportunities can be created for agriculture. The main advantage of C-dots is that they can be also collected from carbon materials by applying simple physical techniques such as extraction methods. This approach is making more lanes in the bio-carbon research area that can overcome the landfilling issues of agricultural residues and can directly support the growth of plant cultivations. In recent years, marine-based resources are gaining more interest in bio-based materials preparation but only restricted literature is available for C-dots using few of the marine sources. Due to their structural aspects and also the presence of multiple components, most of the marine sources can end up in functionalized C-dots in one-step approach. Due to the noticeable development in bioengineering, one can alter the metabolites of the plants (including some marine sources) that can lead to completely different C-dots compared to original plants.

Developing eco-friendly processes with improved quantum yield is the initial task in the use of renewable sources. The next challenge present in this area of research is to prepare the completely bio-based greener CDs without using any other additives with high performances toward the targeted applications. Developing eco-friendly processes for surface passivation/functionalization without harsh synthetic conditions is the next challenge that comes forward in this area. Though most of the bio-derived C-dots were used for sensing purposes due to their fluorescence nature, selective sensor of specific components in the real systems is the main hurdle that presents in nowadays. In general, bio-derived C-dots are non-harmful to human/nature; their lifetime and decay mechanisms have to be studied well that can promote their role more in the biomedicine fields. Finding out the new applications for bio-derived C-dots can create a rapid commercial interest for these materials in the society. Additionally, proper disposable/storage procedure has to be developed for bio-derived C-dots which can enhance the basic understanding as well as future opportunities of these materials. The possibility of reuse for consecutive cycles or derivatization toward specific applications is the alternative approaches to promote the long-term use of bio-derived C-dots. We strongly believe that this documentation will give new directions to the researchers to develop an eco-friendly C-dots from novel renewable sources for the diversified applications.

5 Conclusions

This chapter explores the use of greener renewable resources as an effective source for the preparation of carbon dots by various synthetic methods along with their applications. Generally, C-dots can be prepared in higher quantities by using simple, rapid, and inexpensive processes. The selection of suitable precursors is the most relevant point, and one can tune the properties of C-dots by simply changing the nature of the source and reaction conditions. Renewable source-derived C-dots possess flexible physical and chemical properties which have a huge influence on the possible applications. Their merits such as eco-friendly, low cost, and non-toxic nature create more attention for the development of renewable C-dots with flexible properties. The development of novel applications using bio-derived C-dots is in an emerging stage, which promotes broad opportunities to explore.

Acknowledgements SV acknowledges the University Grants Commission (UGC) for the financial support for this research activity through the Minor Research Project (MRP/UGC-SERO—Proposal No.: 1593). SS acknowledges FONDECYT-CONICYT, Chile, for his postdoctoral fellowship (Project No. 3160392). RN acknowledges Anillo de Investigación en Ciencia y Tecnología GAMBIO Project No ACT172128, and FONDECYT project No 1190769 from CONICYT, Chile.

References

Amjadi M, Abolghasemi-Fakhri Z, Hallaj T (2015) Carbon dots-silver nanoparticles fluorescence resonance energy transfer system as a novel turn-on fluorescent probe for selective determination of cysteine. J Photochem Photobiol, A 309:8–14. https://doi.org/10.1016/j.jphotochem.2015.04.016

Amjadi M, Hallaj T, Asadollahi H, Song Z, de Frutos M, Hildebrandt N (2017) Facile synthesis of carbon quantum dot/silver nanocomposite and its application for colorimetric detection of methimazole. Sens Actuators B: Chem 244:425–432. https://doi.org/10.1016/j.snb.2017.01.003

Arul V, Edison TN, Lee YR, Sethuraman MG (2017) Biological and catalytic applications of green synthesized fluorescent N-doped carbon dots using *Hylocereus undatus*. J Photochem Photobiol B 168:142–148. https://doi.org/10.1016/j.jphotobiol.2017.02.007

Atchudan R, Edison T, Lee YR (2016a) Nitrogen-doped carbon dots originating from unripe peach for fluorescent bioimaging and electrocatalytic oxygen reduction reaction. J Colloid Interface Sci 482:8–18. https://doi.org/10.1016/j.jcis.2016.07.058

Atchudan R, Edison TNJI, Sethuraman MG, Lee YR (2016b) Efficient synthesis of highly fluorescent nitrogen-doped carbon dots for cell imaging using unripe fruit extract of *Prunus mume*. Appl Surf Sci 384:432–441. https://doi.org/10.1016/j.apsusc.2016.05.054

Atchudan R, Edison TNJI, Chakradhar D, Perumal S, Shim J-J, Lee YR (2017) Facile green synthesis of nitrogen-doped carbon dots using *Chionanthus retusus* fruit extract and investigation of their suitability for metal ion sensing and biological applications. Sens Actuators B: Chem 246:497–509. https://doi.org/10.1016/j.snb.2017.02.119

Atchudan R, Edison TNJI, Aseer KR, Perumal S, Lee YR (2018) Hydrothermal conversion of *Magnolia liliiflora* into nitrogen-doped carbon dots as an effective turn-off fluorescence sensing, multi-colour cell imaging and fluorescent ink Colloids Surf B: Biointerfaces 169:321–328. https://doi.org/10.1016/j.colsurfb.2018.05.032

Ahmed KBA, Suresh Kumar P, Veerappan A (2016) A facile method to prepare fluorescent carbon dots and their application in selective colorimetric sensing of silver ion through the formation of silver nanoparticles. J Lumin 177:228–234. https://doi.org/10.1016/j.jlumin.2016.04.053

Baker SN, Baker GA (2010) Luminescent carbon nanodots: emergent nanolights. Angew Chem Int Ed 49:6726–6744. https://doi.org/10.1002/anie.200906623

Bandi R, Gangapuram BR, Dadigala R, Eslavath R, Singh SS, Guttena V (2016) Facile and green synthesis of fluorescent carbon dots from onion waste and their potential applications as sensor and multicolour imaging agents. RSC Adv 6:28633–28639. https://doi.org/10.1039/C6RA01669C

Banhart F, Ajayan P (1996) Carbon onions as nanoscopic pressure cells for diamond formation. Nature 382:433–435. https://doi.org/10.1038/382433a0

Barati A, Shamsipur M, Arkan E, Hosseinzadeh L, Abdollahi H (2015) Synthesis of biocompatible and highly photoluminescent nitrogen doped carbon dots from lime: analytical applications and optimization using response surface methodology. Mater Sci Eng C Mater Biol Appl 47:325–332. https://doi.org/10.1016/j.msec.2014.11.035

Barbosa CDES, Corrêa JR, Medeiros GA, Barreto G, Magalhães KG, de Oliveira AL, Spencer J, Rodrigues MO, Neto BAD (2015) Carbon dots (C-dots) from cow manure with impressive subcellular selectivity tuned by simple chemical modification. Chem—Eur J 21:5055–5060. https://doi.org/10.1002/chem.201406330

Baruah U, Gogoi N, Konwar A, Deka MJ, Chowdhury D, Majumdar G (2014) Carbon dot based sensing of dopamine and ascorbic acid. J Nanoparticles. https://doi.org/10.1155/2014/178518

Bera D, Qian L, Tseng T-K, Holloway PH (2010) Quantum dots and their multimodal applications: a review. Materials 3:2260–2345. https://doi.org/10.3390/ma3042260

Bhaisare ML, Gedda G, Khan MS, Wu H-F (2016) Fluorimetric detection of pathogenic bacteria using magnetic carbon dots. Anal Chim Acta 920:63–71. https://doi.org/10.1016/j.aca.2016.02.025

Bhatt S, Bhatt M, Kumar A, Vyas G, Gajaria T, Paul P (2018) Green route for synthesis of multifunctional fluorescent carbon dots from Tulsi leaves and its application as Cr (VI) sensors, bio-imaging and patterning agents. Colloids Surf B: Biointerfaces 167:126–133

Bhunia SK, Maity AR, Nandi S, Stepensky D, Jelinek R (2016) Imaging cancer cells expressing the folate receptor with carbon dots produced from folic acid. ChemBioChem 17:614–619. https://doi.org/10.1002/cbic.201500694

Bourlinos AB, Trivizas G, Karakassides MA, Baikousi M, Kouloumpis A, Gournis D, Bakandritsos A, Hola K, Kozak O, Zboril R, Papagiannouli I, Aloukos P, Couris S (2015) Green and simple route toward boron doped carbon dots with significantly enhanced non-linear optical properties. Carbon 83:173–179. https://doi.org/10.1016/j.carbon.2014.11.032

Chatzimitakos T, Kasouni A, Sygellou L, Leonardos I, Troganis A, Stalikas C (2018a) Human fingernails as an intriguing precursor for the synthesis of nitrogen and sulfur-doped carbon dots with strong fluorescent properties: analytical and bioimaging applications. Sens Actuators B: Chem 267:494–501. https://doi.org/10.1016/j.snb.2018.04.059

Chatzimitakos TG, Kasouni AI, Troganis AN, Stalikas CD (2018b) Carbonization of human fingernails: toward the sustainable production of multifunctional nitrogen and sulfur codoped carbon nanodots with highly luminescent probing and cell proliferative/migration properties. ACS Appl Mater Interfaces 10:16024–16032. https://doi.org/10.1021/acsami.8b03263

Chen Y, Wu Y, Weng B, Wang B, Li C (2016) Facile synthesis of nitrogen and sulfur co-doped carbon dots and application for Fe(III) ions detection and cell imaging. Sens Actuators B: Chem 223:689–696. https://doi.org/10.1016/j.snb.2015.09.081

Chen J, Liu J, Li J, Xu L, Qiao Y (2017) One-pot synthesis of nitrogen and sulfur co-doped carbon dots and its application for sensor and multicolor cellular imaging. J Colloid Interface Sci 485:167–174. https://doi.org/10.1016/j.jcis.2016.09.040

Chen D, Zhao J, Zhang L, Liu R, Huang Y, Lan C, Zhao S (2018a) Capsicum-derived biomass quantum dots coupled with Alizarin Red S as an inner-filter-mediated illuminant nanosystem for imaging of intracellular calcium ions. Anal Chem 90:13059–13064. https://doi.org/10.1021/acs.analchem.8b04055

Chen Z, Wang S, Yang X (2018b) Phosphorus-doped carbon dots for sensing both Au (III) and l-methionine. J Photochem Photobiol A: Chem 365:178–184. https://doi.org/10.1016/j.jphotochem.2018.08.001

Chowdhury D, Gogoi N, Majumdar G (2012) Fluorescent carbon dots obtained from chitosan gel. RSC Adv 2:12156. https://doi.org/10.1039/c2ra21705h

Costas-Mora I, Romero V, Lavilla I, Bendicho C (2015) situ photochemical synthesis of fluorescent carbon dots for optical sensing of hydrogen peroxide and antioxidants. Talanta 144:1308–1315. https://doi.org/10.1016/j.talanta.2015.07.093

da Silva JCGE, Gonçalves HMR (2011) Analytical and bioanalytical applications of carbon dots. TrAC Trends Anal Chem 30:1327–1336. https://doi.org/10.1016/j.trac.2011.04.009

Demchenko AP, Dekaliuk MO (2013) Novel fluorescent carbonic nanomaterials for sensing and imaging. Methods Appl Fluorescence 1:042001. https://doi.org/10.1088/2050-6120/1/4/042001

Dey D, Bhattacharya T, Majumdar B, Mandani S, Sharma B, Sarma TK (2013) Carbon dot reduced palladium nanoparticles as active catalysts for carbon-carbon bond formation. Dalton Trans 42:13821–13825. https://doi.org/10.1039/C3DT51234G

Dey S, Govindaraj A, Biswas K, Rao CNR (2014) Luminescence properties of boron and nitrogen doped graphene quantum dots prepared from arc-discharge-generated doped graphene samples. Chem Phys Lett 595:203–208. https://doi.org/10.1016/j.cplett.2014.02.012

Ding H, Du F, Liu P, Chen Z, Shen J (2015a) DNA-carbon dots function as fluorescent vehicles for drug delivery. ACS Appl Mater Interfaces 7:6889–6897. https://doi.org/10.1021/acsami.5b00628

Ding H, Yu S-B, Wei J-S, Xiong H-M (2015b) Full-color light-emitting carbon dots with a surface-state-controlled luminescence mechanism. ACS Nano 10:484–491. https://doi.org/10.1021/acsnano.5b05406

Ding B, Huang S, Pang K, Duan Y, Zhang J (2018) Nitrogen-enriched carbon nanofiber aerogels derived from marine chitin for energy storage and environmental remediation. ACS Sustain Chem Eng 6:177–185. https://doi.org/10.1021/acssuschemeng.7b02164

Dong Y, Chen C, Zheng X, Gao L, Cui Z, Yang H, Guo C, Chi Y, Li CM (2012) One-step and high yield simultaneous preparation of single-and multi-layer graphene quantum dots from CX-72 carbon black. J Mater Chem 22:8764–8766. https://doi.org/10.1039/c2jm30658a

D'souza SL, Deshmukh B, Bhamore JR, Rawat KA, Lenka N, Kailasa SK (2016) Synthesis of fluorescent nitrogen-doped carbon dots from dried shrimps for cell imaging and boldine drug delivery system. RSC Adv 6:12169–12179. https://doi.org/10.1039/c5ra24621k

Edison TN, Atchudan R, Sethuraman MG, Shim JJ, Lee YR (2016a) Microwave assisted green synthesis of fluorescent N-doped carbon dots: cytotoxicity and bio-imaging applications. J Photochem Photobiol B 161:154–161. https://doi.org/10.1016/j.jphotobiol.2016.05.017

Edison TN, Atchudan R, Shim JJ, Kalimuthu S, Ahn BC, Lee YR (2016b) Turn-off fluorescence sensor for the detection of ferric ion in water using green synthesized N-doped carbon dots and its bio-imaging. J Photochem Photobiol, B 158:235–242. https://doi.org/10.1016/j.jphotobiol.2016.03.010

Ensafi AA, Sefat HS, Kazemifard N, Rezaei B, Moradi F (2017) A novel one-step and green synthesis of highly fluorescent carbon dots from saffron for cell imaging and sensing of prilocaine. Sens Actuators B: Chem 253:451–460. https://doi.org/10.1016/j.snb.2017.06.163

Eslami A, Borghei SM, Rashidi A, Takdastan A (2018) Preparation of activated carbon dots from sugarcane bagasse for naphthalene removal from aqueous solutions. Sep Sci Technol 53:2536–2549. https://doi.org/10.1080/01496395.2018.1462832

Essner JB, Laber CH, Ravula S, Polo-Paradab L, Baker GA (2016) Pee-dots: biocompatible fluorescent carbon dots derived from the upcycling of urine. Green Chem 18:243–250. https://doi.org/10.1039/c5gc02032h

Fang Y, Guo S, Li D, Zhu C, Ren W, Dong S, Wang E (2012) Easy synthesis and imaging applications of cross-linked green fluorescent hollow carbon nanoparticles. ACS Nano 6:400–409. https://doi.org/10.1021/nn2046373

Fang L, Xu Q, Zheng X, Zhang W, Zheng J, Wu M, Wu W (2016) Soy flour-derived carbon dots: facile preparation, fluorescence enhancement, and sensitive Fe^{3+} detection. J Nanoparticle Res 18:224. https://doi.org/10.1007/s11051-016-3521-z

Feng X, Jiang Y, Zhao J, Miao M, Cao S, Fang J, Shi L (2015a) Easy synthesis of photoluminescent N-doped carbon dots from winter melon for bio-imaging. RSC Adv 5:31250–31254. https://doi.org/10.1039/c5ra02271a

Feng Y, Zhong D, Miao H, Yang X (2015b) Carbon dots derived from rose flowers for tetracycline sensing. Talanta 140:128–133. https://doi.org/10.1016/j.talanta.2015.03.038

Feng J, Wang W-J, Hai X, Yu Y-L, Wang J-H (2016) Green preparation of nitrogen-doped carbon dots derived from silkworm chrysalis for cell imaging. J Mater Chem B 4:387–393. https://doi.org/10.1039/C5TB01999K

Fernando KS et al (2015) Carbon quantum dots and applications in photocatalytic energy conversion. ACS Appl Mater Interfaces 7:8363–8376. https://doi.org/10.1021/acsami.5b00448

Gaddam RR, Mukherjee S, Punugupati N, Vasudevan D, Patra CR, Narayan R, Kothapalli RVSN (2017) Facile synthesis of carbon dot and residual carbon nanobeads: Implications for ion sensing, medicinal and biological applications. Mater Sci Eng C 73:643–652. https://doi.org/10.1016/j.msec.2016.12.095

Gedda G, Lee C-Y, Lin Y-C, Wu H-F (2016) Green synthesis of carbon dots from prawn shells for highly selective and sensitive detection of copper ions. Sens Actuators B: Chem 224:396–403. https://doi.org/10.1016/j.snb.2015.09.065

Geim AK, Novoselov KS (2007) The rise of graphene. Nat Mater 6:183–191. https://doi.org/10.1038/nmat1849

Gholinejad M, Najera C, Hamed F, Seyedhamzeh M, Bahrami M, Kompany-Zareh M (2017) Green synthesis of carbon quantum dots from vanillin for modification of magnetite nanoparticles and formation of palladium nanoparticles: efficient catalyst for Suzuki reaction. Tetrahedron 73:5585–5592. https://doi.org/10.1016/j.tet.2016.11.014

Ghosh B, Gogoi S, Thakur S, Karak N (2016) Bio-based waterborne polyurethane/carbon dot nanocomposite as a surface coating material. Prog Org Coat 90:324–330. https://doi.org/10.1016/j.porgcoat.2015.10.025

Gonçalves H, Jorge PAS, Fernandes JRA, da Silva JCGE (2010) Hg (II) sensing based on functionalized carbon dots obtained by direct laser ablation. Sens Actuators B: Chem 145:702–707. https://doi.org/10.1016/j.snb.2010.01.031

Gong X, Lu W, Paau MC, Hu Q, Wu X, Shuang S, Dong C, Choi MM (2015) Facile synthesis of nitrogen-doped carbon dots for Fe^{3+} sensing and cellular imaging. Anal Chim Acta 861:74–84. https://doi.org/10.1016/j.aca.2014.12.045

Gu D, Shang S, Yu Q, Shen J (2016) Green synthesis of nitrogen-doped carbon dots from lotus root for Hg(II) ions detection and cell imaging. Appl Surf Sci 390:38–42. https://doi.org/10.1016/j.apsusc.2016.08.012

Gu D, Zhang P, Zhang L, Liu H, Pu Z, Shang S (2018) Nitrogen and phosphorus co-doped carbon dots derived from lily bulbs for copper ion sensing and cell imaging. Opt Mater 83:272–278. https://doi.org/10.1016/j.optmat.2018.06.012

Guo L-P, Zhang Y, Li W-C (2017) Sustainable microalgae for the simultaneous synthesis of carbon quantum dots for cellular imaging and porous carbon for CO_2 capture. J Colloid Interface Sci 493:257–264. https://doi.org/10.1016/j.jcis.2017.01.003

Han S, Zhang H, Zhang J, Xie Y, Liu L, Wang H, Li X, Liu W, Tang Y (2014) Fabrication, gradient extraction and surface polarity-dependent photoluminescence of cow milk-derived carbon dots. RSC Adv 4:58084–58089. https://doi.org/10.1039/c4ra09520k

Han S, Zhang H, Xie Y, Liu L, Shan C, Li X, Liu W, Tang Y (2015) Application of cow milk-derived carbon dots/Ag NPs composite as the antibacterial agent. Appl Surf Sci 328:368–373. https://doi.org/10.1016/j.apsusc.2014.12.074

Himaja AL, Karthik PS, Sreedhar B, Singh SP (2014) Synthesis of carbon dots from kitchen waste: conversion of waste to value added product. J Fluorescence 24:1767–1773. https://doi.org/10.1007/s10895-014-1465-1

Himaja AL, Karthik PS, Singh SP (2015) Carbon dots: the newest member of the carbon nanomaterials family. Chem Rec 15:595–615. https://doi.org/10.1002/tcr.201402090

Hou J, Dong J, Zhu H, Teng X, Ai S, Mang M (2015) A simple and sensitive fluorescent sensor for methyl parathion based on L-tyrosine methyl ester functionalized carbon dots. Biosens Bioelectron 68:20–26. https://doi.org/10.1016/j.bios.2014.12.037

Hou J, Dong G, Tian Z, Lu J, Wang Q, Ai S, Wang M (2016) A sensitive fluorescent sensor for selective determination of dichlorvos based on the recovered fluorescence of carbon dots-Cu(II) system Food Chem 202:81–87. https://doi.org/10.1016/j.foodchem.2015.11.134

Hu Y, Yang J, Tian J, Jia L, Yu J-S (2014) Waste frying oil as a precursor for one-step synthesis of sulfur-doped carbon dots with pH-sensitive photoluminescence. Carbon 77:775–782. https://doi.org/10.1016/j.carbon.2014.05.081

Huang C, Dong H, Su Y, Wu Y, Narron R, Yong Q (2019) Synthesis of carbon quantum dot nanoparticles derived from byproducts in bio-refinery process for cell imaging and in vivo bioimaging. Nanomaterials 9:387. https://doi.org/10.3390/nano9030387

Iijima S, Ichihashi T (1993) Single-shell carbon nanotubes of 1-nm diameter. Nature 363:603–605. https://doi.org/10.1038/363603a0

Ilaiyaraja N, Fathima SJ, Khanum F (2018) Quantum dots: a novel fluorescent probe for bioimaging and drug delivery applications. In: Inorganic frameworks as smart nanomedicines. Elsevier, Amsterdam, pp 529–563. https://doi.org/10.1016/b978-0-12-813661-4.00012-2

Jahanbakhshi M, Habibi B (2016) A novel and facile synthesis of carbon quantum dots via salep hydrothermal treatment as the silver nanoparticles support: application to electroanalytical determination of H_2O_2 in fetal bovine serum. Biosens Bioelectron 81:143–150 https://doi.org/10.1016/j.bios.2016.02.064

Janus Ł, Piątkowski M, Radwan-Pragłowska J, Bogdał D, Matysek D (2019) Chitosan-based carbon quantum dots for biomedical applications: synthesis and characterization. Nanomaterials 9:274. https://doi.org/10.3390/nano9020274

Jhonsi MA, Thulasi S (2016) A novel fluorescent carbon dots derived from tamarind. Chem Phys Lett 661:179–184. https://doi.org/10.1016/j.cplett.2016.08.081

Jiang H, Chen F, Lagally MG, Denes FS (2010) New strategy for synthesis and functionalization of carbon nanoparticles. Langmuir 26:1991–1995. https://doi.org/10.1021/la9022163

Jiang D, Zhang Y, Huang M, Liu J, Wan J, Chu H, Chen M (2014a) Carbon nanodots as reductant and stabilizer for one-pot sonochemical synthesis of amorphous carbon-supported silver nanoparticles for electrochemical nonenzymatic H_2O_2 sensing. J Electroanal Chem 728:26–33. https://doi.org/10.1016/j.jelechem.2014.06.017

Jiang C, Wu H, Song X, Ma X, Wang J, Tan M (2014b) Presence of photoluminescent carbon dots in Nescafe® original instant coffee: applications to bioimaging. Talanta 127:68–74. https://doi.org/10.1016/j.talanta.2014.01.046

Jin H, Gui R, Wang Y, Sun J (2017) Carrot-derived carbon dots modified with polyethyleneimine and nile blue for ratiometric two-photon fluorescence turn-on sensing of sulfide anion in biological fluids. Talanta 169:141–148. https://doi.org/10.1016/j.talanta.2017.03.083

Kasibabu BSB, D'Souza SL, Jha S, Kailasa SK (2015a) Imaging of bacterial and fungal cells using fluorescent carbon dots prepared from *Carica papaya* juice. J Fluoresc 25:803–810. https://doi.org/10.1007/s10895-015-1595-0

Kasibabu BSB, D'Souza SL, Jha S, Singhal RK, Basu H, Kailasa SK (2015b) One-step synthesis of fluorescent carbon dots for imaging bacterial and fungal cells. Anal Methods 7:2373–2378. https://doi.org/10.1039/c4ay02737j

Kong D, Yan F, Luo Y, Ye Q, Zhou S, Chen L (2017) Amphiphilic carbon dots for sensitive detection, intracellular imaging of Al(3). Anal Chim Acta 953:63–70. https://doi.org/10.1016/j.aca.2016.11.049

Konwar A, Chowdhury D (2015) Property relationship of alginate and alginate-carbon dot nanocomposites with bivalent and trivalent cross-linker ions. RSC Adv 5:62864–62870. https://doi.org/10.1039/C5RA09887D

Kroto HW, Heath JR, O'Brien SC, Curl RF, Smalley RE (1985) C_{60}: Buckminsterfullerene. Nature 318:162–163. https://doi.org/10.1038/318162a0

Kumar A, Chowdhuri AR, Laha D, Mahto TK, Karmakar P, Sahu SK (2017) Green synthesis of carbon dots from *Ocimum sanctum* for effective fluorescent sensing of Pb^{2+} ions and live cell imaging. Sens Actuators B: Chem 242:679–686. https://doi.org/10.1016/j.snb.2016.11.109

Li H, Kang Z, Liu Y, Lee S-T (2012) Carbon nanodots: synthesis, properties and applications. J Mater Chem 22:24230–24253. https://doi.org/10.1039/c2jm34690g

Li C-L, Ou C-M, Huang C-C, Wu W-C, Chen Y-P, Lin T-E, Ho L-C, Wang C-W, Shih C-C, Zhou H-C, Lee Y-C, Tzeng W-F, Chiou T-J, Chu S-T, Cang J, Chang H-T (2014) Carbon dots prepared from ginger exhibiting efficient inhibition of human hepatocellular carcinoma cells. J Mater Chem B 2:4564. https://doi.org/10.1039/c4tb00216d

Li L, Li L, Wang C, Liu K, Zhu R, Qiang H, Lin Y (2015) Synthesis of nitrogen-doped and amino acid-functionalized graphene quantum dots from glycine, and their application to the fluorometric determination of ferric ion. Microchimica Acta 182:763–770. https://doi.org/10.1007/s00604-014-1383-6

Li Y, Chen T, Ma Y (2016) Nanosized carbon dots from organic matter and biomass. J Wuhan Univ Technol Mater Sci Ed 31:823–826. https://doi.org/10.1007/s11595-016-1452-2

Li K, Xu J, Arsalan M, Cheng N, Sheng Q, Zheng J, Cao W, Yue T (2019) Nitrogen doped carbon dots derived from natural seeds and their application for electrochemical sensing. J Electrochem Soc 166:B56–B62. https://doi.org/10.1149/2.0501902jes

Liang Q, Ma W, Shi Y, Li Z, Yang X (2013) Easy synthesis of highly fluorescent carbon quantum dots from gelatin and their luminescent properties and applications. Carbon 60:421–428. https://doi.org/10.1016/j.carbon.2013.04.055

Liang Z, Zeng L, Cao X, Wang Q, Wang X, Sun R (2014) Sustainable carbon quantum dots from forestry and agricultural biomass with amplified photoluminescence by simple NH_4OH passivation. J Mater Chem C 2:9760–9766. https://doi.org/10.1039/c4tc01714e

Lim SY, Shen W, Gao Z (2015) Carbon quantum dots and their applications. Chem Soc Rev 44:362–381. https://doi.org/10.1039/C4CS00269E

Lin P-Y, Hsieh C-W, Kung M-L, Chu L-Y, Huang H-J, Chen H-T, Wu D-C, Kuo C-H, Hsieh S-L, Hsieh S (2014) Eco-friendly synthesis of shrimp egg-derived carbon dots for fluorescent bioimaging. J Biotechnol 189:114–119. https://doi.org/10.1016/j.jbiotec.2014.08.043

Lin B, Yan Y, Guo M, Cao Y, Yu Y, Zhang T, Huang Y, Wu D (2018) Modification-free carbon dots as turn-on fluorescence probe for detection of organophosphorus pesticides. Food Chem 245:1176–1182. https://doi.org/10.1016/j.foodchem.2017.11.038

Linehan K, Doyle H (2014) Efficient one-pot synthesis of highly monodisperse carbon quantum dots. RSC Adv 4:18–21. https://doi.org/10.1039/C3RA45083J

Liu S, Tian J, Wang L, Zhang Y, Qin X, Luo Y, Asiri AM, Al-Youbi AO, Sun X (2012) Hydrothermal treatment of grass: a low-cost, green route to nitrogen-doped, carbon-rich, photoluminescent polymer nanodots as an effective fluorescent sensing platform for label-free detection of Cu(II) ions. Adv Mater 24:2037–2041. https://doi.org/10.1002/adma.201200164

Liu S-S, Wang C-F, Li C-X, Wang J, Mao L-H, Chen S (2014) Hair-derived carbon dots toward versatile multidimensional fluorescent materials. J Mater Chem C 2:6477–6483. https://doi.org/10.1039/C4TC00636D

Liu R, Zhang J, Gao M, Li Z, Chen J, Wu D, Liu P (2015) A facile microwave-hydrothermal approach towards highly photoluminescent carbon dots from goose feathers. RSC Adv 5:4428–4433. https://doi.org/10.1039/C4RA12077A

Liu G, Chen Z, Jiang X, Feng D-Q, Zhao J, Fan D, Wang W (2016a) In-situ hydrothermal synthesis of molecularly imprinted polymers coated carbon dots for fluorescent detection of bisphenol A. Sens Actuators B: Chem 228:302–307. https://doi.org/10.1016/j.snb.2016.01.010

Liu X, Pang J, Xu F, Zhang X (2016b) Simple approach to synthesize amino-functionalized carbon dots by carbonization of chitosan. Sci Rep 6:31100. https://doi.org/10.1038/srep31100

Liu F, Zhang W, Chen W, Wang J, Yang Q, Zhu W, Wang J (2017a) One-pot synthesis of $NiFe_2O_4$ integrated with EDTA-derived carbon dots for enhanced removal of tetracycline. Chem Eng J 310:187–196. https://doi.org/10.1016/j.cej.2016.10.116

Liu W, Diao H, Chang H, Wang H, Li T, Wei W (2017b) Green synthesis of carbon dots from rose-heart radish and application for Fe^{3+} detection and cell imaging. Sens Actuators B: Chem 241:190–198. https://doi.org/10.1016/j.snb.2016.10.068

Liu W, Li C, Sun X, Pan W, Yu G, Wang J (2017c) Highly crystalline carbon dots from fresh tomato: UV emission and quantum confinement. Nanotechnology 28:485705. https://doi.org/10.1088/1361-6528/aa900b

Liu Y, Zhou Q, Yuan Y, Wu Y (2017d) Hydrothermal synthesis of fluorescent carbon dots from sodium citrate and polyacrylamide and their highly selective detection of lead and pyrophosphate. Carbon 115:550–560. https://doi.org/10.1016/j.carbon.2017.01.035

Lu W, Gong X, Nan M, Liu Y, Shuang S, Dong C (2015) Comparative study for N and S doped carbon dots: synthesis, characterization and applications for Fe^{3+} probe and cellular imaging. Anal Chim Acta 898:116–127. https://doi.org/10.1016/j.aca.2015.09.050

Lu M, Duan Y, Song Y, Tan J, Zhou L (2018) Green preparation of versatile nitrogen-doped carbon quantum dots from watermelon juice for cell imaging, detection of Fe^{3+} ions and cysteine, and optical thermometry. J Mol Liquids 269:766–774. https://doi.org/10.1016/j.molliq.2018.08.101

Ma Z, Ming H, Huang H, Liu Y, Kang Z (2012) One-step ultrasonic synthesis of fluorescent N-doped carbon dots from glucose and their visible-light sensitive photocatalytic ability. New J Chem 36:861–864. https://doi.org/10.1039/C2NJ20942J

Ma Y, Zhang Z, Xu Y, Ma M, Chen B, Wei L, Xiao L (2016) A bright carbon-dot-based fluorescent probe for selective and sensitive detection of mercury ions. Talanta 161:476–481. https://doi.org/10.1016/j.talanta.2016.08.082

Mehta VN, Jha S, Kailasa SK (2014) One-pot green synthesis of carbon dots by using *Saccharum officinarum* juice for fluorescent imaging of bacteria (Escherichia coli) and yeast (*Saccharomyces cerevisiae*) cells. Mater Sci Eng C Mater Biol Appl 38:20–27. https://doi.org/10.1016/j.msec.2014.01.038

Mehta VN, Jha S, Basu H, Singhal RK, Kailasa SK (2015) One-step hydrothermal approach to fabricate carbon dots from apple juice for imaging of mycobacterium and fungal cells. Sens Actuators B: Chem 213:434–443. https://doi.org/10.1016/j.snb.2015.02.104

Meng Y, Zhang Y, Sun W, Wang M, He B, Chen H, Tang Q (2017) Biomass converted carbon quantum dots for all-weather solar cells. Electrochim Acta 257:259–266. https://doi.org/10.1016/j.electacta.2017.10.086

Mewada A, Pandey S, Shinde S, Mishra N, Oza G, Thakur M, Sharon M, Sharon M (2013) Green synthesis of biocompatible carbon dots using aqueous extract of *Trapa bispinosa* peel. Mater Sci Eng C 33:2914–2917. https://doi.org/10.1016/j.msec.2013.03.018

Miao H, Wang L, Zhuo Y, Zhou Z, Yang X (2016) Label-free fluorimetric detection of CEA using carbon dots derived from tomato juice. Biosens Bioelectron 86:83–89. https://doi.org/10.1016/j.bios.2016.06.043

Mishra V, Patil A, Thakur S, Kesharwani P (2018) Carbon dots: emerging theranostic nanoarchitectures. Drug Discovery Today 23:1219–1232. https://doi.org/10.1016/j.drudis.2018.01.006

Mochalin VN, Shenderova O, Ho D, Gogotsi Y (2012) The properties and applications of nanodiamonds. Nat Nanotechnol 7:11–23. https://doi.org/10.1038/nnano.2011.209

Murugan N, Sundramoorthy AK (2018) Green synthesis of fluorescent carbon dots from *Borassus flabellifer* flowers for label-free highly selective and sensitive detection of Fe^{3+} ions. New J Chem 42:13297–13307. https://doi.org/10.1039/c8nj01894d

Muthulingam S, Bae KB, Khan R, Lee I-H, Uthirakumar P (2016) Carbon quantum dots decorated N-doped ZnO: synthesis and enhanced photocatalytic activity on UV, visible and daylight sources with suppressed photocorrosion. J Environ Chem Eng 4:1148–1155. https://doi.org/10.1016/j.jece.2015.06.029

Nabid MR, Bide Y, Fereidouni N (2016) Boron and nitrogen co-doped carbon dots as a metal-free catalyst for hydrogen generation from sodium borohydride. New J Chem 40:8823–8828. https://doi.org/10.1039/C6NJ01650B

Oza G, Oza K, Pandey S, Shinde S, Mewada A, Thakur M, Sharon M, Sharon M (2015) A green route towards highly photoluminescent and cytocompatible carbon dot synthesis and its separation using sucrose density gradient centrifugation. J Fluoresc 25:9–14. https://doi.org/10.1007/s10895-014-1477-x

Oza G, Ravichandran M, Merupo VI, Shinde S, Mewada A, Ramirez JT, Velumani S, Sharon M, Sharon M (2016) Camphor-mediated synthesis of carbon nanoparticles, graphitic shell encapsulated carbon nanocubes and carbon dots for bioimaging. Sci Rep 6:21286. https://doi.org/10.1038/srep21286

Pakkath SAR, Chetty SS, Selvarasu P, Murugan, AV, Kumar Y, Periyasamy L, Santhakumar M, Sadras SR, Santhakumar K (2018) Transition metal ion (Mn^{2+}, Fe^{2+}, Co^{2+}, and Ni^{2+})-doped carbon dots synthesized via microwave-assisted pyrolysis: a potential nanoprobe for magneto-fluorescent dual-modality bioimaging. ACS Biomater Sci Eng 4:2582–2596. https://doi.org/10.1021/acsbiomaterials.7b00943

Park SY, Lee HU, Park ES, Lee SC, Lee JW, Jeong SW, Kim CH, Lee YC, Huh YS, Lee J (2014) Photoluminescent green carbon nanodots from food-waste-derived sources: large-scale synthesis, properties, and biomedical applications. ACS Appl Mater Interfaces 6:3365–3370. https://doi.org/10.1021/am500159p

Pei S, Zhang J, Gao M, Wu D, Yang Y, Liu R (2015) A facile hydrothermal approach towards photoluminescent carbon dots from amino acids. J Colloid Interface Sci 439:129–133. https://doi.org/10.1016/j.jcis.2014.10.030

Phadke C, Mewada A, Dharmatti R, Thakur M, Pandey S, Sharon M (2015) Biogenic synthesis of fluorescent carbon dots at ambient temperature using *Azadirachta indica* (Neem) gum. J Fluorescence 25:1103–1107. https://doi.org/10.1007/s10895-015-1598-x

Pires NR, Santos CMW, Sousa RR, de Paula RCM, Cunha PLR, Feitosa JPA (2015) Novel and fast microwave-assisted synthesis of carbon quantum dots from raw cashew gum. J Braz Chem Soc 26:1274–1282. https://doi.org/10.5935/0103-5053.20150094

Pradeep T (2007) Nano: the essentials. Understanding nanoscience and nanotechnology. Tata McGraw-Hill Education. ISBN: 9780071548298

Prasannan A, Imae T (2013) One-pot synthesis of fluorescent carbon dots from orange waste peels. Ind Eng Chem Res 52:15673–15678. https://doi.org/10.1021/ie402421s

Pudza MY, Abidin ZZ, Abdul-Rashid Suraya, Md Yassin F, Noor ASM, Abdullah M (2019) Synthesis and characterization of fluorescent carbon dots from tapioca. Chem Select 4: 4140–4146. https://doi.org/10.1002/slct.201900836

Purbia R, Paria S (2016) A simple turn on fluorescent sensor for the selective detection of thiamine using coconut water derived luminescent carbon dots. Biosens Bioelectron 79:467–475. https://doi.org/10.1016/j.bios.2015.12.087

Qiao Z-A, Wang Y, Gao Y, Li H, Dai T, Liu Y, Huo Q (2010) Commercially activated carbon as the source for producing multicolor photoluminescent carbon dots by chemical oxidation. Chem Commun 46:8812–8814. https://doi.org/10.1039/C0CC02724C

Qu D, Zheng M, Zhang L, Zhao H, Xie Z, Jing X, Haddad RE, Fan H, Sun Z (2014) Formation mechanism and optimization of highly luminescent N-doped graphene quantum dots. Sci Rep 4:5294. https://doi.org/10.1038/srep05294

Ramezani Z, Qorbanpour M, Rahbar N (2018) Green synthesis of carbon quantum dots using quince fruit (*Cydonia oblonga*) powder as carbon precursor: application in cell imaging and As^{3+} determination. Colloids Surf A: Physicochem Eng Aspects 549:58–66. https://doi.org/10.1016/j.colsurfa.2018.04.006

Ren G, Zhang Q, Li S, Fu S, Chai F, Wang C, Qu F (2017) One pot synthesis of highly fluorescent N doped C-dots and used as fluorescent probe detection for Hg^{2+} and Ag^{+} in aqueous solution. Sens Actuators B: Chem 243:244–253. https://doi.org/10.1016/j.snb.2016.11.149

Rodrigues CV, Correa JR, Aiube CM, Andrade LP, Galvão PM, Costa PA, Campos AL, Pereira AJ, Ghesti GF, Felix JF, Weber IT, Neto BA, Rodrigues MO (2015) Down- and up-conversion photoluminescence of carbon-dots from brewing industry waste: application in live cell-imaging experiments. J Braz Chem Soc 26:2623–2628. https://doi.org/10.5935/0103-5053.20150291

Romero V, Vila V, de la Calle I, Lavilla I, Bendicho C (2019) Turn–on fluorescent sensor for the detection of periodate anion following photochemical synthesis of nitrogen and sulphur co–doped carbon dots from vegetables. Sens Actuators B: Chem 280:290–297. https://doi.org/10.1016/j.snb.2018.10.064

Rong M-C, Zhang K-X, Wang Y-R, Chen X (2017) The synthesis of B, N-carbon dots by a combustion method and the application of fluorescence detection for Cu^{2+}. Chin Chem Lett 28:1119–1124. https://doi.org/10.1016/j.cclet.2016.12.009

Roy P, Periasamy AP, Chuang C, Liou Y-R, Chen Y-F, Joly J, Liang C-T, Chang H-T (2014) Plant leaf-derived graphene quantum dots and applications for white LEDs. New J Chem 38:4946–4951. https://doi.org/10.1039/C4NJ01185F

Sabet M, Mahdavi K (2019) Green synthesis of high photoluminescence nitrogen-doped carbon quantum dots from grass via a simple hydrothermal method for removing organic and inorganic water pollutions. Appl Surf Sci 463:283–291. https://doi.org/10.1016/j.apsusc.2018.08.223

Sachdev A, Gopinath P (2015) Green synthesis of multifunctional carbon dots from coriander leaves and their potential application as antioxidants, sensors and bioimaging agents. Analyst 140:4260–4269. https://doi.org/10.1039/c5an00454c

Sachdev A, Matai I, Kumar SU, Bhushan B, Dubey P, Gopinath P (2013) A novel one-step synthesis of PEG passivated multicolour fluorescent carbon dots for potential biolabeling application. RSC Adv 3:16958–16961. https://doi.org/10.1039/c3ra42415d

Sahu S, Behera B, Maiti TK, Mohapatra S (2012) Simple one-step synthesis of highly luminescent carbon dots from orange juice: application as excellent bio-imaging agents. Chem Commun 48:8835–8837. https://doi.org/10.1039/C2CC33796G

Sankaranarayanan S, Vishnukumar P, Hariram M, Vivekanandhan S, Camus C, Buschmann AH, Navia R (2019) Hydrothermal synthesis, characterization and seed germination effects of green-emitting graphene oxide-carbon dot composite using brown macroalgal bio-oil as precursor. J Chem Technol Biot 94:3269–3275. https://doi.org/10.1002/jctb.6137

Sharma M, Mondal D, Singh N, Upadhyay K, Rawat A, Devkar RV, Sequeira RA, Prasad K (2017) Seaweed-derived nontoxic functionalized graphene sheets as sustainable materials for the efficient removal of fluoride from high fluoride containing drinking water. ACS Sustain Chem Eng 5:3488–3498. https://doi.org/10.1021/acssuschemeng.7b00198

Shasha P, Kim JH, Park SJ (2019) Celery stalk-derived carbon dots for detection of copper ions. J Nanosci Nanotechnol 19:6077–6082. https://doi.org/10.1166/jnn.2019.17022

Shchipunov YA, Khlebnikov ON, Silant'ev VE (2015) Carbon quantum dots hydrothermally synthesized from chitin. Polymer Sci Ser B 57:16–22. https://doi.org/10.1134/S1560090415010121

Shen L-M, Liu J (2016) New development in carbon quantum dots technical applications. Talanta 156–157:245–256. https://doi.org/10.1016/j.talanta.2016.05.028

Shen L, Chen M, Hu L, Chen X, Wang J (2013) Growth and stabilization of silver nanoparticles on carbon dots and sensing application. Langmuir: The ACS J Surf Colloids 29:16135–16140. https://doi.org/10.1021/la404270w

Shen J, Shang S, Chen X, Wang D, Cai Y (2017) Facile synthesis of fluorescence carbon dots from sweet potato for Fe^{3+} sensing and cell imaging. Mater Sci Eng C 76:856–864. https://doi.org/10.1016/j.msec.2017.03.178

Shi W, Fan H, Ai S, Zhu L (2015) Preparation of fluorescent graphene quantum dots from humic acid for bioimaging application. New J Chem 39:7054–7059. https://doi.org/10.1039/C5NJ00760G

Shi L, Zhao B, Li X, Zhang G, Zhang Y, Dong C, Shuang S (2016) Eco-friendly synthesis of nitrogen-doped carbon nanodots from wool for multicolor cell imaging, patterning, and biosensing. Sens Actuators B: Chem 235:316–324. https://doi.org/10.1016/j.snb.2016.05.094

Sun Y-P, Zhou B, Lin Y, Wang W, Fernando KAS, Pathak P, Meziani MJ, Harruff BA, Wang X, Wang H (2006) Quantum-sized carbon dots for bright and colorful photoluminescence. J Am Chem Soc 128:7756–7757. https://doi.org/10.1021/ja062677d

Sun D, Ban R, Zhang P-H, Wu G-H, Zhang J-R, Zhu J-J (2013) Hair fiber as a precursor for synthesizing of sulfur- and nitrogen-co-doped carbon dots with tunable luminescence properties. Carbon 64:424–434. https://doi.org/10.1016/j.carbon.2013.07.095

Sung-Soo K, Jeong-Yong C, Kwan K, Byeong-Hyeok S (2012) Large area tunable arrays of graphene nanodots fabricated using diblock copolymer micelles. Nanotechnology 23:125301. https://doi.org/10.1088/0957-4484/23/12/125301

Suvarnaphaet P, Tiwary CS, Wetcharungsri J, Porntheeraphat S, Hoonsawat R, Ajayan PM, Tang IM, Asanithi P (2016) Blue photoluminescent carbon nanodots from limeade. Mater Sci Eng C Mater Biol Appl 69:914–921. https://doi.org/10.1016/j.msec.2016.07.075

Tan M, Zhang L, Tang R, Song X, Li Y, Wu H, Wang Y, Lv G, Liu W, Ma X (2013) Enhanced photoluminescence and characterization of multicolor carbon dots using plant soot as a carbon source. Talanta 115:950–956. https://doi.org/10.1016/j.talanta.2013.06.061

Tan XW, Romainor ANB, Chin SF, Ng SM (2014) Carbon dots production via pyrolysis of sago waste as potential probe for metal ions sensing. J Anal Appl Pyrol 105:157–165. https://doi.org/10.1016/j.jaap.2013.11.001

Teng X, Ma C, Ge C, Yan M, Yang J, Zhang Y, Morais PC, Bi H (2014) Green synthesis of nitrogen-doped carbon dots from konjac flour with "off–on" fluorescence by Fe^{3+} and L-lysine for bioimaging. J Mater Chem B 2:4631. https://doi.org/10.1039/c4tb00368c

Thambiraj S, Shankaran DR (2016) Green synthesis of highly fluorescent carbon quantum dots from sugarcane bagasse pulp. Appl Surf Sci 390:435–443. https://doi.org/10.1016/j.apsusc.2016.08.106

Tripathi KM, Tran TS, Tung TT, Losic D, Kim T (2017) Water soluble fluorescent carbon nanodots from biosource for cells imaging. J Nanomater 2017:10. https://doi.org/10.1155/2017/7029731

Tyagi A, Tripathi KM, Singh N, Choudhary S, Gupta RK (2016) Green synthesis of carbon quantum dots from lemon peel waste: applications in sensing and photocatalysis. RSC Adv 6:72423–72432. https://doi.org/10.1039/C6RA10488F

Vandarkuzhali SAA, Natarajan S, Jeyabalan S, Sivaraman G, Singaravadivel S, Muthusubramanian S, Viswanathan B (2018) Pineapple peel-derived carbon dots: applications as sensor, molecular keypad lock, and memory device. ACS Omega 3:12584–12592. https://doi.org/10.1021/acsomega.8b01146

Vivekanandhan S, Schreiber M, Muthuramkumar S, Misra M, Mohanty AK (2017) Carbon nanotubes from renewable feedstocks: a move toward sustainable nanofabrication. J Appl Polym Sci 134:44255. https://doi.org/10.1002/app.44255

Wang Y, Hu A (2014) Carbon quantum dots: synthesis, properties and applications. J Mater Chem C 2:6921–6939. https://doi.org/10.1039/c4tc00988f

Wang L, Zhou HS (2014) Green synthesis of luminescent nitrogen-doped carbon dots from milk and its imaging application. Anal Chem 86:8902–8905. https://doi.org/10.1021/ac502646x

Wang F, Pang S, Wang L, Li Q, Kreiter M, Liu C-Y (2010) One-step synthesis of highly luminescent carbon dots in noncoordinating solvents. Chem Mater 22:4528–4530. https://doi.org/10.1021/cm101350u

Wang Q, Zheng H, Long Y, Zhang L, Gao M, Bai W (2011) Microwave–hydrothermal synthesis of fluorescent carbon dots from graphite oxide. Carbon 49:3134–3140. https://doi.org/10.1016/j.carbon.2011.03.041

Wang J, Wang C-F, Chen S (2012) Amphiphilic egg-derived carbon dots: rapid plasma fabrication, pyrolysis process, and multicolor printing patterns. Angew Chem Int Ed 51:9297–9301. https://doi.org/10.1002/anie.201204381

Wang D, Wang X, Guo Y, Liu W, Qin W (2014a) Luminescent properties of milk carbon dots and their sulphur and nitrogen doped analogues. RSC Adv 4:51658–51665. https://doi.org/10.1039/C4RA11158C

Wang J, Ng YH, Lim Y-F, Ho GW (2014b) Vegetable-extracted carbon dots and their nanocomposites for enhanced photocatalytic H_2 production. RSC Adv 4:44117–44123. https://doi.org/10.1039/c4ra07290a

Wang L, Bi Y, Hou J, Li H, Xu Y, Wang B, Ding H, Ding L (2016a) Facile, green and clean one-step synthesis of carbon dots from wool: application as a sensor for glyphosate detection based on the inner filter effect. Talanta 160:268–275. https://doi.org/10.1016/j.talanta.2016.07.020

Wang L, Li B, Xu F, Shi X, Feng D, Wei D, Li Y, Feng Y, Wang Y, Jia D, Zhou Y (2016b) High-yield synthesis of strong photoluminescent N-doped carbon nanodots derived from hydrosoluble chitosan for mercury ion sensing via smartphone APP. Biosens Bioelectron 79:1–8. https://doi.org/10.1016/j.bios.2015.11.085

Wang L, Li W, Wu B, Li Z, Wang S, Liu Y, Pan D, Wu M (2016c) Facile synthesis of fluorescent graphene quantum dots from coffee grounds for bioimaging and sensing. Chem Eng J 300:75–82. https://doi.org/10.1016/j.cej.2016.04.123

Wang N, Wang Y, Guo T, Yang T, Chen M, Wang J (2016d) Green preparation of carbon dots with papaya as carbon source for effective fluorescent sensing of Iron (III) and *Escherichia coli*. Biosens Bioelectron 85:68–75. https://doi.org/10.1016/j.bios.2016.04.089

Wang ZH, Yu JB, Gui RJ, Jin H, Xia YZ (2016c) Carbon nanomaterials-based electrochemical aptasensors. Biosens Bioelectron 79:136–149. https://doi.org/10.1016/j.bios.2015.11.093

Wang X, Yang P, Feng Q, Meng T, Wei J, Xu C, Han J (2019) Green preparation of fluorescent carbon quantum dots from cyanobacteria for biological imaging polymers 11:616. https://doi.org/10.3390/polym11040616

Weng C-I, Chang H-T, Lin C-H, Shen Y-W, Unnikrishnan B, Li Y-J, Huang C-C (2015) One-step synthesis of biofunctional carbon quantum dots for bacterial labeling. Biosens Bioelectron 68:1–6. https://doi.org/10.1016/j.bios.2014.12.028

Wu H, Mi C, Huang H, Han B, Li J, Xu S (2012) Solvothermal synthesis of green-fluorescent carbon nanoparticles and their application. J Lumin 132:1603–1607. https://doi.org/10.1016/j.jlumin.2011.12.077

Wu ZL, Zhang P, Gao MX, Liu CF, Wang W, Leng F, Huang CZ (2013) One-pot hydrothermal synthesis of highly luminescent nitrogen-doped amphoteric carbon dots for bioimaging from *Bombyx mori* silk—natural proteins. J Mater Chem B 1: 2868–2873. https://doi.org/10.1039/c3tb20418a

Xiao D, Yuan D, He H, Lu J (2013) Microwave-assisted one-step green synthesis of amino-functionalized fluorescent carbon nitride dots from chitosan. Luminescence 28:612–615. https://doi.org/10.1002/bio.2486

Xu X, Ray R, Gu Y, Ploehn HJ, Gearheart L, Raker K, Scrivens WA (2004) Electrophoretic analysis and purification of fluorescent single-walled carbon nanotube fragments. J Am Chem Soc 126:12736–12737. https://doi.org/10.1021/ja040082h

Xu Y, Tang C-J, Huang H, Sun C-Q, Zhang Y-K, Ye Q-F, Wang A-J (2014) Green synthesis of fluorescent carbon quantum dots for detection of Hg^{2+}. Chin J Anal Chem 42:1252–1258. https://doi.org/10.1016/s1872-2040(14)60765-9

Xu M, Gao Z, Zhou Q, Lin Y, Lu M, Tang D (2016a) Terbium ion-coordinated carbon dots for fluorescent aptasensing of adenosine 5′-triphosphate with unmodified gold nanoparticles. Biosens Bioelectron 86:978–984. https://doi.org/10.1016/j.bios.2016.07.105

Xu M, Huang Q, Sun R, Wang X (2016b) Simultaneously obtaining fluorescent carbon dots and porous active carbon for supercapacitors from biomass. RSC Adv 6:88674–88682. https://doi.org/10.1039/C6RA18725K

Xue M, Zhan Z, Zou M, Zhang L, Zhao S (2016) Green synthesis of stable and biocompatible fluorescent carbon dots from peanut shells for multicolor living cell imaging. New J Chem 40:1698–1703. https://doi.org/10.1039/C5NJ02181B

Yan Z, Zhang Z, Chen J (2016) Biomass-based carbon dots: synthesis and application in imatinib determination. Sens Actuators B: Chem 225:469–473. https://doi.org/10.1016/j.snb.2015.10.107

Yang X, Zhuo Y, Zhu S, Luo Y, Feng Y, Dou Y (2014a) Novel and green synthesis of high-fluorescent carbon dots originated from honey for sensing and imaging. Biosens Bioelectron 60:292–298. https://doi.org/10.1016/j.bios.2014.04.046

Yang Z, Xu M, Liu Y, He F, Gao F, Su Y, Wei H, Zhang Y (2014b) Nitrogen-doped, carbon-rich, highly photoluminescent carbon dots from ammonium citrate. Nanoscale 6:1890–1895. https://doi.org/10.1039/c3nr05380f

Yang W, Huang TT, Zhao M, Luo F, Weng W, Wei Q, Lin Z, Chen G (2017a) High peroxidase-like activity of iron and nitrogen co-doped carbon dots and its application in immunosorbent assay. Talanta 164:1–6. https://doi.org/10.1016/j.talanta.2016.10.099

Yang X, Wang Y, Shen X, Su C, Yang J, Piao M, Jia F, Gao G, Zhang L, Lin Q (2017b) One-step synthesis of photoluminescent carbon dots with excitation-independent emission for selective bioimaging and gene delivery. J Colloid Interface Sci 492:1–7. https://doi.org/10.1016/j.jcis.2016.12.057

Yang Q, Duan J, Yang W, Li X, Mo J, Yang P, Tang Q (2018) Nitrogen-doped carbon quantum dots from biomass via simple one-pot method and exploration of their application. Appl Surf Sci 434:1079–1085. https://doi.org/10.1016/j.apsusc.2017.11.040

Yao Y-Y, Gedda G, Girma WM, Yen C-L, Ling Y-C, Chang J-Y (2017) Magnetofluorescent carbon dots derived from crab shell for targeted dual-modality bioimaging and drug delivery. ACS Appl Mater Interfaces 9:13887–13899. https://doi.org/10.1021/acsami.7b01599

Ye Q, Yan F, Luo Y, Wang Y, Zhou X, Chen L (2017) Formation of N, S-codoped fluorescent carbon dots from biomass and their application for the selective detection of mercury and iron ion. Spectrochim Acta A Mol Biomol Spectrosc 173:854–862. https://doi.org/10.1016/j.saa.2016.10.039

Yu H, Li X, Zeng X, Lu Y (2015a) Preparation of carbon dots by non-focusing pulsed laser irradiation in toluene. Chem Commun 52:819–822. https://doi.org/10.1039/C5CC08384B

Yu J, Song N, Zhang Y-K, Zhong S-X, Wang A-J, Chen J (2015b) Green preparation of carbon dots by Jinhua bergamot for sensitive and selective fluorescent detection of Hg^{2+} and Fe^{3+}. Sens Actuators B: Chem 214:29–35. https://doi.org/10.1016/j.snb.2015.03.006

Yu C, Xuan T, Chen Y, Zhao Z, Liu X, Lian G, Li H (2016) Gadolinium-doped carbon dots with high quantum yield as an effective fluorescence and magnetic resonance bimodal imaging probe. J Alloy Compd 688:611–619. https://doi.org/10.1016/j.jallcom.2016.07.226

Yuan M, Zhong R, Gao H, Li W, Yun X, Liu J, Zhao X, Zhao G, Zhang F (2015) One-step, green, and economic synthesis of water-soluble photoluminescent carbon dots by hydrothermal treatment of wheat straw, and their bio-applications in labeling, imaging, and sensing. Appl Surf Sci 355:1136–1144. https://doi.org/10.1016/j.apsusc.2015.07.095

Yue X, Chun-Jing T, Hong H, Chao-Qun S, Ya-Kun Z, Qun-Feng Y, Ai-Jun W (2014) Green synthesis of fluorescent carbon quantum dots for detection of Hg^{2+}. Chinese J Anal Chem 42:1252–1258

Zhai H, Zheng B, Yang F, Wang M, Xiao D (2018) Synthesis of water-soluble fluorescent carbon dots from *Setcreasea purpurea boom* and its application for Br_2 detection. Anal Methods 10:151–157. https://doi.org/10.1039/c7ay02631e

Zhang Y-Q, Ma D-K, Zhuang Y, Zhang X, Chen W, Hong L-L, Yan Q-X, Yu K, Huang S-M (2012) One-pot synthesis of N-doped carbon dots with tunable luminescence properties. J Mater Chem 22:16714–16718. https://doi.org/10.1039/C2JM32973E

Zhang L, Aboagye A, Kelkar A, Lai C, Fong H (2014) A review: carbon nanofibers from electrospun polyacrylonitrile and their applications. J Mater Sci 49:463–480. https://doi.org/10.1007/s10853-013-7705-y

Zhang J, Dong L, Yu S-H (2015a) A selective sensor for cyanide ion (CN^-) based on the inner filter effect of metal nanoparticles with photoluminescent carbon dots as the fluorophore. Sci Bull 60:785–791. https://doi.org/10.1007/s11434-015-0764-5

Zhang J, Yuan Y, Liang G, Yu S-H (2015b) Scale-up synthesis of fragrant nitrogen-doped carbon dots from bee pollens for bioimaging and catalysis advanced. Science 2:1500002. https://doi.org/10.1002/advs.201500002

Zhang Z, Sun W, Wu P (2015c) Highly photoluminescent carbon dots derived from egg white: facile and green synthesis, photoluminescence properties, and multiple applications. ACS Sustain Chem Eng 3:1412–1418. https://doi.org/10.1021/acssuschemeng.5b00156

Zhang C, Xiao Y, Ma Y, Li B, Liu Z, Lu C, Liu X, Wei Y, Zhu Z, Zhang Y (2017) Algae biomass as a precursor for synthesis of nitrogen-and sulfur-co-doped carbon dots: a better probe in Arabidopsis guard cells and root tissues. J Photochem Photobiol, B 174:315–322. https://doi.org/10.1016/j.jphotobiol.2017.06.024

Zhang X, Jiang M, Niu N, Chen Z, Li S, Liu S, Li J (2018) Natural-product-derived carbon dots: from natural products to functional materials. Chemsuschem 11:11–24. https://doi.org/10.1002/cssc.201701847

Zhao Q-L, Zhang Z-L, Huang B-H, Peng J, Zhang M, Pang D-W (2008) Facile preparation of low cytotoxicity fluorescent carbon nanocrystals by electrooxidation of graphite. Chem Commun. 5116–5118. https://doi.org/10.1039/b812420e

Zhao XJ, Zhang WL, Zhou ZQ (2014) Sodium hydroxide-mediated hydrogel of citrus pectin for preparation of fluorescent carbon dots for bioimaging. Colloids Surf B Biointerfaces 123:493–497. https://doi.org/10.1016/j.colsurfb.2014.09.048

Zhao S, Lan M, Zhu X, Xue H, Ng T-W, Meng X, Lee C-S, Wang P, Zhang W (2015) Green synthesis of bifunctional fluorescent carbon dots from garlic for cellular imaging and free radical scavenging. ACS Appl Mater Interfaces 7:17054–17060. https://doi.org/10.1021/acsami.5b03228

Zhao C, Jiao Y, Hu F, Yang Y (2018) Green synthesis of carbon dots from pork and application as nanosensors for uric acid detection. Spectrochim Acta Part A Mol Biomol Spectrosc 190:360–367. https://doi.org/10.1016/j.saa.2017.09.037

Zheng L, Chi Y, Dong Y, Lin J, Wang B (2009) Electrochemiluminescence of water-soluble carbon nanocrystals released electrochemically from graphite. J Am Chem Soc 131:4564–4565. https://doi.org/10.1021/ja809073f

Zhong D, Zhuo Y, Feng Y, Yang X (2015) Employing carbon dots modified with vancomycin for assaying Gram-positive bacteria like *Staphylococcus aureus*. Biosens Bioelectron 74:546–553. https://doi.org/10.1016/j.bios.2015.07.015

Zhou J, Booker C, Li R, Zhou X, Sham T-K, Sun X, Ding Z (2007) An electrochemical avenue to blue luminescent nanocrystals from multiwalled carbon nanotubes (MWCNTs). J Am Chem Soc 129:744–745. https://doi.org/10.1021/ja0669070

Zhou J, Sheng Z, Han H, Zou M, Li C (2012) Facile synthesis of fluorescent carbon dots using watermelon peel as a carbon source. Mater Lett 66:222–224. https://doi.org/10.1016/j.matlet.2011.08.081

Zhu C, Zhai J, Dong S (2012) Bifunctional fluorescent carbon nanodots: green synthesis via soy milk and application as metal-free electrocatalysts for oxygen reduction. Chem commun 48:9367–9369. https://doi.org/10.1039/C2CC33844K

Zhu L, Yin Y, Wang C-F, Chen S (2013) Plant leaf-derived fluorescent carbon dots for sensing, patterning and coding. J Mater Chem C 1:4925–4932

Zong J, Zhu Y, Yang X, Shen J, Li C (2011) Synthesis of photoluminescent carbogenic dots using mesoporous silica spheres as nanoreactors. Chem Commun 47:764–766. https://doi.org/10.1039/C0CC03092A

Zou S, Hou C, Fa H, Zhang L, Ma Y, Dong L, Li D, Huo D, Yang M (2017) An efficient fluorescent probe for fluazinam using N, S co-doped carbon dots from L-cysteine. Sens Actuators B: Chem 239:1033–1041. https://doi.org/10.1016/j.snb.2016.07.169

Zuo P, Lu X, Sun Z, Guo Y, He H (2016a) A review on syntheses, properties, characterization and bioanalytical applications of fluorescent carbon dots. Microchimica Acta 183:519–542

Zuo PL, Lu XH, Sun ZG, Guo YH, He H (2016b) A review on syntheses, properties, characterization and bioanalytical applications of fluorescent carbon dots Microchimica Acta 183:519–542. https://doi.org/10.1007/s00604-015-1705-3

Advances with Synthesis and Applications of Green Bionanomaterials

Aswathy Jayakumar, K. V. Heera, Thoniparambil Sunil Sumi, Meritta Joseph and E. K. Radhakrishnan

Abstract Increasing environmental concern over the synthetic materials generates much interest on biopolymer-based greener alternatives. Major advantages with biopolymers include its biodegradability, renewability, cost-effectiveness and eco-friendly nature. Due to the same, biopolymers have been widely exploited for the nanotechnological applications in various areas including biomedical, water management and food industry. Nanotechnological advances with biopolymer-based materials indicate its revolutionary applications. Hence, the current chapter is designed to include advances with the design of biopolymer-based nanomaterials and its diverse applications. The chapter specifically includes synthesis, processing and characterization of biopolymers-based nanomaterials developed from cellulose, starch, alginate and chitosan. This chapter also gives an overview about the applications of biopolymers in medicine and food packaging.

Keywords Biopolymer · Nanoparticles · Chitosan · Starch · Alginate · Cellulose

1 Introduction

In recent years, there is a tremendous raise in research on biopolymers especially to modify them to novel biopolymer composites with superior qualities. Enhanced accumulation of nondegradable plastics and its low recycling rates has increased the demand for biopolymers. Also, polymers from renewable resources have been considered to have superior features than fossil-fuel-derived polymers. Biopolymers include covalently bonded monomeric units produced by living organisms. They are biocompatible, biodegradable, nontoxic, highly stable and safe for use. Also, biopolymers can easily be degraded by microorganisms present in soil, compost or marine sediment. Based on the chemical nature of their repeating units, biopolymers

A. Jayakumar · K. V. Heera · T. S. Sumi · M. Joseph · E. K. Radhakrishnan (✉)
School of Biosciences, Mahatma Gandhi University, P.D Hills (P.O), Kottayam, Kerala 686560, India
e-mail: radhakrishnanek@mgu.ac.in

S. Ahmed and W. Ali (eds.), *Green Nanomaterials*,
Advanced Structured Materials 126,
https://doi.org/10.1007/978-981-15-3560-4_8

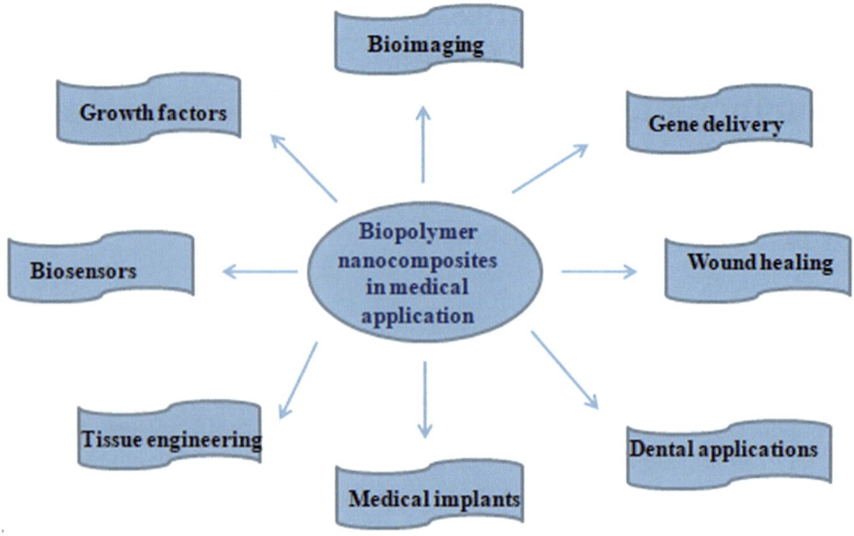

Fig. 1 Applications of biopolymer nanocomposites in medicine

can be polysaccharides, proteins and nucleic acids. They can have significant applications in field of medicine, food industry, cloth industries, water treatment and as biosensors (Figs. 1 and 2).

Among the biopolymers, polysaccharides comprise of repeated monosaccharide units connected by gylcosidic linkages. They include starch, cellulose, alginate, pullulan, heparin, chitosan, hyaluronic acid and dextran. They are preferred in diverse applications due to their high stability, biocompatibility and biodegradable nature. Because of same, their nanotechnological promises can be tremendous. Most interestingly, these have various derivable moieties in its structure which can be simply engineered biochemically or chemically. According to the mechanism of formation, these can be categorized as self assembled, cross linked by chemically or physically and polyionic complex nanoparticles.

2 Cellulose

Cellulose is the most abundant naturally occurring carbohydrate polymer on the earth which is composed of over 3000 glucose residues. It is the prominent scaffolding polysaccharide in cell wall of plant and composed of 33% of all vegetable matter. In plants, it is found as microfibrils (2–20 nm diameter and 100–40,000 nm length) and forms the structural frame work of cell wall. Some bacteria (Rojas et al. 2015), algae (Mihranyan 2011) and some sea animals (George and Sabapathi 2015) have also demonstrated to produce cellulose in a highly hydrated form. Cellulose is an insoluble molecule and is nondigestible by human due to the lack of appropriate

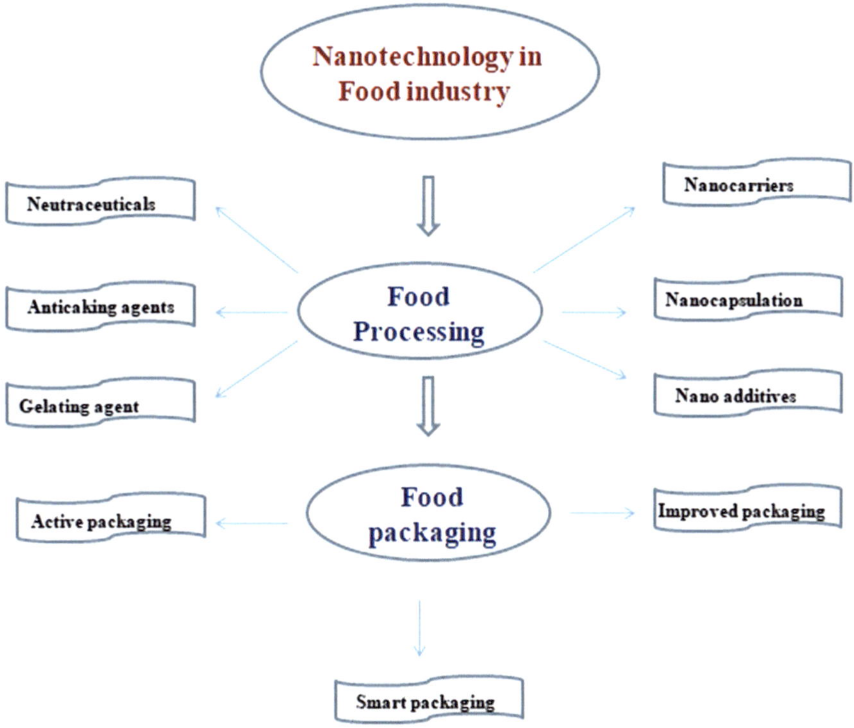

Fig. 2 Application of nanotechnology in food industry

enzymes to breakdown the beta acetal linkages. They are used for the production of papers, fiber plastics, photographic films and rayon after chemical modification. Many derivatives of cellulose are commonly used as adhesing agent, food thickening agents and as explosives.

Cellulose nanofibers (CNFs) and cellulose nanocrystals (CNCs) are the major cellulose nanoparticles (CNCs) and are prepared by acid hydrolysis, enzymatic treatment and mechanical disintegration. Cellulose nanocrystals which are primarily extracted from cellulose fibers are hydrophilic in nature but could be processed for the production of hydrophobic polymer matrices. Highly crystalline cellulose microfibrils can be obtained by chemical, mechanical and enzymatic actions (Domingues et al. 2014). High-intensity ultrasonic treatments, cryocrushing and high-pressure homogenizations are the mechanical processes used for extracting cellulose microfibrils. The shear forces applied during these mechanical action help to split the cellulose fibers. Chemical methods are most effective in the conversion of cellulose microfibrils. These methods reduce the energy consumption and can result in short nanocrystals having improved crystalline property. The removal of amorphous domains is accomplished by strong acid hydrolysis which can penetrate easily in the amorphous regions without affecting crystalline regions. These nanocrystals also have different names such as nanoparticles, whiskers, nanofibers and microcrystallites.

Cellulose nanofibers have diameter of less than 100 nm or can be microfibers with cross section as nanodimension (Isogai et al. 2011). This makes nanocellulose fibers to have significant application as precursor in a diversity of novel functional biomaterials. Cellulose nanofibers take various forms depending on the mode of synthesis and include cellulose microfibril (4 nm), cellulose microfibril bundles (10–20 nm), cellulose nanofiber (5–20 nm) and finally microfibrillar cellulose (5–50 nm). The complication in extraction process restricts the usage of CNF from plant origin even though it possesses excellent features. Bacterial CNF overcome this restriction due to its extracellular secretion which makes the extraction easier. General extraction of CNF from plant source consists of variety of physical, chemical and enzymatic methods. CNFs have various applications in biomedical (Czaja et al. 2007), bioimaging (Dong and Roman 2007), as a part of nanocomposites (Juntaro et al. 2008; Liu et al. 2010), and gas barrier in food packaging films (Fukuzumi et al. 2009).

2.1 Preparation of Cellulose Nanoparticles

2.1.1 Acid Hydrolysis

It is the commonly employed method for preparing cellulose nanocrystal (CNC) from microcrystalline cellulose (Sofla et al. 2016), bacterial cellulose (Vasconcelos et al. 2017), cotton (Sun et al. 2015), sisal (Mariano et al. 2016), pineapple leaves (Santos et al. 2013), fibers of coconut husk (Espino et al. 2014), bananas and barks of mulberry (Jonoobi et al. 2015). The process is carried out using an acid solution (usually employed is H_2SO_4/HCl) at a temperature of 45 °C for 60 min. Water is further added followed by centrifugation at 12,000 rpm for 15 min for removing the acid content and also for stopping ongoing hydrolysis reaction. To further remove acid, dialysis is carried out until the pH neutrality is reached. The solution will further be sonicated for 10 min in an ice bath. The length of CNC depends on cellulose microfibrils and time and temperature in acid hydrolysis treatment (George and Sabapathi 2015).

2.1.2 Chemo-mechanical Treatment

In the mechanical method, the fibers treated with deionized water are homogenized for 20 min at 10,000 rpm followed by the use of microfluidizer to prepare cellulose nanofibers. Chemico-mechanical method utilizes a combination of TEMPO oxidation and microfluidizer for the preparation of cellulose nanofibers. The pretreated fibers are added to a solution of 0.000125 wt% of TEMPO (2, 2, 6, 6-tetramethyl-1-piperidinyloxy) and 0.00125 wt% sodium bromide. Addition of 0.01 wt% sodium hypochlorite initiates the reaction and is maintained at pH 10.5 using sodium hydroxide. The reaction is assumed to be finished when there is no more requirement of sodium hydroxide for maintaining the pH at 10.5. Using hydrochloric acid, the pH will then be adjusted to 7 at room temperature. The whole mixture will finally be allowed to pass through a microfluidizer. For removing the residual salts, the produced cellulose nanofibers can be transferred into dialysis membrane tube.

3 Applications

3.1 In Food Packaging

At a concentration of <2%, nanocellulose appears as gel like and transparent which makes it to be suitable for the production of biodegradable and environmentally safe films (Vartiainen et al. 2011). Nanocellulose films have been prepared from variety of sources. The combination of nanocellulose, polyvinyl alcohol and montmorillonite nanoclay in nanocomposite film making has been described to have interesting features. High-quality nanocellulose films are also produced by semi-industrial roll. Cellulose nanofibrils are widely used in thermoplastics and thermoset plastics.

3.2 Biomedical Applications

Nanocellulose has significant applications in wound dressing, tissue repairing, medical implanting, wound healing and for recovering periodontal tissue. This reveals recent trend to explore nanocellulose as a new-generation biomaterial. It is also used in cartilage repair (Gao et al. 2016; Pretzel et al. 2013), bone regeneration and nervous tissue reconstruction (Kowalska-Ludwicka et al. 2013). Other uses of nanocellulose include tissue scaffolds and drug delivery, and it is also demonstrated to have use in immobilization and recognition of protein, replacement of blood vessel, ligament, meniscus, cartilage and pulposus (Lin and Dufresne 2014). CNF has been considered to have greater thermal stability and stronger reinforcement than the CNC (Sofla et al. 2016) polymer-based matrix.

4 Chitosan

The natural polymer chitosan is derived from crustacean shells by segmental alkaline deacetylation of chitin. The structure of chitosan consists of units of N-acetylglucosamine and glucosamine (Badawy and Rabea 2011). They are also found in some yeast and fungi (Krishna sailaja et al. 2011). This polymer is considered to have beneficial physicochemical properties such as high biocompatibility, nontoxicity, biodegradability and water solubility (Singh Dhillon et al. 2013). Its outstanding features make them a promising material with application in medicine (Rampino et al. 2013). It can be chemically tailored to improve suitability for diverse medical applications. The amino and hydroxyl groups present in them provide an attention to functionally improving the properties. Major techniques used for the modification of chitosan include crosslinking, methylation of carboxyl group, graft copolymerization, etherification and esterification (Rafique et al. 2016).

4.1 Chitosan Nanoparticles

They are biocompatible, nonhazardous and environment friendly having cationic nature (Keawchaoon and Yoksan 2011; Zeng 2011). Because of these, they are acknowledged for drug delivery (Patel et al. 2012; Prabaharan 2015; Agnihotri et al. 2004). However, its application depends on factors like type and shape of material (Ren et al. 2009) concentration (Kim et al. 2003) hydrophobicity, functional groups and chemical modifications (Nel 2006).

4.2 Preparation of Chitosan Nanoparticles (ChNps)

ChNP are formed by various techniques like ionotropic gelation, microemulsion, emulsification, solvent diffusion and complexation using polyelectrolyte.

4.2.1 Ionotropic Gelation Method

This method is a simple technique with mild preparation in the aqueous environment. The mechanism involve the electrostatic interaction of amine group of chitosan with polyanion having negative charge. In this technique, chitosan is dissolved in acetic acid, and polyanion or anionic polymers were added subsequently dropwise under constant mechanical stirring. Stabilizing agents like poloxamer can also be added to the solution before or after adding polyanion (Krishna sailaja et al. 2011).

4.2.2 Microemulsion Method

Here, nanoparticle formation takes place in aqueous core of reverse micellar droplets and is subsequently linked through glutaraldehyde. For stability, surfactant is dissolved in N-hexane and is added to the solution. The system is stirred overnight to complete the crosslinking of free amine group with glutaraldehyde. By evaporation under low pressure, organic solvent is removed and excess surfactant retained with the ChNp can be eliminated by precipitation with calcium chloride. The precipitant can further be removed by centrifugation, and from that nanoparticles will be obtained by further dialysis and lyophilization. The size distribution of the nanoparticles can be made less than 100 nm by this technique. The size of the nanoparticle can be further controlled by changing the amount of glutaraldehyde. However, it has many disadvantages as it is time consuming, uses organic solvents and an involvement of complex washing step (Krishna sailaja et al. 2011).

4.2.3 Emulsification

This makes use of partial miscibility of organic solvents with water content. Emulsion of oil in water is prepared by injecting organic phase into the chitosan containing solution with stabilizing agent like poloxamer under continuous stirring followed by homogenization. Emulsion thus obtained is then diluted with large amount of water to overcome the miscibility of organic solvent with water. The diffusion of organic solvent into water results in the formation of the nanoparticles. Advantages of this technique involve high drug entrapment and suitability for hydrophobic drug, and the drawbacks are the harsh processing conditions and the usage of organic solvents.

4.2.4 Polyelectrolyte Complexation (PEC)

Polyelectrolyte complexes are formed via the interaction between oppositively charged polyions or their corresponding salts. The electrostatic interaction between the ionic groups of polyelectrolyte is responsible for the formation of PEC. Other interactions such as hydrophobic and hydrogen bonds also contribute to complex formation. During the formation of chitosan-PEC, aggregate formation occurs as a result of charge neutralization. Here, the polyelectrolyte solutions must be diluted, and one of the polyions must be used in excess. The difference in preparation methods will result in the formation of nanoparticle, such as nanosphere, nanoaggregate and nanocapsule, and can be determined by water solubility of active agent and the polyanion used.

4.2.5 Electrospinning

Chitosan nanofibers can be prepared by electrospinning technique. This technique makes use of high-voltage electric current at room temperature (Sill and von Recum 2008). Fibrous structures of chitosan can be formed by the electrospinning of chitosan solution in aqueous acetic acid or trifluoroacetic acid (Ohkawa et al. 2004). Chitosan is often mixed with other polymers like PVA in acetic acid, PEO in aqueous acid and PLA in trifluoroacetic acid electrospinning (Lee et al. 2009).

5 Applications

5.1 In Medicine

Chitosan nanoparticles serve as an effective candidate for drug carrier system as it has small size and hydrophilic nature. The small size of chitosan nanoparticles makes them capable of evading the reticuloendothelial system and thereby circulates in the blood for a considerable period. Because of the less penetrable nature of antibiotics, chitosan nanoparticle-based antibiotic carrier can be used as an alternative. Chitosan nanoparticle is a promising candidate in the oral administration of labile drugs which otherwise may become susceptible to degradation by an enzyme in gastrointestinal tract. Many studies have demonstrated chitosan nanoparticles as an effective material for ocular administration of drug. Nanofibers of chitosan find many applications such as in the filtration composite reinforcement (Kim et al. 2005), multifunctional tissue engineering scaffolds (Ma et al. 2005), wound dressing, artificial organs and vascular grafts (Katti et al. 2004).

5.2 In Food Packaging

After harvesting, fruits and vegetables can be coated with chitosan for better shelf life and protecting it from microbial spoilage (Ravi Kumar 2000). Chitosan-based food packaging films are now being fabricated utilizing its antimicrobial property. Numerous strategies like casting, coating and dipping have been developed for preparing chitosan-based films (Wang et al. 2018).

6 Starch

Starch is one of the most abundant polysaccharide present in nature. They are produced by plants as a stored form of energy. The majority of it is found in plant roots, stalks, crop seeds and staple crops such as rice, corn, wheat, tapioca and potato. The major monomer is glucose, which is in two forms consisting 20–30% of amylose and 70–80% amylopectin, which varies from plants to plants. Amylose is linear molecules of glucose units linked by (1–4) α-D glycoside bonds and slightly branched by α (1–6) linkages. Amylopectins are highly branched polymer with linkages and interlinkages at α-D (1–4) glycopyranose and α-D (1–6) glycosidic bonds, respectively. Wet grinding, sieving and drying are the simple methods used by industries for native starch extraction. The different chemical and physical properties of this polysaccharide can be efficiently utilized to replace many toxin compounds used in our day-to-day life. Therefore, the polysaccharide can be modified by different physical and chemical methods to nanoparticles and nanofibers. These compounds have much more efficiency than native starch with variety of industrial and biomedical applications.

6.1 Starch-Based Nanoparticles

Nanoparticles prepared from natural polymers have the advantages like nontoxicity, biocompatibility, biodegradability, renewability and availability. Starch nanoparticles (SNPs) have been prepared using a variety of methods, including acid hydrolysis, extrusion, high-pressure homogenization and emulsification and nanoprecipitation. Nanostarch has high specific area, surface energy, strong absorptive capacity, good biocompatibility and stability. Methods like ultrasonic radiation, air flow crumbling and ball milling which involve friction, collision, impingement, or shear can break down the long chains of polysaccharide and thereby improve efficiency. However, the expected size and shape cannot be acquired accurately. The chemical treatment has the disadvantages of longer time and low yield. Starch nanocrystals are present as water insoluble semi-crystalline granules at normal temperature, and hence, its potential use as a reinforcing phase in polymeric matrix and barrier agent in packaging materials has been highlighted several publications. The crystalline property of the starch is due to A-, B-, C- and V-type starches present in different plants differentiated by X-ray diffraction patterns. The A and B are relatively less sensitive to temperature, whereas C-type starch maintains their crystalline structures at high temperature even being polymorph of A- and B- type starches. A-type crystalline structure is a characteristic of cereal starches. B-type starch is found in potato, and here, the amylopectin forms a closed circle with water molecules in the cavity. The tuber and legume starch consists of C-type polymorph. Unlike other starches, V-type pattern is observed in fully gelatinized starches.

6.2 Preparation of Starch Nanoparticles

6.2.1 Acid Hydrolysis

Acid hydrolysis is the simplest method used for the semi-crystalline polysaccharide nanoparticle preparation. On treatment with mild acids like sulfuric acid and hydrochloric acid, the amorphous regions erode and will facilitate the isolation of resistant crystalline moieties. Šárka and Dvořáček (2017) have reported the response of waxy maize starch toward hydrochloric acid in a submerged reactor under controlled conditions. From this, the acid-treatment process has shown to result in the starch granules with nanoscale size range of 30–80 nm.

6.2.2 Enzymatic Treatment

The highly branched amylopecin gets debranched in the presence of enzyme pullulanase. The debranching of the amylopectin is strongly related to its crystalline structure. The pullulanase hydrolytic study in the gelatinized suspension of maize starch has shown to result in starch nanoparticles with irregular shapes and size range of 15–45 nm.

6.2.3 Ultrasonication

For this, piezoelectric converter which works at high-voltage pulses of energy at frequency like 20 kHz is used. It causes rapid vibration in liquid samples, resulting in violent collapse of microscopic bubbles called cavitation. Its major advantage when compared to acid hydrolysis is the absence of chemical treatment. Hence, the effect of ultrasound treatment on the physical properties of starch suspensions has been studied by various groups. Izidoro et al. (2011) have showed an improvement in starch solubility, swelling power and water absorption capacity due to the ultrasound treatment for 1 h. This was in accordance with the previous reports where the ultrasound treatment has been demonstrated to cause decrease in the viscosity consistency index and enthalpy of gelatinization (Luo et al. 2008; Jambrak et al. 2010). Due to this, ultrasound treatment can results in the distortion of the crystalline region favoring the swelling power of the treated starch. Czechowska-Biskup et al. (2005) described the effect of ultrasound at 360 kHz to cause a drop in the molecular weight of starch due to chain scission of starch macromolecules. Many previous reports also demonstrated the impact of ultrasound treatment on physical degradation of starch granules (Gallant et al. 1997; Degrois et al. 1974). When waxy maize starch was treated with high-intensity frequency of 24 kHz at a low temperature of 8–10 °C, the elementary size of the nanoparticles formed ranged from 20 to 80 nm. Also with the increasing time of ultrasonication, the surface of the granules appeared to be progressively broken down with the release of particles with 20–100 nm size.

6.2.4 Nanoprecipitation

Nanoprecipitation is the most simple and easily reproducible single-step process used to produce starch-based synthetic polymer nanoparticles using polylactic acid (PLA) or polylactic-co-glycolic acid (PLGA). Chin et al. (2011) have synthesized starch nanoparticles with absolute ethanol under controlled precipitation conditions. The shape and size of the particle were maintained by the use of surfactants. In presence of increased ethanol volume, the particle size varied from fibrous and elongated fibrous to spherical shapes. On SEM characterization, the spherical particle size was found to range from 300 to 400 nm.

6.2.5 Emulsions

Starch microspheres are produced by traditional method of using crosslinking agent in water–oil emulsion. It is produced in larger-sized aggregates which hinder its use in drug delivery system. As the chemical and physical properties of starch allow it to be used at room temperature, successful preparation of nanoparticle has been reported.

6.2.6 High-pressure Homogenization and Mini-emulsion Crosslinking

Crosslinking is a chemical process for the production of starch nanoparticles and nanospheres. The enclosed chain structure of the starch is opened and dispersed by alkali, acid and enzymes. As the starch, gelatin and their derivatives are hydrophilic, it can easily form aggregates after dispersing in an aqueous solution of water, oil or emulsifiers. Crosslinking technique requires a low chemical treatment (Liu et al. 2009). They are stable molecules and the shape and size are maintained by the emulsions within the dispersed phase. The unstable mini-emulsion are smaller than normal molecules, so they need energy input through chemical devices or mechanical devices like rotor-stator emulsifiers, sonicators or high-pressure homogenizers. Nanoparticle preparation by high-pressure homogenizer from mini-emulsion is much effective. Shi et al. (2011) generated nanoparticle with water soluble starch using high-pressure homegenizer from water–oil emulsion in presence of sodium trimetaphosphate as crosslinking agent. The purified nanoparticle showed promising sherecity with a particle size range 50–250 nm as revealed by TEM analysis.

7 Applications

7.1 In Packaging

The barrier properties of starch enable it to be used in food packaging. Further, mechanical and physical properties can be improved by the use of nanofillers which help to increase the shelf life of food by protecting it from humidity, oxidation reactions and other deleterious effects. The bioactive compound-encapsulated nanofillers can permit the controllable release of compounds to enhance the shelf life of food without damage.

7.2 In Medicine

Nowadays, systemic delivery of bioactive compounds into the body is gaining much attention. This method overcomes most of the limitations of the protein drugs and antibiotics like poor permeability. Even though raw, native starch has various limitations like poor solubility and high viscosity, it can be modified. Various studies have demonstrated the effective promise of nanoscale drug delivers in biological system like stomach, intestine and blood. Once the starch structure is opened and dispersed by the alkali, acid or enzyme treatment, the formation of nanoparticles can be favoured and accelerated. These starch nanoparticles can be used to encapsulate and deliver the hydrophilic drugs to targeted spots. This approach of encapsulation of active drug in starch nanoparticles can minimize the undesired release of the drug, thereby improving the drug delivering efficiency and its prolonged performance and also reduces the side effects related to drug.

7.3 Alginate

Alginate is an anionic polysaccharide consisting of a chain of β-D-mannuronic acid (M) and α-L-guluronic acid (G) and can be extracted from cell wall of marine brown algae or from bacteria. It is safe, biodegradable, less cost and easily available biopolymer which was found to be biocompatible, mucoadhesive and nonimmunogenic in nature. It is produced as exopolysaccharide during the growth phase and is well known for its gel-forming properties. The gelling is due to the interactions of carboxylic acid and bivalent counter ions.

7.4 Preparation of Alginate Nanoparticles

7.4.1 Ionic Gelation

This is the most common method for preparing alginate-based nanoparticles, microparticles, beads and hydrogels due to the nontoxic procedure and simplicity. This method employs the capacity of crosslinking of polyelectrolytes with counter ions. Here, the incorporation of compounds takes place via the crosslinking of alginate in aqueous solutions ionotropically. This gelation method is otherwise known as egg-box model and can be of internal and external method. Difference between these two geletation processes is the type of calcium salt used. Internal gelation method (internal setting) typically uses aqueous solution of insoluble calcium sources which is characterized by release of calcium ions in a controlled manner. Change in pH can limit the solubility of calcium, and chelating agents present in them determine the controlled release of calcium. External gelation method or diffusion method usually uses dropping of particle-loaded alginate solution into aqueous soluble calcium salt.

7.4.2 Desolvation Technique

It is a thermodynamically driven, self-assembly process which is employed for preparing nanoparticles from polymers. This process typically involves dissolution, aggregation and deaggregation. By using crosslinking agents like glutaraldehyde and sodium sulfate, the aggregated nanoparticles are crosslinked. Desolvating agents like acetone, isopropanol or n-butanol are commonly used for dissolution. The addition of agents is optimized by nephlometer. This techniques are used for entrapment of both lipophilic and hydrophilic drugs. The size of nanoparticles depends on pH, ionic strength, amount of desolvating agent, protein content, agitating speed and concentration of crosslinking agent.

7.4.3 Emulsification

Alginate microparticles and nanoparticles are usually obtained by emulsification and are followed by gelation. These are comparatively simple to perform, as this requires the dispersion of fluid into another immiscible fluid. The alginate nanosphere formation occurs via the combination of emulsification and external gelation. The alginate containing solution with drug or enzyme is emulsified in an oil phase which occurs and forms water–oil emulsion. Then, addition of crosslinking agent causes the alginate emulsion droplets to gelate followed by the demixing of emulsion.

7.4.4 Bubble Bursting

Bubble bursting is a powerful natural phenomenon occurs via the breaking of wave in marine environment which produces nanosized and micronsized droplets. The bubble bursting mechanism involves bubble film disintegration and jetting. Numerous microfluidic devices are used for preparation of microbubble. T-junctions devices are usually used for monodispersed microbubbles of narrow bubble size. Here, the mechanism behind nanoparticle formation depends on volatile solvent evaporation from the microbubble shell coating.

8 Applications

8.1 *In Medicine*

Alginate nanoparticles have remarkable role in medical applications as drug delivery carriers, and also it can boost the bioavailability of drugs. The mucoadhesive properties enable to adhere to the mucosal membranes. The pH-sensitive property of alginate-based nanoparticles can be explored for controlled release of entrapped drug from vehicle. The study conducted by Ahmad et al. (2006) used alginate nanoparticle to deliver pyrazinamide, isoniazid, rifampicin and ethambutol for the treatment of tuberculosis (Ahmad et al. 2006). Cefepime, gentamicin, ceftazidime (Balaure et al. 2013), ciprofloxacin (Ghaffari et al. 2012) and gatifloxacin (Motwani et al. 2007) are the other antibiotics delivered through alginate-based drug delivery. Anticancer drugs such as doxorubicin (Rajaonarivony et al. 1993) (Le et al. 2018) and 5-fluorouracil (Nagarwal et al. 2012) can also delivered by this. Other application involve the delivery of rhodamine, vitamin D3, 5-aminolevulinic acid and quinapyramine as reported from various studies (Le et al. 2018).

9 Conclusions

The increasing awareness for sustainable development has paved the way for replacing synthetic polymers with biopolymers. Naturally occurring biopolymers are ubiquitous and are produced by all living organisms. Biocompatibility of these polymers along with its biodegradability makes it to be used in various applications. The desired structure and function of biopolymers can be derived through physical and chemical modifications. Successful exploitation of these biopolymers in different tissues and organs suggests their promising future for application in repair and regeneration. The nanofiller incorporation in biopolymers based films can help in improving the mechanical and barrier properties which enables it to be used in food packaging applications.

References

Agnihotri SA, Mallikarjuna NN, Aminabhavi TM (2004) Recent advances on chitosan-based micro- and nanoparticles in drug delivery. J Control Release 100(1):5–28. https://doi.org/10.1016/j.jconrel.2004.08.010

Ahmad Z, Pandey R, Sharma S, Khuller GK (2006) Alginate nanoparticles as antituberculosis drug carriers: formulation development, pharmacokinetics and therapeutic potential. Indian J Chest Dis & Allied Sci 48(3):171–176

Badawy MEI, Rabea EI (2011) A biopolymer chitosan and its derivatives as promising antimicrobial agents against plant pathogens and their applications in crop protection. Int J Carbohydr Chem 2011:1–29. https://doi.org/10.1155/2011/460381

Balaure PC, Andronescu E, Grumezescu AM, Ficai A, Huang K-S, Yang C-H, Chifiriuc CM, Lin Y-S (2013) Fabrication, characterization and in vitro profile based interaction with eukaryotic and prokaryotic cells of alginate–chitosan–silica biocomposite. Int J Pharm 441(1–2):555–561. https://doi.org/10.1016/j.ijpharm.2012.10.045

Chin SF, Pang SC, Tay SH (2011) Size controlled synthesis of starch nanoparticles by a simple nanoprecipitation method. Carbohydr Polym 86(4):1817–1819. https://doi.org/10.1016/j.carbpol.2011.07.012

Czaja WK, Young DJ, Kawecki M, Brown RM (2007) The future prospects of microbial cellulose in biomedical applications. Biomacromol 8(1):1–12. https://doi.org/10.1021/bm060620d

Czechowska-Biskup R, Rokita B, Lotfy S, Ulanski P, Rosiak JM (2005) Degradation of chitosan and starch by 360-kHz ultrasound. Carbohydr Polym 60(2):175–184. https://doi.org/10.1016/j.carbpol.2004.12.001

Degrois M, Gallant D, Baldo P, Guilbot A (1974) The effects of ultrasound on starch grains. Ultrasonics 12(3):129–131. https://doi.org/10.1016/0041-624x(74)90070-5

Domingues RMA, Gomes ME, Reis RL (2014) The potential of cellulose nanocrystals in tissue engineering strategies. Biomacromol 15(7):2327–2346. https://doi.org/10.1021/bm500524s

Dong S, Roman M (2007) Fluorescently Labeled cellulose nanocrystals for bioimaging applications. J Am Chem Soc 129(45):13810–13811. https://doi.org/10.1021/ja076196l

Espino E, Cakir M, Domenek S, Román-Gutiérrez AD, Belgacem N, Bras J (2014) Isolation and characterization of cellulose nanocrystals from industrial by-products of Agave tequilana and barley. Ind Crop Prod 62:552–559. https://doi.org/10.1016/j.indcrop.2014.09.017

Fukuzumi H, Saito T, Iwata T, Kumamoto Y, Isogai A (2009) Transparent and high gas barrier films of cellulose nanofibers prepared by TEMPO-mediated oxidation. Biomacromol 10(1):162–165. https://doi.org/10.1021/bm801065u

Gallant DJ, Bouchet B, Baldwin PM (1997) Microscopy of starch: evidence of a new level of granule organization. Carbohydr Polym 32(3–4):177–191. https://doi.org/10.1016/s0144-8617(97)00008-8

Gao X, Shi Z, Liu C, Yang G, Silberschmidt VV (2016) Fracture behaviour of bacterial cellulose hydrogel: microstructural effect. Procedia Struct Integr 2:1237–1243. https://doi.org/10.1016/j.prostr.2016.06.158

George J, Sabapathi SN (2015) Cellulose nanocrystals: synthesis, functional properties, and applications. Nanotechnol, Sci Appl 45. https://doi.org/10.2147/nsa.s64386

Ghaffari S, Varshosaz J, Haririan I, Khoshayand MR, Azarmi S, Gazori T (2012) Ciprofloxacin loaded alginate/chitosan and solid lipid nanoparticles, preparation, and characterization. J Dispers Sci Technol 33(5):685–689. https://doi.org/10.1080/01932691.2011.579831

Isogai A, Saito T, Fukuzumi H (2011) TEMPO-oxidized cellulose nanofibers. Nanoscale 3(1):71–85. https://doi.org/10.1039/c0nr00583e

Izidoro DR, Sierakowski M-R, Haminiuk CWI, de Souza CF, Scheer AdP (2011) Physical and chemical properties of ultrasonically, spray-dried green banana (Musa cavendish) starch. J Food Eng 104(4):639–648. https://doi.org/10.1016/j.jfoodeng.2011.02.002

Jambrak AR, Herceg Z, Šubarić D, Babić J, Brnčić M, Brnčić SR, Bosiljkov T, Čvek D, Tripalo B, Gelo J (2010) Ultrasound effect on physical properties of corn starch. Carbohydr Polym 79(1):91–100. https://doi.org/10.1016/j.carbpol.2009.07.051

Jonoobi M, Oladi R, Davoudpour Y, Oksman K, Dufresne A, Hamzeh Y, Davoodi R (2015) Different preparation methods and properties of nanostructured cellulose from various natural resources and residues: a review. Cellulose 22(2):935–969. https://doi.org/10.1007/s10570-015-0551-0

Juntaro J, Pommet M, Kalinka G, Mantalaris A, Shaffer MSP, Bismarck A (2008) Creating hierarchical structures in renewable composites by attaching bacterial cellulose onto sisal fibers. Adv Mater 20(16):3122–3126. https://doi.org/10.1002/adma.200703176

Katti DS, Robinson KW, Ko FK, Laurencin CT (2004) Bioresorbable nanofiber-based systems for wound healing and drug delivery: optimization of fabrication parameters. J Biomed Mater Res Part B, Appl Biomater 70(2):286–296. https://doi.org/10.1002/jbm.b.30041

Keawchaoon L, Yoksan R (2011) Preparation, characterization and in vitro release study of carvacrol-loaded chitosan nanoparticles. Colloids Surf B: B 84(1):163–171. https://doi.org/10.1016/j.colsurfb.2010.12.031

Kim B, Kim D, Cho D, Cho S (2003) Bactericidal effect of TiO_2 photocatalyst on selected food-borne pathogenic bacteria. Chemosphere 52(1):277–281. https://doi.org/10.1016/s0045-6535(03)00051-1

Kim C-W, Frey MW, Marquez M, Joo YL (2005) Preparation of submicron-scale, electrospun cellulose fibers via direct dissolution. J Polym Sci, Part B: Polym Phys 43(13):1673–1683. https://doi.org/10.1002/polb.20475

Kowalska-Ludwicka K, Cala J, Grobelski B, Sygut D, Jesionek-Kupnicka D, Kolodziejczyk M, Bielecki S, Pasieka Z (2013) New methods modified bacterial cellulose tubes for regeneration of damaged peripheral nerves. Arch Med Sci 3:527–534. https://doi.org/10.5114/aoms.2013.33433

Krishna sailaja A, Amareshwar P, Chakravarty P (2011) Different technique used for the preparation of nanoparticles using natural polymers and their application, vol 3

Le TTH, Bui TQ, Ha TMT, Le MH, Pham HN, Ha PT (2018) Optimizing the alginate coating layer of doxorubicin-loaded iron oxide nanoparticles for cancer hyperthermia and chemotherapy. J Mater Sci 53(19):13826–13842. https://doi.org/10.1007/s10853-018-2574-z

Lee KY, Jeong L, Kang YO, Lee SJ, Park WH (2009) Electrospinning of polysaccharides for regenerative medicine. Adv Drug Deliv Rev 61(12):1020–1032. https://doi.org/10.1016/j.addr.2009.07.006

Lin N, Dufresne A (2014) Nanocellulose in biomedicine: Current status and future prospect. Eur Polym J 59:302–325. https://doi.org/10.1016/j.eurpolymj.2014.07.025

Liu D, Wu Q, Chen H, Chang PR (2009) Transitional properties of starch colloid with particle size reduction from micro- to nanometer. J Colloid Interface Sci 339(1):117–124. https://doi.org/10.1016/j.jcis.2009.07.035

Liu H, Liu D, Yao F, Wu Q (2010) Fabrication and properties of transparent polymethylmethacrylate/cellulose nanocrystals composites. Bioresour Technol 101(14):5685–5692. https://doi.org/10.1016/j.biortech.2010.02.045

Luo Z, Fu X, He X, Luo F, Gao Q, Yu S (2008) Effect of ultrasonic treatment on the physicochemical properties of maize starches differing in amylose content. Starch - Stärke 60(11):646–653. https://doi.org/10.1002/star.200800014

Ma Z, Kotaki M, Inai R, Ramakrishna S (2005) Potential of nanofiber matrix as tissue-engineering scaffolds. Tissue Eng 11(1–2):101–109. https://doi.org/10.1089/ten.2005.11.101

Mariano M, Cercená R, Soldi V (2016) Thermal characterization of cellulose nanocrystals isolated from sisal fibers using acid hydrolysis. Ind Crop Prod 94:454–462. https://doi.org/10.1016/j.indcrop.2016.09.011

Mihranyan A (2011) Cellulose from cladophorales green algae: from environmental problem to high-tech composite materials. J Appl Polym Sci 119(4):2449–2460. https://doi.org/10.1002/app.32959

Motwani S, Chopra S, Talegaonkar S, Kohli K, Ahmad F, Khar R (2007) Chitosan–sodium alginate nanoparticles as submicroscopic reservoirs for ocular delivery: Formulation, optimisation and in vitro characterisation. Eur J Pharm Biopharm. https://doi.org/10.1016/j.ejpb.2007.09.009

Nagarwal RC, Kumar R, Pandit JK (2012) Chitosan coated sodium alginate-chitosan nanoparticles loaded with 5-FU for ocular delivery: in vitro characterization and in vivo study in rabbit eye. Eur J Pharm Sci: Off J Eur Fed Pharm Sci 47(4):678–685. https://doi.org/10.1016/j.ejps.2012.08.008

Nel A (2006) Toxic potential of materials at the nanolevel. Science 311(5761):622–627. https://doi.org/10.1126/science.1114397

Ohkawa K, Cha D, Kim H, Nishida A, Yamamoto H (2004) Electrospinning of chitosan. Macromol Rapid Commun 25(18):1600–1605. https://doi.org/10.1002/marc.200400253

Patel T, Zhou J, Piepmeier JM, Saltzman WM (2012) Polymeric nanoparticles for drug delivery to the central nervous system. Adv Drug Deliv Rev 64(7):701–705. https://doi.org/10.1016/j.addr.2011.12.006

Prabaharan M (2015) Chitosan-based nanoparticles for tumor-targeted drug delivery. Int J Biol Macromol 72:1313–1322. https://doi.org/10.1016/j.ijbiomac.2014.10.052

Pretzel D, Linss S, Ahrem H, Endres M, Kaps C, Klemm D, Kinne RW (2013) A novel in vitro bovine cartilage punch model for assessing the regeneration of focal cartilage defects with biocompatible bacterial nanocellulose. Arthritis Res & Ther 15(3):R59. https://doi.org/10.1186/ar4231

Rafique A, Mahmood Zia K, Zuber M, Tabasum S, Rehman S (2016) Chitosan functionalized poly(vinyl alcohol) for prospects biomedical and industrial applications: a review. Int J Biol Macromol 87:141–154. https://doi.org/10.1016/j.ijbiomac.2016.02.035

Rajaonarivony M, Vauthier C, Couarraze G, Puisieux F, Couvreur P (1993) Development of a new drug carrier made from alginate. J Pharm Sci 82(9):912–917

Rampino A, Borgogna M, Blasi P, Bellich B, Cesàro A (2013) Chitosan nanoparticles: preparation, size evolution and stability. Int J Pharm 455(1–2):219–228. https://doi.org/10.1016/j.ijpharm.2013.07.034

Ravi Kumar MNV (2000) A review of chitin and chitosan applications. React Funct Polym 46(1):1–27. https://doi.org/10.1016/s1381-5148(00)00038-9

Ren G, Hu D, Cheng EWC, Vargas-Reus MA, Reip P, Allaker RP (2009) Characterisation of copper oxide nanoparticles for antimicrobial applications. Int J Antimicrob Agents 33(6):587–590. https://doi.org/10.1016/j.ijantimicag.2008.12.004

Rojas J, Bedoya M, Ciro Y (2015) Current trends in the production of cellulose nanoparticles and nanocomposites for biomedical applications. https://doi.org/10.5772/61334

Santos RMd, Flauzino Neto WP, Silvério HA, Martins DF, Dantas NO, Pasquini D (2013) Cellulose nanocrystals from pineapple leaf, a new approach for the reuse of this agro-waste. Ind Crop Prod 50:707–714. https://doi.org/10.1016/j.indcrop.2013.08.049

Šárka E, Dvořáček V (2017) New processing and applications of waxy starch (a review). J Food Eng 206:77–87. https://doi.org/10.1016/j.jfoodeng.2017.03.006

Shi A-m, Li D, Wang L-j, Li B-z, Adhikari B (2011) Preparation of starch-based nanoparticles through high-pressure homogenization and miniemulsion cross-linking: influence of various process parameters on particle size and stability. Carbohydr Polym 83(4):1604–1610. https://doi.org/10.1016/j.carbpol.2010.10.011

Sill TJ, von Recum HA (2008) Electrospinning: applications in drug delivery and tissue engineering. Biomaterials 29(13):1989–2006. https://doi.org/10.1016/j.biomaterials.2008.01.011

Singh Dhillon G, Kaur S, Jyoti Sarma S, Kaur Brar S, Verma M, Yadagiri Surampalli R (2013) Recent development in applications of important biopolymer chitosan in biomedicine, pharmaceuticals and personal care products. Curr Tissue Eng 2(1):20–40. https://doi.org/10.2174/2211542011302010004

Sofla MRK, Brown RJ, Tsuzuki T, Rainey TJ (2016) A comparison of cellulose nanocrystals and cellulose nanofibres extracted from bagasse using acid and ball milling methods. Adv Nat Sci: Nanosci Nanotechnol 7(3):035004

Sun B, Zhang M, Hou Q, Liu R, Wu T, Si C (2015) Further characterization of cellulose nanocrystal (CNC) preparation from sulfuric acid hydrolysis of cotton fibers. Cellulose 23(1):439–450. https://doi.org/10.1007/s10570-015-0803-z

Vartiainen J, Pöhler T, Sirola K, Pylkkänen L, Alenius H, Hokkinen J, Tapper U, Lahtinen P, Kapanen A, Putkisto K, Hiekkataipale P, Eronen P, Ruokolainen J, Laukkanen A (2011) Health and environmental safety aspects of friction grinding and spray drying of microfibrillated cellulose. Cellulose 18(3):775–786. https://doi.org/10.1007/s10570-011-9501-7

Vasconcelos NF, Feitosa JPA, da Gama FMP, Morais JPS, Andrade FK, de Souza Filho MdSM, Rosa MdF (2017) Bacterial cellulose nanocrystals produced under different hydrolysis conditions: Properties and morphological features. Carbohyd Polym 155:425–431. https://doi.org/10.1016/j.carbpol.2016.08.090

Wang RH, Xin JH, Tao XM (2005) UV-blocking property of dumbbell-shaped ZnO crystallites on cotton fabrics. Inorg Chem 44(11):3926–3930. https://doi.org/10.1021/ic0503176

Wang H, Qian J, Ding F (2018) Emerging chitosan-based films for food packaging applications. J Agric Food Chem 66(2):395–413. https://doi.org/10.1021/acs.jafc.7b04528

Zeng Z (2011) Recent advances of chitosan nanoparticles as drug carriers. Int J Nanomedicine 765. https://doi.org/10.2147/ijn.s17296

Green Nanomaterials for Wastewater Treatment

Poonam Singh, Sanjay Kumar Yadav and Mohammed Kuddus

Abstract Nanomaterials are very small-sized particles that have nanoscale dimension approximately from 1 to 100 nm. The green nanomaterials or nanoparticles are synthesized by biological methods. During the synthesis of green nanoparticles, natural materials such as plants, microorganisms and organic polymers like carbohydrates, proteins and fats are actively involved. Due to the very small size, nanomaterials are playing a significant role in wastewater treatment system and give up possibilities for targeting very specific points. Different types of biosynthesized nanomaterials are used in the controlling of water pollution, such as zinc oxide nanoparticles, phytogenic magnetic nanoparticles, biopolymer-coated metal nanoparticles, silver-impregnated cyclodextrin nanocomposites, silver nanoparticles in *Aloe vera* plant extract and *Photocatalytic titania* nanoparticles. Green nanoparticles are providing an alternative way for elimination of pollutants from the water bodies. The application of green nanoparticles provides an economical, convenient and eco-friendly way of wastewater treatment.

Keywords Nanoparticle · Metal · Pollutant · Eco-friendly · Biological · Phytogenic

1 Introduction

Without water, life cannot be possible on earth; approximately 70% of the total earth's surface is covered by water, and nearly, 97.0% of the total water is found in the sea. The sea water contains a large amount of salt concentration, which is not fit for drinking before desalinization. In the whole world, about one-sixth of population are suffering from lack of clean drinking water (Amin et al. 2014). The clean water demand is continuously rising due to extended drought and population

P. Singh · S. K. Yadav
Department of Molecular and Cellular Engineering, SHUATS, Prayagraj (Allahabad), India

M. Kuddus (✉)
Department of Biochemistry, University of Hail, Hail, Saudi Arabia
e-mail: mkuddus@gmail.com

S. Ahmed and W. Ali (eds.), *Green Nanomaterials*,
Advanced Structured Materials 126,
https://doi.org/10.1007/978-981-15-3560-4_9

growth. Currently, the wastewater is treated by the conventional method including physical, chemical and biological ways, now urgent needs are to adopt an advance technology for wastewater treatments to providing the requirements of clean water. The nanotechnology is a promising field which effectively provides an alternative wastewater treatment process. Due to the high surface area-to-mass ratios of the nanomaterials/nanoparticles, they significantly increase the adsorption ability of sorbent substances (DeFriend et al. 2003). The composition of the wastewater depends on the source from which it is generated, and the common constituents of wastewater are inorganic substances like solutes, heavy metals, metal ions, ammonia along with gases, complex organic compounds such as excreta, plants material, food, protein, and biological pathogens like coliform organisms, faecal streptococci, protozoan cysts and virus particles. Green nanomaterial-based technologies (adsorbents, catalysts, etc.) can create eco-friendly, efficient and affordable solutions for wastewater treatment. Nanomaterials are particles or chemical substances which have nanoscale dimensions (1–100 nm), and the procedure applied to synthesis and manipulate such materials is known as nanotechnology. Nanomaterials show special properties such as increased strength, chemical reactivity and conductivity as compared to non-nanoscale materials. Green nanomaterials are synthesized by using the natural ingredients, which is eco-friendly and harmless for surroundings and human being. In the synthesis of nanoparticles, natural ingredients such as biopolymers, vitamins, proteins, peptides, sugars and plant extracts behave as reducing agents. Combination of green chemistry and nanotechnology is caused by the synthesis of green nanomaterials. Commonly used green nanoparticles for water and wastewater treatment system are metals oxide nanoparticles, viz. TiO_2, ZnO, Fe_3O_4 and silver nanoparticles. The promising elements like aluminium, gold, zinc, carbon, titanium, palladium, iron and copper have been usually used in the formation of nanoparticles. The fullerenes (C60) are used as pollution tracers, and the magnetic nanoparticles successfully adsorbed the metals ion and natural compounds. The nanocatalyst effectively reduces the concentration of oxidized pollutants (Rickerby and Morrison 2007). Live organism plays a vital role in the synthesis of metal green nanoparticles; among all organisms, the plants are found to be the most appropriate for the manufacturing of the green nanoparticles at the commercial level. As compared to the microorganism, plants are produced more reliable and faster rate of green nanoparticles, plant-based nanoparticles are varied in shape and size as compared to other organisms. Some advanced microscopic technique is involved in the morphological study of nanoparticles during the synthesis like scanning electron microscopy, transmission electron microscopy and atomic force microscopy. The synthesized green nanoparticles are characterizing by UV-VIS, TEM, HR-TEM, EDX, XRD, SAED and FTIR (Vetrivel et al. 2015). Due to the small size of nanomaterials, their properties like mechanical, electrical, optical and magnetic properties are quite different from other conventional materials. The strong adsorption capacities and high reactivity of nanomaterials are successfully removed the water pollutants such as heavy metals, organic pollutants, inorganic pollutants and pathogenic microorganism during wastewater treatment.

The silver, Zn and TiO_2 nanoparticles are very effective against pathogenic microorganism like *Escherichia coli, Bacillus subtilis, Vibria cholera, Pseudomonas aeruginosa, Syphillis typhus* and *Staphylococcus aureus* (Baruah et al. 2012). The green nanoparticles having higher surface areas facilitate the attachments of varied chemical functional groups; the various functional groups increase their affinity for target contaminants and successfully remove from water bodies (Chatterjee et al. 2016). Nanomaterials are used by different forms in wastewater treatment system such as nanofiltration membranes, as a nanocatalyst, magnetic nanoparticles and nanosensors for the detection of contaminants. The nanoscale filtration membrane is developed by nanoparticles which separate the pollutant from wastewater through adsorption technique and contaminant breakdown by nanoparticles catalyst. The nanoabsorbents like nanoclays, zeolites, metal oxide nanoparticles, nanoporous carbon fibres and polymeric adsorbents actively participate in the removal of a particular contaminant from wastewater. The advantage of green nanotechnology in the wastewater treatment system is that it reduces the number of steps, cost-effective, energy efficient and successfully removes the contaminants from wastewater. The application of green nanotechnology other than water treatment system is successfully used in healthcare products, electronic items, cosmetics product, textiles industry, information technology and environmental protection (Patidar and Jain 2017).

2 Green Nanomaterials

Green nanomaterials are synthesized by using the natural ingredients, which is eco-friendly and safe to the environment and human health. Nanoparticles are not a new thing; it is very common in nature; life is depending upon many nanoscale matters, including proteins, enzymes and DNA. Green nanotechnology is the study of the synthesis of nanomaterials through natural ingredients which are the benefit to the environment (Ahmed et al. 2016). Green syntheses of nanomaterials mean that raw materials participate in the synthesis of nanomaterials which are natural, biocompatible and biodegradable in nature. Some numerous naturally occurring substances serve as appropriate reducing agents for the formation of green nanoparticles such as plant extracts, biopolymers, vitamins, proteins, peptides and various carbohydrates. The green nanotechnology is a suitable combination of green chemistry and nanotechnology to developing a sustainable technique for water and water treatment system.

2.1 Synthesis of Green Nanomaterials

The basic approach in the synthesis of green nanomaterials is the reduction/oxidation process. The natural substances and organisms are having antioxidant or reducing properties. In the synthesis of green nanomaterials, microorganisms like bacteria, fungi, yeast and actinomycetes are acting as reducing or protecting agents, and in case of the higher plant, phytochemicals are a very good reducing agent and playing important role in the preparation of metals and metal oxide nanoparticles, while nanodimensional natural ingredients like biopolymers, vitamins, proteins, peptides and sugars work as reducing agent in synthesis nanomaterials. Some promising metals including silver, aluminium, gold, zinc, carbon, titanium, palladium, iron and copper are normally used for the formation of green nanoparticles (Sundrarajan and Gowri 2011).

2.1.1 Metal Nanomaterials

The main characteristics of metals nanomaterials are larger surface area-to-volume ratio, large surface energy and transition between molecular and metallic states. Transition state of metals provides a particular electronic arrangement for synthesis of metals nanoparticles. The unique electronic arrangements, optical activity and special mechanical, chemical and magnetic properties of following metals such as silver, gold, platinum and palladium, and non-metallic, inorganic oxides like the zinc oxide and titanium oxide are commonly used in the formation of metal nanoparticles (Kelly et al. 2003).

Green Synthesis of Silver Nanomaterials

Some techniques like ion sputtering, chemical reduction, sol-gel, etc., are using in the formation of silver nanoparticles, but all these methods are producing hazardous chemicals. The silver nanoparticles produced by the green synthesis method are easy, one step, low cost, eco-friendly and stable. Naturally occurring materials like carbohydrates, biodegradable organic polymers, plant extracts matter and microorganisms are using as a reducing and capping agents in the formation of the silver nanoparticles through greener ways (Fatimah 2016). The microorganism and higher plants are reliable sources of the green formation of silver nanoparticles, and in case of microorganism rate of synthesis, nanoparticles are slower as compared to plants.

Synthesis of Silver Nanoparticles Using Microorganism

Microorganisms can synthesize the nanoparticles by intracellular and extracellular forms, *Pseudomonas stutzeri* AG259 strain was the first known bacterial strain which has used in the synthesis of silver nanoparticles. The bacterial strain can grow in a higher concentration of metals ions because they have the ability to develop the adaptation against the specific metals ions. Silver nitrate is used as metallic sources for the in vitro production of silver nanoparticles (Xiangqian et al. 2011). In the normal metabolic process, bacterial cells secrete NADPH-dependent enzymes nitrate reductase. During the process of silver ion reduction, enzyme nitrate reductase reduced nitrate ions to nitrite ions and transferred electron is received by silver ions, the silver ions now reduced to silver (Ag^+ to Ag). Some common fungal strains are used in silver nanomaterial synthesis, including *Fusarium oxysporum, Aspergillus fumigates, Fusarium semitectum, Aspergillus flavus, Trichoderma* species and *Verticillium* species. The benefit of the formation of silver nanoparticles by means of microorganisms has the potential to obtain various shapes and sizes of nanoparticles. The process of extracellular biosynthesis of the silver nanoparticles, the selected bacterial cells are grown in nutrient media, after full growth in the medium bacterial cells mass are separated by supernatant through centrifugation. The supernatant is now properly mixed with the filter-sterilized silver nitrate solution (1 mM). During the intracellular biosynthesis of silver nanoparticles, 2 g of biomass re-suspended in 100 ml (1 mM) aqueous solution of silver nitrate, and now the reaction mixtures are incubates for 72 h in light and shaking condition (Das et al. 2014). Biosynthesized silver nanoparticles are characterized on the basis of size, charges and morphology of nanoparticles using some sophisticated microscopic system like Atomic Force Microscopy (AFM), Scanning Electron Microscopy (SEM) and Transmission Electron Microscopy (TEM).

Synthesis of Silver Nanoparticles Using Plant Extract

Most of the biosynthesis method of nanomaterials is facing problems like stability, aggregation of the nanoparticles, control of crystal growth, and morphology of nanoparticles, size and size distribution. The metal nanoparticles biosynthesized by using plants extract are more stable than microorganism, because plant extract more rapidly reduced metal ions as compare to microorganism and synthesized nanoparticles are extra diverse in morphology (Sadowski et al. 2008). The phytochemicals of some common higher plants species like *Pelargonium graveolens, Cinnammum camphora, Eclipta, Cycas, Ziziphora tenuior, Abutilon indicum, Solanum tricobatum, Erythrin aindica, Ocimum tenuiflorum, Spirogyra varians, Meliadubia, Acalypha indica* and *Sesuvium portulacastrum* are having antioxidant or reducing properties. The process of synthesis of green silver nanoparticles, collect the selected plants materials and washed with sterile water to remove cells debris. The washed plant materials are now dried in shady place and powdered through grinding. The 10% w/v aqueous mixture of powder in deionised distilled water is prepared and carefully boiled. The boiled mixture is filtered carefully up to no insoluble materials show in the broth. For the reduction of silver ions, plant extract mixed with 0.001 M $AgNO_3$ solution in the ratio of 1:9. The silver nanoparticle biosynthesis is determined by using UV-VIZ spectrophotometer (absorption peak around 450 nm). The metallic nature properties of the synthesized nanoparticles are studies by the using of X-ray diffraction (XRD) technique and the crystalline nature of silver nanoparticles are determined by XRD pattern. The binding pattern of linked organic material with nanoparticles is determined by using the Fourier transform infrared (FTIR) spectroscopy which is a chemical analytical technique (Sarvanan et al. 2017).

Green Synthesis of Gold Nanoparticles

The gold nanoparticles are conventionally synthesized by the method of chemical and physical treatments using sodium borohydride, hydrazinium hydroxide, cetyltriethylammnonium bromide and solvents. The chemical method of gold nanoparticles synthesis is not good for the environment; they generate hazard material during synthesis. The gold nanoparticles synthesized by microorganism and plant extract using green chemistry and biosynthesis methods are eco-friendly for environment (Jayaseelan et al. 2013).

Synthesis of Gold Nanoparticles Using Microorganism

The gold nanoparticle is synthesized by the microorganism either intracellular or extracellular ways, and during the intracellular biosynthesis of gold nanoparticles, positively charged Au^{3+} ions move into negatively charged cell wall by electrostatic attraction (Menon et al. 2017). The cell enzymes and various metabolites reduced the gold metal ions into gold nanoparticles. The extracellular way of gold nanoparticles biosynthesis is to use nitrate enzyme, nitrate reductase or hydroquinone synthesized by the reduction mechanism, and during such reduction mechanism, Au^{3+} ions change into Au^{0} nanoparticles. Lactic acid bacteria present in buttermilk are commonly used in the synthesis of gold nanoparticle. Some bacterial species, viz. *Rhodopseudomonas capsulata*, *Pseudomonas aeruginosa* and *Pseudomonas denitrificans*is, are commonly used for biosynthesized gold nanoparticles (He et al. 2007). The yeast *Hansenula anomala* are having the potential to reduce the Au^{3+} ions to Au^{0} nanoparticles. The green biosynthesizes gold nanoparticles are non-toxic in nature.

Synthesis of Gold Nanoparticles Using Plant Extract

The biosynthesis of gold nanoparticle using plant extract is superior over the other biosynthesis methods. The plant species like *Terminalia catappa, Allium cepa,* tea and lemongrass actively involved in the biosynthesis of gold nanoparticles (Lee et al. 2016). Process of biosynthesis of gold nanoparticles is to initially prepare the aqueous mixture (2.5% w/v) of plants materials (fine powder) and boil at 60 °C for 30 min. Now crude extracts are filtered using filter paper, and 20 ml of plants extract is reacted with 10 mM of tetrachloroaurate at room temperature under static conditions. Within an hour during the reaction, mixture colour is changed from pale brownish to purple colour (cherry yellowish red colour in case of *Allium cepa* extract) indicating the biosynthesis of gold nanoparticles (Parida et al. 2011). Prepared gold nanoparticles are stabilized by adding 2% of aqueous chitosan solution. The reducing agent present in plant extracts such as vitamin C or other metabolites reduced the Au^{3+} to Au^{0}. The prepared gold nanoparticles are analysed by UV-visible spectra, a specific absorbance is observed at 540 nm wavelength which evidently indicates the excitation of surface plasmon vibrations in the gold nanoparticles, the size of the nanoparticle is by SEM, and the structure of the gold nanoparticles is examined by using XRD and TEM.

2.1.2 Metal Oxide Nanoparticles

The various metal oxides are capable to form nanoparticles because they show specific physical and chemical characters due to their unique size and density. The electrostatic character of the metal oxide is related to its small size, and due to low surface free energy, the nanoparticles are showing higher mechanical stability. Some common metal oxide is used in wastewater treatments including titanium dioxide (TiO_2), zinc oxide (ZnO), cerium oxide (CeO_2), magnesium oxide (MgO), CuO, NiO and ZrO_2 nanoparticles (Kaushik et al. 2010; Slavin et al. 2017).

Green Synthesis of Titanium Dioxide Nanoparticles (TiO_2 NPs)

Titanium dioxide (TiO_2) is one of the most widely used metal oxide for nanoparticles synthesis, due to its high thermal stability and polymorph formation named as anatase, rutile and brookite (Khadar et al. 2016). Green biosynthesis of titanium dioxide nanoparticles is produced from different plant material extracts, and before preparation of extract, plant materials are washed carefully by sterile water, after drying it is grinded as fine powder. Prepare the 5% w/v aqueous mixture of fine powder in distilled water and boiled at 60 °C for 30 min, and an extract is filtered with filter paper. Now prepare the 1 N TiO_2 solution in distilled water and add the plant extract drop-wise under constant stirring condition to achieve pH which becomes 7 of the solution. The mixture kept as such stirring condition up to 4 h; during the stirring condition, nanoparticles are formed. Afterwards, nanoparticles are separated by filtration using Whatman filter paper; material is washed with distilled water to eliminate the by-product (Rao et al. 2015). The synthesized nanoparticles are dry at 100 °C temperature for overnight and calcined at 500 °C temperature for 4 h. The synthesized titanium nanoparticles of the crystal structure and the average crystalline size are studied by X-ray diffractometer, the average particle size is analysed by particle size analyser, and shape and elemental analysis of the nanoparticles by scanning electron microscope and structure arrangement are observed by transmission electron microscope (Patidar and Jain 2017).

Green Synthesis of Zinc Oxide Nanoparticles (ZnO NPs)

Green synthesis of zinc oxide nanoparticles is eco-friendly and cost-effective as compared to other processes. Various phytochemicals are found in plant extract, which is helping the biosynthesis of zinc oxide nanoparticle through oxidation and reduction processes. The various amines and alkenes (terpenoids, flavonoids and alkaloids) are present in the plant metabolites; they contained functional groups which are facilitating the biosynthesis of nanoparticles (Agarwal et al. 2017). The process of biosynthesis of zinc oxide nanoparticles through plant extract, in the beginning the plant extract is prepared through boiling of the fine powder plant materials in distilled water. Now prepare the 50 ml (1 mM) zinc acetate solution under stirring condition, and then, 20 ml of NaOH solution and 25 ml of prepared plants extract are added in zinc acetate solution under stirring condition. The change in colour of the reaction mixture is observed after 1 h of incubation periods, and nearly after 3 h of mixing, yellow colour appears to indicate the zinc oxide nanoparticle biosynthesis. Now, the precipitates are separated by centrifugation (8000 rpm, 15 min) of the reaction mixture at 60 °C, and the collected pellets were dried at 80 °C for 2 h (Snthoshkumar et al. 2017). The biosynthesized zinc oxide nanoparticles are measured by UV-visible spectroscopy, and the maximum absorbance spectra are reported as 380 nm. The crystal density, purity and size of the zinc oxide nanoparticles are analysed by the application of X-ray crystallography, and with the help of SEM, the shape and size of nanoparticles are analysed. The elemental dispersion analysis of X-ray (EDAX) and atomic force microscopy (AFM) determine the elemental compositions and topography of synthesized zinc oxide nanoparticles, respectively.

3 Characterization of Green Nanomaterials

The nanomaterials' characterization must be required before application to find out behaviours and properties of synthesizes nanomaterials. The nanomaterials are characterized in terms of size, shape, state of dispersion, surface area, physical and chemical properties. The advanced modern techniques like atomic force microscopy, scanning electron microscopy and transmission electron microscopy are used to analyse the size, shape and surface charge of nanomaterials (Powers et al. 2006). The metallic properties of green nanoparticles are characterized by the X-ray diffraction technique, and the character of an attached organic molecule of plant extract to the nanoparticles is analysed by Fourier transform infrared spectroscopy. UV-Vis spectroscopy is indicating the presence of specific nanomaterials in the reaction mixture, by exhibiting characteristic peaks at different absorptions (Sapsford et al. 2011; Hall et al. 2007).

4 Application of Green Nanomaterials for Wastewater Treatment

Different forms of green nanomaterials like nanowires, nanotubes, films, particles, quantum dots and colloids are applied in wastewater treatment system. The pollutants are removed from industrial effluents, surface water, ground water and drinking water with the help of nanomaterial with eco-friendly and cost-effective manners. Traditionally, the three types of the membrane separation system were applied in water purification system, which was comprised as reverse osmosis (RO), ultrafiltration (UF) and microfiltration (MF). The membrane filtration system is very common in the purification of drinking water from wastewater, seawater and surface water.

Currently, nanomaterials' incorporated filtration membrane is very popular in the field of water treatment system. The nanoparticles performed a varied function in water treatment system including pollutant materials degradation, inorganic materials remediation, antimicrobial activity and disinfectant. Numbers of commonly known nanoparticles such as silver nanoparticles, gold nanoparticles, zinc oxide nanoparticles, titanium dioxide nanoparticles, magnesium oxide and carbon nanotubes efficiently are utilized in disinfection of water and wastewater (Li et al. 2008; Elmi et al. 2014). The zinc oxide nanoparticles are having the strong capability to kill the pathogenic microorganism, which is a substitute of UV disinfection to improve water quality. Pd nanoparticles and silver-based nanocatalyst have enhanced the biodegradation of halogenated organic compounds and degraded the organic dyes (Zhang et al. 2012a, b; Wu et al. 2010). Magnetic nanoparticles, carbon nanotubes, metals nanoparticles, quantum dots and dye-doped nanoparticles are extensively employed in the tracing of the pathogen (Yan et al. 2007). The special feature of nanomaterials is a very small size and large surface area, and possesses excellent adsorption properties and strong reducing ability; these notable characteristics contribute to the removal of contaminants from the wastewater (Theron et al. 2008). The application technique of green nanomaterials for water and wastewater treatment is categorized into three groups: (1) nanoscale filtration technique, (2) adsorption of the pollutant on nanoparticles and (3) breakdown of the contaminant by nanoparticle catalyst.

4.1 Nanoscale Filtration Technique

The nanomembranes are made up of nanofibers which employ to remove the micro-sized contaminant from water bodies to reduce the fouling tendency (Srivastava et al. 2004). Multifunctional advance membranes are prepared by the combination of nanomaterials, and an inorganic membrane is called as nanocomposite membranes (Dotzauer et al. 2006). Nanomembranes are employed as a pretreatment measure in ultrafiltration or reverse osmosis. The nanoparticles including alumina, silica, zeolite and TiO_2 are incorporated into polymeric ultrafiltration membranes, the nanoparticles enhanced the membranes efficiency term of surface hydrophilicity, water permeability and fouling resistance. The titanium dioxide nanoparticles containing inorganic membranes are employed for the reductive degradation of chlorinated compounds (Zhang et al. 2008). Silver nanoparticles containing polymeric membranes effectively suppressed the bacterial biofilm formation on the membrane which are broadly used for water disinfection.

4.2 Adsorption of Pollutant on Nanoparticles

Nanomaterials have the ability to adsorb the contaminant from water, the surface of adsorbing material containing specific functional groups which interact with ionic groups of contaminant. The adsorbents are made up of nanomaterials or nanostructured materials which have very high specific surface areas, possibly combined with functional groups, low diffusion distances, definable pore sizes and surface chemistry (Gubin et al. 2005). The commonly used nanoparticles for adsorption of contaminant are manganese oxide, zinc oxide, magnesium oxide, titanium oxide and ferric oxides and carbon nanotubes (Gupta et al. 2015).

4.3 Breakdown of the Contaminant by Nanocatalysts

Due to some unique characters, the nanocatalysts are effective in the removal of a contaminant from wastewater. The catalytic nanoparticles not only remove pollutants but also they can chemically degrade (Dutta et al. 2014). The water contaminants including organochlorine-based pesticides, halogenated herbicides, Azo dyes, polychlorinated biphenyls and nitro-aromatics compounds are readily degraded by catalytic nanoparticles. The nanocatalytic materials including zerovalent metallic nanoparticles, semiconductor materials nanoparticles and bimetallic nanoparticles are generally used in wastewater treatment system. Magnetic nanoparticles are specially used for the elimination of metallic salt and heavy metals and other organic contaminants from water bodies (Ma et al. 2015). Nanosensors can detect single cell of a pathogen and even atoms of contaminants wastewater (Chaturvedi et al. 2012). Magnetic nanosorbents are broadly used in removing the organic contaminants from water. Nanofiber mats having tremendous potential for elimination of heavy metal ions from water bodies due to its heterogeneous adsorbent properties, specific surface area, high porosity and small pore size. The microbial biofilm formation on nanofibril significantly supported the nanofiber structures.

4.4 Water Pollutant Removal Using Different Green Nanomaterials

4.4.1 Disinfection

If the microorganisms, natural organic matter and biological toxins are present in the excess amount in water, it is considered as contaminants. There are various types of green nanomaterials like silver nanoparticles, TiO_2 nanomaterials and zinc nanoparticles used as disinfecting agents for disease-causing microorganism in water body. The nanoparticles also show antibacterial properties due to their smallest size, high specific surface area and high charge carrying capacity (Srinivasan et al. 2008). Some potential green nanomaterials including TiO_2 photo nanocatalyst, metallic nanomaterials and metal oxide nanoparticles are extensively shown as effective antimicrobial character. The average size of silver nanoparticles is 1–40 nm, which is act as antibiotic, antimicrobial and antifungal agents. Bactericidal activity of silver nanoparticles, they initially make free radicals which are adhered on the bacterial cell wall and damage it. After damaging the cell wall they interconnected with bacterial DNA and altering the cell membrane properties and denature the enzymes. The significant role of silver nanoparticles for wastewater treatment is efficiently inactivating the microbial cells and also reducing membrane biofouling (Jain and Pradeep 2005). TiO_2 nanoparticles are the most significant photocatalysts for the water cleansing. The semiconductor-based photocatalysts such as TiO_2 nanoparticles produced the extremely reactive oxidants like OH radicals which are showing the disinfection properties against pathogenic microorganisms. Nitrogen-doped TiO_2 nanoparticle catalysts and TiO_2 nanocomposites with multi walled carbon nanotubes effectively degraded the microbial contaminants in water. The nanostructured TiO_2 films and membranes work as effective disinfecting agents even in presence of UV and visible-light irradiation (Escobar et al. 2001).

4.4.2 Removal of Heavy Metals and Ions

Metal-based nanomaterials effectively eliminated the heavy metal salt from wastewater. The TiO_2 nanoparticles and nanosized magnetite are adsorbing the arsenic from water. The TiO_2 nanoparticles when attached with graphene sheet reduced the chromium (VI) to chromium (III) in presence of sunlight (Zhang et al. 2012a, b). The iron oxide nanomaterials (Fe_2O_3 and Fe_3O_4) are capable of eliminating the heavy metals such as arsenic (Takafuji et al. 2004). The polymer-grafted Fe_2O_3 nanocomposite is effectively used for removal of divalent heavy metal ions such as copper, nickel and cobalt at broad pH range from pH3 to pH7. The three-dimensional nanostructures of CeO_2 are applied as excellent adsorbents for the removal of As and Cr heavy metals (Zhong et al. 2007).

4.4.3 Removal of Organic Contaminants

Natural pollutants are made up of various hydrophilic and hydrophobic organic compounds such as polychlorinated biphenyls (PCBs), benzenes and chlorinated alkanes (Kabra et al. 2004; Lu and Su 2007). TiO_2 nanomaterials in addition to CeO_2 are applied as a catalyst for the whole degradation of natural contaminants along with ozonation processes (Nawrocki and Kasprzyk-Hordern 2010; Orge et al. 2011). Doping of the noble metals into TiO_2 nanoparticles enhanced the decomposition of an organic compound, due to increase of the hydroxyl radical production. TiO_2 nanocrystals modified with the noble metal, which are capable for the decomposition of methylene blue in the presence of light. TiO_2 nanocomposites with mesoporous silica are capable to eliminate the aromatic contaminants from wastewater. Nanocatalysts contained semiconductor materials, zerovalence metals and bimetallic nanoparticles are having the capability to decompose the organic pollutants like

PCBs, pesticides and azo dyes (Qiao et al. 2003). Iron oxide nanomaterials show excellent properties for the removal of organic contaminants as compared to the bulk materials. The chlorinated organic compounds and PCBs are transformed successfully by zerovalent iron nanomaterials. Fe_2O_3 nanoparticles effectively removed the coloured humic acids from wastewater (Zhao et al. 2011).

4.4.4 Desalination

Thin nanocomposite membranes have silver nanoparticle, and TiO_2 nanoparticles show excellent salt elimination properties (Lee et al. 2007). Iron oxide nanoparticles (Fe_2O_3) are coated on TiO_2/Al_2O_3 (aluminium oxide) composite ceramic membranes which exhibit membrane permeability and salt elimination (Lee et al. 2008). Nanocomposite membranes are suitable for the desalination process, during the process which increases the water permeability and a higher amount of salt elimination.

5 Summary

Water pollution, the concentration of one or more substances, is increased up to a limit when they cause the problems for live organisms and environments, the substances are called as pollutants, and the phenomenon is called as water pollution (Owa 2014). Water bodies naturally can clean up a certain amount of pollution by microbial degradation or by dispersing it harmlessly, but concentration of pollutant is so high as it is not clean up naturally. The life of all the plants, animals and humans depends on the water; life cannot survive without water. The main causes of water pollution are industrial liquid waste, domestic sewage, mining activities, marine dumping, accidental oil leakage, agricultural chemical fertilizers and pesticides. Green nanoparticles having the vast possibilities to be applying in wastewater treatment process, the small size, high surface area and charge provided unique characteristics for nanomaterials. The green nanomaterials efficiently removed the toxic heavy metals ion, pathogenic microorganism, inorganic matters and organic contaminant from wastewater (Alrumman et al. 2016). The varied groups of nanomaterials are applying in water treatment system including metal-based nanoparticles (silver nanoparticles and gold nanoparticles), metal oxide nanoparticles (ZnO, TiO_2 and Fe_2O_3), carbonaceous nanomaterials and zeolites. The mode of applications of green nanomaterials in wastewater is divided into major three categories, including treatment and remediation process, sensing and detection process, and pollution prevention mechanism. The nanomaterials are nanoscale size (smaller than 100 nm). Nanoparticles, nanomembrane and nanopowder are broadly applied for the tracing and elimination of chemicals, organic materials and biological materials, heavy metals ions including arsenic, cadmium, copper, lead, mercury and nickel. The application of green nanomaterials in the field of wastewater treatment is playing a vital role in water security (Tiwari et al. 2008).

6 Future Perspective of Nanomaterial for Wastewater Treatment

In future, there are needs to a strong technique for removal of the contaminant from water bodies; the nanotechnology is fulfilling the requirement of the improved technique, who ensures to provide a high-quality drinking water. In future, the nanoparticles will provide the effective alternative ways for wastewater treatment process. Due to its small size, the nanoparticles are having higher interacting, absorbing and reacting capabilities. In near future, some new and special types of nanoparticles will be developing, which is lighter, stronger, more effective, better characterized and easy to adjust in the existing system (Berekaa 2016).

References

Agarwal H, Kumar SV, Rajeshkumar S (2017) A review on green synthesis of zinc oxide. Res Efficient Technol 3(4):406–413

Ahmed S, Ahmad SM, Swami BL, Ikram S (2016) Green synthesis of silver nanoparticles using *Azadirachta indica* aqueous leaf extract. J Radiat Res Appl Sci 9(1):1–7

Alrumman SA, El-kott EF, Keshk SMAS (2016) Water pollution: source & treatment. Am J Environ Eng 6(3):88–98

Amin MT, Alazba AA, Manzoor U (2014) A review of removal of pollutants from water/wastewater using different types of nanomaterials. Adv Mater Sci Eng 2014:1–24

Baruah S, Pal SK, Dutta J (2012) Nanostructured Zinc oxide for water treatment. Nanosci Nanotechnol Asia 2(2):90–102

Berekaa MM (2016) Nanotechnology in wastewater treatment; influence of nanomaterials on microbial systems. Int J Curr Microbiol App Sci 5(1):713–726

Chatterjee A, Nishanthini D, Sandhiya N, Abraham J (2016) Biosynthesis of titanium dioxide nanoparticles using *Vigna radiate*. Asian J Pharm Clin Res 9(4):85–88

Chaturvedi S, Dave PN, Shah NK (2012) Applications of nanocatalyst in new era. J Saudi Chem Soc 16:307–325

Das VL, Thomus R, Varghese RT, Soniya EV, Mathew J, Radhakrishnan EK (2014) Extracellular synthesis of silver nanoparticles by the *Bacillus* strain CS 11 isolated from industrialized area. 3 Biotech 4(2):121–126

DeFriend KA, Wiesner MR, Barron AR (2003) Alumina and aluminate ultrafiltration membranes derived from alumina nanoparticles. J Membr Sci 224(1–2):11–28

Dotzauer DM, Dai J, Sun L, Bruening ML (2006) Catalytic membranes prepared using layer-by-layer adsorption of polyelectrolyte/metal nanoparticle films in porous supports. Nano Lett 6:2268–2272

Dutta AK, Maji SK, Adhikary B (2014) C-Fe2O3 nanoparticles: an easily recoverable effective photo-catalyst for the degradation of rose bengal and methylene blue dyes in the waste-water treatment plant. Mater Res Bull 49:28–34

Elmi F, Alinezhad H, Moulana Z, Salehian F, Tavakkoli SM, Asgharpour F, Fallah H, Elmi MM (2014) The use of antibacterial activity of ZnO nanoparticles in the treatment of municipal wastewater. Water Sci Technol 70(5):763–770

Escobar IC, Randall AA, Taylor JS (2001) Bacterial growth in distribution systems: effect of assimilable organic carbon and biodegradable dissolved organic carbon. Environ Sci Technol 35(17):3442–3447

Fatimah I (2016) Green synthesis of silver nanoparticles using extract of *ParkiaspeciosaHassk* pods assisted by microwave irradiation. J Adv Res 7(6):961–969

Gubin SP, Koksharov YA, Khomutov GB, Yurkov GYE (2005) Magnetic nanoparticles: preparation, structure and properties. Russ Chem Rev 74:489–520

Gupta VK, Tyagi I, Sadegh H, Shahryari-Ghoshekand R, Makhlouf ASH, Maazinejad B (2015) Nanoparticles as adsorbent; a positive approach for removal of noxious metal ions: a review. Sci Technol Dev 34:195

Hall JB, Dobrovolskaia MA, Patri AK, McNeil SE (2007) Characterization of nanoparticles for therapeutics. Nanomedicine (London) 2:789–803

He S, Guo Z, Zhang Y, Wang J, Jhang S, Gu N (2007) Biosynthesis of gold nanoparticles using the bacteria *Rhodopseudomonas capsulata*. Mater Lett 62(18):3984–3987

Jain P, Pradeep T (2005) Potential of silver nanoparticle-coated polyurethane foam as an antibacterial water filter. Biotechnol Bioeng 90(1):59–63

Jayaseelan C, Rajkumar R, Rahuman AA, Perumal P (2013) Green synthesis of gold nanoparticles using seed aqueous extract of *Abelmoschus esculentus* and its antifungal activity. Ind Crops Prod 45:423–429

Kabra K, Chaudhary R, Sawhney RL (2004) Treatment of hazardous organic and inorganic compounds through aqueous-phase photocatalysis: a review. Ind Eng Chem Res 43(24):7683–7696

Kaushik N, Thakur MS, Snehit S, Mhatre MS, Rajesh Y, Parikh MS (2010) Biological synthesis of metallic nanoparticles. Nanomed Nanotechnol Biol Med 6(2):257–262

Kelly KL, Coronado E, Zhao LL, Schatz GC (2003) The optical properties of metal nanoparticles: the influence of size, shape, and dielectric environment. J Phys Chem B 107(3):668–677

Khadar A, Behara DK, Kumar MK (2016) Synthesis and characterization of controlled size TiO_2 nanoparticles via green route using *Aloe vera* extract. Int J Sci Res 11(5):1913–1916

Lee HS, Im SJ, Kim JH, Kim HJ, Kim JP, Min BR (2008) Polyamide thin-film nanofiltration membranes containing TiO_2 nanoparticles. Desalination 219(1–3):48–56

Lee KX, Shameli K, Miyake M, Kuwano N, Khairudin NBBA, Mohamad SEB, Yew YP (2016) Green synthesis of gold nanoparticles using aqueous extract of Garciniamangostana fruit peels. J Nanomaterials 2016:1–7

Lee SY, Kim HJ, Patel R, Im SJ, Kim JH, Min BR (2007) Silver nanoparticles immobilized on thin film composite polyamide membrane: characterization, nanofiltration, antifouling properties. Polym Adv Technol 18(7):562–568

Li D, Lyon DY, Li Q, Alvarez PJJ (2008) Effect of soil sorption and aquatic natural organic matter on the antibacterial activity of a fullerene water suspension. Environ Toxicol Chem 27(9):1888–1894

Lu C, Su F (2007) Adsorption of natural organic matter by carbon nanotubes. Sep Purif Technol 58(1):113–121

Ma H, Wang H, Na C (2015) Microwave-assisted optimization of platinum-nickel nanoalloys for catalytic water treatment. Appl Catal B Environ 163:198–204

Menon S, Rajkumar S, Venkat KS (2017) A review on biogenic synthesis of gold nanoparticles, characterization, and its applications. Res Efficient Technol 3(4):516–527

Nawrocki J, Kasprzyk-Hordern B (2010) The efficiency and mechanisms of catalytic ozonation. Appl Catal B Environ 99(1–2):27–42

Orge CA, Orfao JMM, Pereira MMFR, Duarte de Farias AM, Neto RCR, Fraga MA (2011) Ozonation of model organic compounds catalysed by nanostructured cerium oxides. Appl Catal B Environ 103(1–2):190–199

Owa FW (2014) Water pollution: sources, effects, control and management. Int Lett Nat Sci 3:1–6

Parida UK, Bindhani BK, Nayak P (2011) Green synthesis and characterization of gold nanoparticles using Onion (*Allium cepa*) extract. World J Nano Sci Eng 1:93–98

Patidar V, Jain P (2017) Green synthesis of TiO_2 nanoparticle using *Moringaoleifera* leaf extract. Int Res J Eng Technol (IRJET) 04(3):470–473

Powers KW, Brown SC, Krishna VB, Wasdo SC, Moudgil BM, Roberts SM (2006) Research strategies for safety evaluation of nanomaterials. Part VI characterization of nanoscale particles for toxicological evaluation. Toxicol Sci 90:296–303

Qiao S, Sun DD, Tay JH, Easton C (2003) Photocatalytic oxidation technology for humic acid removal using a nano structured TiO_2/Fe_2O_3 catalyst. Water Sci Technol 47(1):211–217

Rao KG, Ashok CH, Rao KV, Chakra CHS, Tambur P (2015) Green synthesis of TiO_2 nanoparticles using *Aloe Vera* extract. Int J Adv Res Phys Sci (IJARPS) 2(1A):28–34

Rickerby DG, Morrison M (2007) Nanotechnology and the environment: a European perspective. Sci Technol Adv Mater 8(1–2):19–24

Sadowski Z, Maliszewska IH, Grochowalska B, Polowczyk I, Kozlecki T (2008) Synthesis of silver nanoparticles using microorganisms. Mater Sci Poland 26(2):219–424

Sapsford KE, Tyner KM, Dair BJ, Deschamps JR, Medintz IL (2011) Analyzing nanomaterial bioconjugates: a review of current and emerging purification and characterization techniques. Anal Chem 83:4453–4488

Sarvanan C, Rajesh R, Kaviarasan T, Muthukumar K, Kavitaki D, Shetty PH (2017) Synthesis of silver nanoparticles using bacterial exopolysaccharide and its application for degradation of azo-dyes. Biotechnol Rep 15:33–40

Slavin YN, Asnis J, Hafeli UO, Horacio Bach H (2017) Metal nanoparticles: understanding the mechanisms behind antibacterial activity. J Nanobiotechnology 15:65

Santhoshkumar J, Kumar SV, Rajeshkumar S (2017) Synthesis of zinc oxide nanoparticles using plant leaf extract against urinary tract infection pathogen. Res Efficient Technol 3(4):459–465

Srinivasan S, Harrington GW, Xagoraraki I, Goel R (2008) Factors affecting bulk to total bacteria ratio in drinking water distribution systems. Water Res 42(13):3393–3404

Srivastava A, Srivastava ON, Talapatra S, Vajtai R, Ajayan PM (2004) Carbon nanotube filters. Nature Mater 3:610–614

Sundrarajan M, Gowri S (2011) Green synthesis of titanium dioxide nanoparticles by *Nyctanthes arbor-tristis*leaves extract. Chalcogenide Lett 8(8):447–451

Takafuji M, Ide S, Ihara H, Xu Z (2004) Preparation of poly (1-vinylimidazole)-grafted magnetic nanoparticles and their application for removal of metal ions. Chem Mater 16(10):1977–1983

Theron J, Walker JA, Cloete TE (2008) Nanotechnology and water treatment: applications and emerging opportunities. Crit Rev Microbiol 34:43–69

Tiwari DK, Behari J, Sen P (2008) Application of nanoparticles in waste water treatment. World Appl Sci J 3(3):417–433

Vetrivel V, Rajendran K, Kalaiselvi V (2015) Synthesis and characterization of pure titanium dioxide nanoparticles by Sol-gel method. Int J Chem Tech Res 7(3):1090–1097

Wu Z, Zhang Y, Tao T, Zhang T, Fong H (2010) Silver nanoparticles on amidoxime fibers for photo-catalytic degradation of organic dyes in waste water. Appl Surf Sci 257(3):1092–1097

Xiangqian L, Huizhong X, Zhe-Sheng C, Guofang C (2011) Biosynthesis of nanoparticles by microorganisms and their applications. J Nanomaterials 2011:1–16

Yan JL, Estevez MC, Smith JE, Wang KM, He XX, Wang L, Tan WH (2007) Dye-doped nanoparticles for bioanalysis. Nano Today 2(3):44–50

Zhang C, Sui J, Li J, Tang Y, Cai W (2012a) Efficient removal of heavy metal ions by thiol-functionalized super paramagnetic carbon nanotubes. Chem Eng J 210:45–52

Zhang K, Kemp KC, Chandra V (2012b) Homogeneous anchoring of TiO_2 nanoparticles on graphene sheets for waste water treatment. Mater Lett 81:127–130

Zhang X, Du AJ, Lee P, Sun DD, Leckie JO (2008) TiO_2 nanowire membrane for concurrent filtration and photocatalytic oxidation for humic acid in water. J Membr Sci 313:44–51

Zhao X, Lv L, Pan B, Zhang W, Zhang S, Zhang Q (2011) Polymer-supported nanocomposites for environmental application: a review. Chem Eng J 170(2–3):381–394

Zhong LS, Hu JS, Cao AM, Liu Q, Song WG, Wan LJ (2007) 3D flowerlike ceria micro/nanocomposite structure and its application for water treatment and CO removal. Chem Mater 19(7):1648–1655

Bionanomaterials from Agricultural Wastes

Manpreet Kaur, Akshita Mehta, Kamal Kumar Bhardwaj and Reena Gupta

Abstract Agricultural biomass is the organic material left as a by-product after agricultural activities, particularly in developing countries. Present-day scenario witnesses a declined supply of raw materials which is a cause for concern. Natural fiber possesses properties that make them a suitable alternative material to be used in local timber industries for the production of value-added products. Agricultural cellulose offers many advantages because of its renewable and biodegradable properties. Cellulose fibers exhibit a unique structural hierarchy derived from their biological origin. Cellulose nanofibers (CNFs) owing to their morphology and physical properties have been proven a promising material not only in the fields of cosmetics, medicine, biocomposites and health care but are also progressing immensely to many other unlimited applications including high gas barrier packaging material, filter material and electronic devices. Depletion of natural resources, growing population and environmental concerns has increased the attention for the extreme development and use of nanomaterials from biomass. Marvelous and complex structure of bionanoparticles is helpful while understanding the chemical applications, and as a result the bionanomaterials can serve as the filler/reinforcement in polymer composites. The goal of this chapter is to discuss the properties of agricultural wastes along with its use as a bionanomaterial and its potential applications.

Keywords Cellulose · Bionanomaterials · Agricultural biomass · Biocomposites · Biopolymers

---

M. Kaur · A. Mehta · K. K. Bhardwaj · R. Gupta (✉)
Department of Biotechnology, Himachal Pradesh University, Summerhill, Shimla 171005, India
e-mail: reenagupta_2001@yahoo.com

M. Kaur
e-mail: mkaur0314@gmail.com

A. Mehta
e-mail: aksmeh1@gmail.com

K. K. Bhardwaj
e-mail: kamalbhardwaj595@gmail.com

S. Ahmed and W. Ali (eds.), *Green Nanomaterials*,
Advanced Structured Materials 126,
https://doi.org/10.1007/978-981-15-3560-4_10

1 Introduction

Agricultural biomass refers to the organic materials that are produced as by-products from agriculture activities. Plenty of agricultural biomass is produced due to the increased activity in the modern agricultural sector, which poses a tremendous threat to the environment. Million metric tonnes of biomass are produced and unutilized every year. Future materials having higher performance are obtained from agricultural biomass which serves as a raw biomaterial (Robles et al. 2015). Agricultural production systems produce large quantities of wastes as a result adding majorly to total annual biomass agricultural waste production. Agricultural biomass serves as the most abundant renewable resource. However, the use of farming-based (energy from) wood and plant material has been based on the paradigm of a fossil resource-based society, as a result of which it is critically important to establish a sustainable production and utilization system for agricultural supplies, mainly in regions near to the equator/hot and humid areas where there is plenty of useful, valuable supplies and biodiversity.

Widespread concerns over the declined supply of raw materials have stimulated tremendous interest in the use of natural fibers as an alternative to timber industry for the production of value-added biocomposites (Kaur et al. 2018). The main fiber waste produced from the activity of agriculture is called as cellulose fiber (CF) which has the potential as reinforcing material due to abundant availability, low weight, renewable, degradable, cheaper and low abrasive property (Jawaid and Abdul Khalil 2011). Biomass-derived fiber such as cellulose can be broken down into small-sized particles until the size up to micrometer or nanometer scale is achieved (Pickering 2008). Hierarchical structure of cellulose fiber is a result of its biological origin. Nanofiber assemblies with a diameter ranging from 2 to 20 nm and a length of more than a few micrometers are used to form cellulose fibers. Morphological and physical properties of cellulose nanofibers (CNFs) make them a potential biomaterial to be used in various sectors including electronics, food, cosmetics, medicine, biocomposites and health care. It can also be used as a high gas barrier packaging material and filler material.

Recent research showed that cereal by-products and agricultural crop waste can be used for nanocellulose production. CNF can be obtained from biomaterial residues, produced by the agricultural industry. Features like larger crystalline region and higher specific surface area are required for the development of nanocomposites, and a study reported that CNF obtained from biomass possesses these features. Nanocomposites in which CNF is used as a filler material improve the barrier properties, and hence, these nanocomposites can be used for the packaging of food items (Durán et al. 2012; Lavoine et al. 2012).

In this chapter, the fundamental properties of different agricultural wastes as future materials are discussed. Polymer nanocomposites and their formation by using the technology to produce bionanomaterials are also described. Some points related to applications of nanocomposites in various fields are also discussed.

2 Fundamental Properties of Agricultural Wastes

To evaluate various properties of agricultural waste, many studies have been carried on its isolation and characterization. To assess the suitability of various agricultural waste-based fibers for future applications, the physical, mechanical and chemical properties must be examined. The fundamental properties will not only help to open up a new avenue for these fibers but will also emphasize the importance of natural fibers from agricultural waste as a future biomaterial.

2.1 Types of Agriculture Waste

Biomaterials which are the primary products of any industrial crop and agricultural biomass or plant residues left after harvesting serve as an environment-friendly and abundant source of raw materials. Income and employment in developing countries are contributed by the use of agricultural crop residues and lignocellulosic materials for the production of component materials (Abdul Khalil et al. 2013). Biomaterials of various types are available. The basis of these different types of biomaterials depends upon the agricultural biomass. Various biomaterials which are obtained out of agricultural biomass are available. A wide range of products can be yielded from industrial crops. Filamentous matter produced out of bast tissue or any other plant/crop parts can be processed for numerous applications in different industries. Primary and secondary wastes are produced out of agricultural activities. Primary wastes are produced by direct agriculture activities, while industrial processes which deal with agricultural activities and their by-products constitute the secondary waste. Kenaf, jute and tertiary wastes also come under agricultural-based biomaterials, which mainly constitute the palm oil industry. Another term "forest biomass" refers to the wooden materials produced by industrial processes particularly by timber industries and by the logging residues of trees. Forest biomass and forestry residues serve as raw material for the biocomposites including fireboard, particleboard and oriented strand board (Wang et al. 2007).

Secondary wastes serve as a potential raw material source, and this in turn will increase the application of agricultural biomass in numerous fields (Namvar et al. 2014). In the future, the contribution of biomass coming from flora and agricultural crop residues is likely to be increased as another source of biomaterial for the sustainability of raw materials (Akil et al. 2011). The classification of agricultural biomass as a biomaterial for biocomposite components according to their origin is presented in Fig. 1.

Residues left after harvesting of any plantation include fruits, leaves, seed coverings, grasses, flowers, small stems, tree tops and branches, and all come under different types of agricultural wastes. Figure 2 shows the classification of agricultural biomass waste on the basis of the conversion of cellulosic agricultural biomass waste to biomaterials and their products.

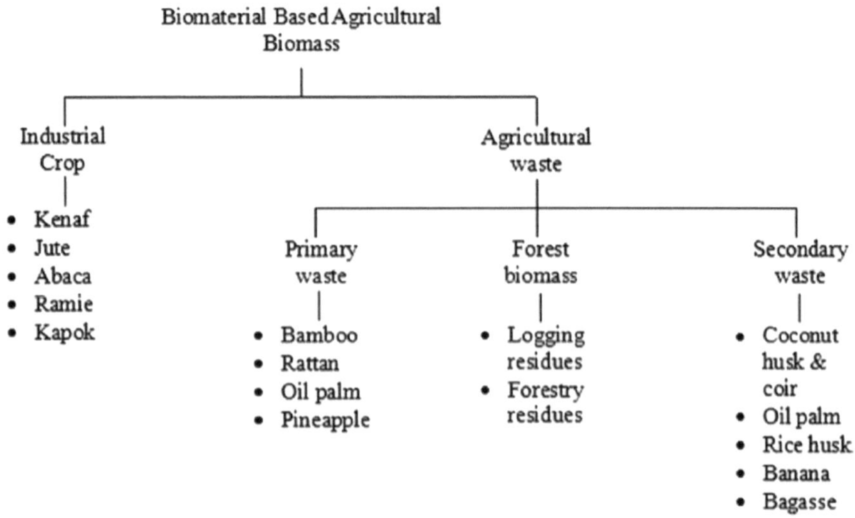

Fig. 1 Classification of agricultural biomass

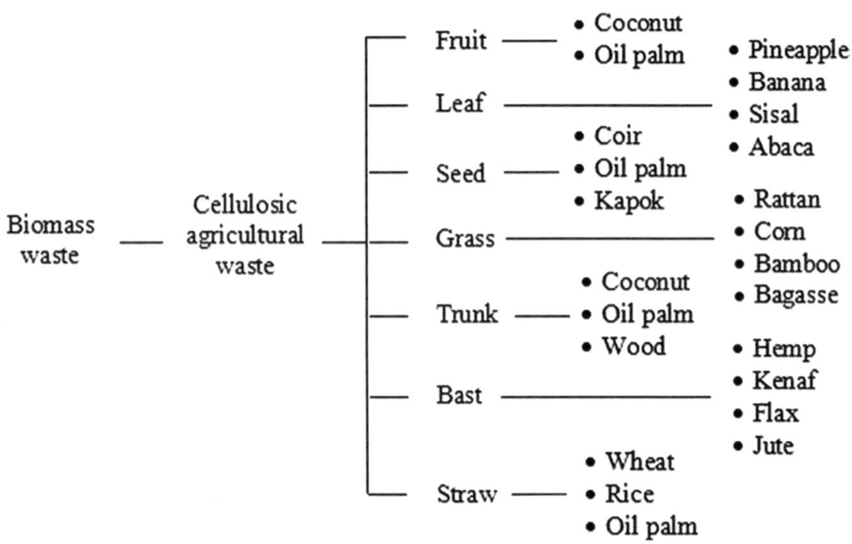

Fig. 2 Classification of agricultural waste as a biomaterial

2.2 Structure and Chemical Composition

Primary cell wall, secondary cell wall, middle lamella and lumen constitute the structure of a natural fiber cell (Kalia et al. 2011). The cell wall is embedded in a hemicellulose–lignin matrix. Hemicellulose present in the natural fibers is known to be a compatibilizer between cellulose and lignin (Thomas et al. 2011). The cell wall in a fiber is not a homogenous membrane (Akil et al. 2011). Primary wall is the first layer that gets deposited during cell growth and encircles a secondary wall, and these together make up the complex layered structure of fiber. Three layers make up the secondary wall, and a thick middle layer determines the mechanical properties of the fiber. A series of helically wound cellular microfibrils formed from long-chain cellulose molecules make up the middle layer. Depending on fiber structure, fiber cell dimensions, microfibril angle and chemical composition, the agricultural waste fibers are seen to possess high variability in properties. When we talk in context to the strength properties of natural fiber-based fiber composites, then the fiber length and width are very important points which need to be considered (Rowell et al. 2000). A wide range of fiber length, width and thickness properties are indicated while considering the characteristics of agricultural waste. Various chemical substances mainly cellulose, hemicellulose and lignin with a minor amount of pectin constitute a complex natural composite, i.e., plant fiber (Azwa et al. 2013). Plant fibers are also called as cellulosic or lignocellulosic fibers, because of such chemical composition (Siqueira et al. 2010). The composition of different chemicals in the different fibers varies depending upon the origin of fiber, climatic conditions during plantation and harvesting of crop. Extraction method also affects the chemical composition of the cell wall layers. As a result of all these factors, the chemical and physical properties of fiber are influenced (Rowell et al. 2000).

2.3 *Physical and Mechanical Properties*

The growing condition, extraction methods, chemical composition and the ratio of fiber highly affect the physical and mechanical properties of the fiber (Huang et al. 2013). The mechanical properties of the fiber are determined by the thickness of the middle layer of the secondary cell walls (Azwa et al. 2013). Tensile strength and Young's modulus are the mechanical properties which were reported due to the better orientation of crystalline cellulose caused by the high water absorb capability. High mechanical properties are shown by the fibers possessing a higher degree of polymerization of cellulose and longer cell length. Fiber with higher cellulose content is reported to possess better tensile strength and Young's modulus. Microfibrillar angles also affect the mechanical properties of the fiber. Smaller angles are responsible for high strength and stiff fiber, whereas larger angles provide higher ductility. Fibers show ductility if microfibrils possess a spiral orientation to the fiber axis, while when oriented parallel to the fiber axis, then they show rigidity and inflexibility.

3 Agricultural Waste to Bionanomaterials

Introduction of cellulose nanofibrils and nanoparticles has led to increased production of bionanomaterials from lignocellulosic and agricultural biomass. Owing to low density, optically transparent properties, strengthened mechanical properties, larger surface area (aspect ratio), flexibility, specific barrier properties, low thermal expansion, biodegradability and environment-friendly nature, the nanocellulose materials can be used in various applications (Moon et al. 2011). Mechanical, chemical and chemo-mechanical treatment processes can be used for the production of nanocellulose (Abdul Khalil et al. 2012). Applications of bio-based polymers for the benefit of numerous industries including transparent films (Zhu et al. 2014), strength enhancers in paper (Tang et al. 2014), reinforcements for polymer composite (Miao and Hamad 2013), emulsions and elemental oxygen barrier films for the packaging of plastics and food items, etc., (Lin and Dufresne 2014) can be enhanced with the aid of appropriate modifications and characterizations of nanocellulose. Research showed the application of lignocellulosic biomass as filler/reinforcement in the composites of polymers. These include the use of coir fiber in polypropylene (PP) composite (Ayrilmis et al. 2011), bagasse fiber in thermoplastic composites (Bajwa et al. 2011), nanocellulose sisal fiber as reinforcement in polyolefin composites (Chand et al. 2012), carbonized jatropha seed coverings as filler in vinyl ester biocomposites, nanocellulose fiber char in poly(vinyl acetate) formed composites, oil palm shell (OPS) nanoparticles as filler in polyester hybrid composites, etc. Nanosized cellulose imparts higher stiffness to the nanocomposites if incorporated in polymer matrices. Nanosized cellulose is the result of H-bonded interconnected network structures. The nanosized cellulose obtained can be used as a reinforcement agent in the composites of polymers. Key properties of nanocellulose include morphology, dimension, crystallinity and surface chemistry. A variety of cellulosic biomass including oil palm, kenaf, bamboo, bagasse, rice husk, wheat husk, rice straw, etc., are available for the extraction of nanocellulose (Chaker et al. 2014). Availability of fiber in any geographical area, the chemical composition of fiber and the economic considerations are the key points while selecting the cellulosic biomass. Though cellulose is the most abundant and easily available biomaterial on earth, its processing into types of nanostructures has received considerable attention recently. Over the past few decades, a great deal of scientists and researchers has been attracted by lignocellulosic biomass, in a manner to isolate nanocellulose from biomaterials so as to design a variety of functional and useful products.

3.1 Properties and Characterization of CNFs from Agricultural Waste

A large range of nanoscaled fillers out of cellulose have been obtained after the isolation of CNFs from biomass. Inherent properties of agricultural biomass and their structure limit its digestibility for the extraction of cellulose. Stronger lignin layers, low accessibility of cellulose to chemicals and increased crystallinity of cellulose are some of the structural properties of agricultural waste which are responsible for the limited extraction of cellulose. Nanocellulosic content in lignocellulosic biomass can be deconstructed with the help of some biorefinery processes, while maintaining cellulose product for further hydrolysis into nanocellulose material (Stenstad et al. 2008). Supramolecular structure of cell wall structure is broken down as the first requirement of lignocellulosic plants which result in the increased accessibility to the polysaccharide components of the raw lignocellulose. In the next step, the accessibility to nanocellulose microfibrils is increased by the breakdown of cross-linked elements of raw material (Siro and Placket 2010). Nanofibrillated cellulose (NFC) and cellulose nanocrystal (CNC) are the types of nanocellulose extracted by processes from different sources of cellulose.

Transmission electron microscopy (TEM) has revealed the nanoscale structure of nanocellulose. Transverse cleavage of microfibril fibers into free and individualized nanocellulose has been suggested by TEM images of the nanocellulose (CNC and NFC) (Fig. 3a, b).

Various reports are available on non-conventional sources of nanocellulose from agricultural waste. These sources include pineapple leaf fibers (Sheltami et al. 2012), grass (Saritha et al. 2013), rice husk (Johar et al. 2012), rice straw, empty fruit bunches, etc. In these types of nanocellulose, tailor-made nanofibrils with specific morphology and surface chemistry are produced, after applying various pretreatments. Nanocomposites of varied reinforcements can be achieved with the help of nanocellulose possessing different properties. Being cheaper to produce, safe to deal and completely renewable, the physical and chemical properties of nanocellulosic materials are also exceptional. Applications of nanofibers include the improved performance of flame retardant in furniture by increased manufacture of carbon

(a) (b)

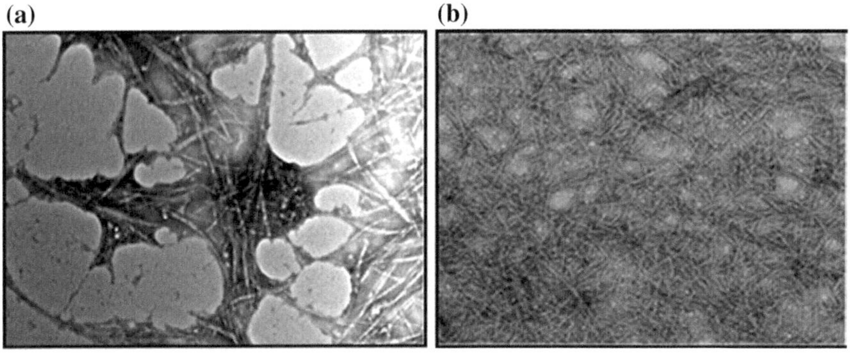

Fig. 3 TEM image of nanocellulose from agricultural waste **a** TEM image of nanocrystal/nanowhiskers/CNC; **b** TEM image of NFC (Source of **a**: Saurabh et al. (2016) with permission; Source of **b**: Fatah et al. (2014) with permission)

nanofibers (Seo et al. 2015). There are many methods to isolate cellulose fibers. Cellulose chains are packed in an ordered manner to form compact microfibrils, which are stabilized by both inter- and intramolecular hydrogen bonding (Alemdar and Sain 2008). Chemical treatment of CNCs forms separated single crystals because of the polar nature of CNCs and attraction caused by H-bonding. Steam explosion method, (Cherian et al. 2010), high-intensity ultrasonication combined with chemical pretreatments (Chen et al. 2011), 2,2,6,6-tetramethylpiperidine-1-oxyl radical (TEMPO) method (Lavoine et al. 2012), acid treatment, ultrasonication method and enzymatic hydrolysis (Aprilia et al. 2016) can be used to isolate CNFs. Microorganisms have been strongly beneficial for biological pretreatment. Numerous advantages including low toxicity, biodegradable properties and factors like reduced inhibition are provided when the CNFs are formed by using microorganisms as tools. Some fungi, bacteria and yeast by using their enzymatic mechanism are capable of degrading cellulosic contents in plants; thereafter, the resultant components can be utilized for the production of nanofibril materials (Ahmed and Saiqa Ikram 2015). In the future, the efficient pretreatment of agricultural biomaterial for the degradation of cellulose will be served by the microorganisms. This, in turn, will enhance the production of CNF.

3.2 Properties and Characterization of Nanoparticles from Agricultural Waste

Particles that measure one of the dimensions equal to 100 nm or less are considered as nanoparticles. In other words, nanoparticles are the collection/gathering of atoms bonded together to form a structural radius of less than 100 nm. When nanoparticles combine to form conventional materials, in certain cases, their properties get changed. Since nanoparticles have a greater surface area per weight than larger particles, they are more reactive to some other molecules. Like proteins exist in almost all biological systems, nanoparticles are very ordinary in nature. Fullerenes, metal clusters, giant molecules like proteins and H-bonded structures of water molecules, existing at ambient temperatures, all come under the category of nanoparticles. Metallic nanoparticles are important because of the large surface area to volume ratio. Synthesis of silver nanoparticles from renewable plant materials is beneficial and serves as an alternative for the development of metal nanoparticles (Nisha et al. 2014). Cellulose nanoparticles can be obtained from agricultural wastes such as sesame husk (Purkait et al. 2011), OPS (Dungani et al. 2013), oil palm ash (Abdul Khalil et al. 2011) and coconut shell (CS) (Sulaeman et al. 2016). A variety of chemical and physical methods can be employed for the production of nanoparticles form agricultural wastes. The effect of chemical treatment is enhanced with the high-energy planetary ball-milling process, and this in turn serves as an effective method for the synthesis of nanoparticles. This method works on the principle of removal of lignin and hemicelluloses by chemical treatment and then the grinding of material into fine powder or cellulose nanoparticles by the ball-milling process. Size, shape and surface structure and processing are the principal properties of nanoparticles which tend to introduce surface imperfections. Overall physicochemical properties of nanoparticle surface are significantly affected by surface imperfections (Jiang et al. 2009).

Polymeric materials containing different amounts of oil constitute the lignocellulosic biomass. The very first challenge is the presence of oil left in the lignocellulosic biomass including oil palm, coconut and jatropha. Irreversible adsorption, aggregation of nanoparticles and loss of significant biomaterial due to purification process are considered as the second challenge (Andersen et al. 2012). The effect of this biomass in advanced applications such as composites (Bi et al. 2008), pulp and paper production, etc., may be lowered by the presence of any impurities/contaminations of the bionanoparticles. The process of oil removal is necessary for the elimination of this problem which will be beneficial in further exploitation of biomaterials for numerous applications. Methods used for the separation of lignocellulosic biomass include mechanical pressing, supercritical fluid extraction and extraction by solvents (Yahaya et al. 2016). The successful extraction of oil removal from lignocellulosic biomass nanoparticles has been reported (Sulaeman et al. 2016). Physical and mechanical properties along with the size of particle and size distribution play an important role in the property characterization of nanoparticles. As an example in a study, it was shown that CS nanoparticles having the size in the range of nanometer scale were thermally more stable as compared to the raw CS (Sulaeman et al. 2016). Hence, it is important to reduce the size of the particles up to nanometer scale in order to prevail the new features of these nanoparticles (Akbari et al. 2011). Nanoparticles significantly differ from the bulk material having the same chemical composition, in terms of functional properties. Size and size distribution properties of nanoparticles can be determined by using the microscopic techniques, which include the relationship between size and behavior of different particles. Numerous commercially available instruments, such as TEM, SEM, dynamic light scattering (DLS), X-ray diffraction (XRD), photon correlation spectroscopy (PCS) and Brunauer–Emmett–Teller (BET), can be used to determine the particle size and size distribution of nanoparticles. Research is being carried out on the different instruments, used for particle sizing of nanoparticles. As an example, an investigation on OPS size determination by TEM showed the range of OPS particle between 10 and 50 nm, and the average size of particle was approximately 50 nm. An average size of OPS nanoparticle was found to be 31.75 nm when calculated by using X-ray diffraction peaks of XRD images (Dungani et al. 2013). Extra information related to the nanoparticles can be gathered by comparing the results from different instruments. Figure 4 shows the DLS technique used to determine the average size and particle size distribution of nanoparticles. Nanoparticles possess different morphologies, with their names being characterized by their different shapes.

Carbon-derived nanoparticles of medlar wood possess homogeneous, smooth and highly dense morphology with a large number of pores and cavities in different sizes and shapes. Size exhibited by adsorbent is in the mesoporous domain with an average pore diameter lower than 10 nm (Ghaedi et al. 2015). Surface morphology of the

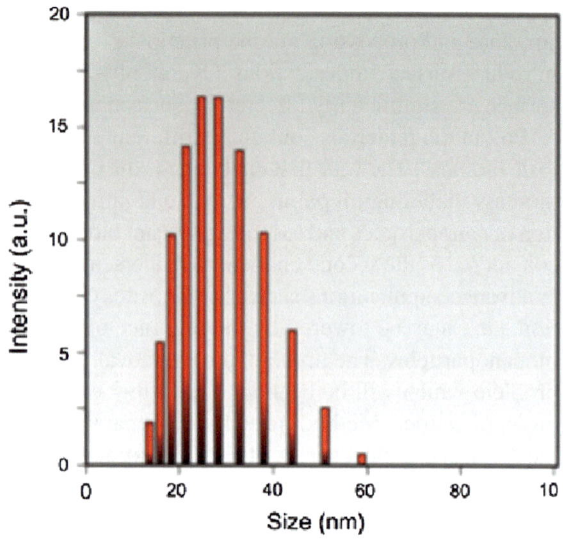

Fig. 4 DLS image showing the particle size distribution of the nanoparticles (*Source* Mollick et al. (2016); with permission)

particles highly affects the functional properties of nanoparticles; reliable characterization of the nanoparticle's properties is possible with the aid of precise measurements of a particle's morphology. In order to exploit the properties of nanoparticles, it is important to control their morphology. The size and shape of nanoparticles can be visualized by SEM analysis. From the SEM analysis, it was found that silica nanoparticles from agricultural waste have their own shape and size arrangement (Vaibhav et al. 2015). Silica powder is obtained by sintering the waste materials to ash at 900 °C and then characterized by SEM showing bamboo leaf-like particles. However, the groundnut shell ash showed uniform spherical particles. It was observed that rice husk ash has a fiber-like appearance; whereas, sugarcane bagasse ash was highly porous in its structure. Studies reported that nanosilica could be obtained from a waste material called as bagasse ash (Hariharan and Sivakumar 2013). Nanosilica with various sizes and prismatic and spherical geometry was also reported.

Surface chemistry can be examined by using Fourier transform infrared spectroscopy (FTIR), because of the attached organic functional groups to the surface of nanoparticles. Rice husk extracted silica nanoparticles showed the presence of organic functional groups at a combustion temperature of approx. 600–700 °C (Chen et al. 2013; Ghorbani et al. 2015).

4 Various Applications of Bionanomaterial

New generation of products and new market segments can be developed in the coming years with the involvement of nanotechnology, since it presents an opportunity to wooden and plant-based materials for improved performance and functionality.

Research is being done on nanocellulose and its development for commercial and sustainable utilization of biomass, mainly in Japan, North America and Europe. The demand for sustainable resource management and the utilization of these resources in industrial applications is growing. Nanocellulose can be used as a prime candidate for the sustainable and recyclable materials in packaging. Nanocellulose derived from selected biomass and fibers with improved characteristics can be used for the production of composite-based products. Advanced materials including high-performance nanocomposite materials and films can be formed from surface-modified nanofibrillar cellulose fibers and extraction technologies, as well as modification and characterization (Regan et al. 2005). Demand for materials having a higher strength-to-weight ratio is continuously growing, and as a result the market for nanoparticles and their utilization in various products including structural components is also expanding at a very fast rate. Components with such properties can be used for the manufacture of aircraft parts. Considerable efforts are being made for the development of bionanomaterials. Novel bionanomaterials with new, multifunctional properties can be used for packaging, automotive components, medical devices and other applications. The following section describes, in brief, the applications of bionanomaterials obtained from agriculture waste in several sectors.

4.1 As a Reinforcing Composite Material

Over the last twenty years, interest is increasing in the field of bionanomaterials and their applications as the reinforcing agent for polymer composite materials. Cellulose fibers and particles possessing required properties are used in these bionanomaterials. Synthetic fibers like glass fibers have been replaced by the cellulose fibers because of many advantages of CFs over the artificial fibers (Wambua et al. 2003). Hydrophobic polymer matrix incompatibility and formation of aggregates during processing and poor resistance to moisture are some of the drawbacks responsible for the limited use of these fibers as a reinforcement agent in thermoplastic matrices. Fiber compatibility and interfacial adhesion can be improved by certain physical treatment strategies (Chen et al. 2013; Vaccaro et al. 2013). Fibers can be modified by chemical methods in order to reduce the polarity and hydrophobicity (Li et al. 2007). A wide range of biomaterials are available which can be used as reinforcers or fillers. Certain properties of any material are enhanced by the addition of nanoparticles to a nanocomposite matrix (Leblanc 2002). Nanoscale additives including soot and silica nanoparticles are frequently used polymer reinforcing agents (Ooi et al. 2013). Use of this material in several fields has received much attention because of the properties of nanocomposites which attracted researchers and companies. Various industrial applications can be accomplished with the aid of bionanomaterial from agriculture waste. Bionanomaterials are reinforced into polymers, such as in thermoset and thermoplastic polymers, which can be used in various applications such as electronic devices (Masoodi et al. 2012). In the context of electronic devices, CNFs possess certain properties as a result of which they become highly valuable in many fields (Liu et al. 2015). Nowadays, composites made by reinforcing silica into polymer are prepared by various methods, which have many applications (Masoodi et al. 2012). Electronic substrates, thin film substrates, electrical insulators, thermal insulators and humidity sensors, etc., make use of silica nanoparticles (Tabatabaei et al. 2006). The size and size distribution of the silica particles highly affect the quality of products.

4.2 Packaging Applications

Development of research in the field of nanotechnology and concern to reduce environmental impact and development of alternative packaging materials based on cellulosic materials are gaining attention. Research studies show the use of natural cellulose-based nanocomposites for the packaging of products or items. It can also be used for the packaging of non-food materials including medical and pharmaceutical items. In packaging structure, the functional ingredients are integrated in order to provide a positive impact on consumer health, and hence, packaging is being designed as a health system (Chen 2014). Gas barrier properties and heat stability of nanocellulosic-based polymer composites are enhanced by the addition of nanocellulose to polymeric composites. Reinforcing structures and nanocellulose like matrices can be incorporated to reduce poor mechanical and barrier properties of bio-based food packaging materials (Azeredo 2013). Research has been done to show that cellulosic materials enhance the mechanical and barrier properties of packaging materials (Kalia et al. 2011). Properties like food protection, enhanced quality of food and release of active compounds onto the surface of food must be covered under active food packaging systems (Azeredo 2013).

4.3 Medical Applications

Advancements of biocomposites have been supported by the production of biofibers, microfibrillated or nanosize fibers. Biofiber-based nanocomposites formed by combining CNFs with biodegradable polymers have been proved to be used in a wide range of applications including cardiac devices, scaffolds for tissue engineering, repair of articular cartilage, vascular grafts, urethral catheters, mammary prostheses, blood bags, penile prostheses, adhesion barriers and artificial skin (John and Thomas 2008). The commercial prospects of biofibers for medical applications have been increased by the development of composites from these biofibers. Many studies have been conducted to show the development of biomaterials from pineapple leaf fiber type of agricultural waste (Cherian et al. 2010). The strength and stiffness of fiber were increased by supplementation of polyurethane with 5% of cellulose nanofibrils. It has been reported that the polyvinyl alcohol-reinforced bacterial cellulose fibers will form biocompatible nanocomposites similar to that of cardiovascular tissues

(Millon and Wan 2006). Numerous uses of bionanomaterials have been reported in cosmetics and medical product industries, including pharmaceutical, medical, veterinary medicine and dental applications. Moreover, they can also be used in the delivery of drugs, medical implants, wound healing dressings, tissue engineering and cell culture (Hua et al. 2014).

4.4 Automotive Industry Application

To meet the consumer needs and regulatory requirements in the automotive industry, the improvement of automobiles and their performance is necessary. Nanotechnology and production of nanomaterials are receiving considerable attention in the automobile industry to fulfill the demands of this industry (Kiziltas et al. 2013). Thermoplastic matrix of the automobile sector is increasingly using nanocellulose-based materials over the last several years. Front ends, bumper beams, dashboards and underbody shields of automobiles use these plastic composites reinforced with cellulose fiber. Cellulose-based materials, such as natural fiber, are emerging as a realistic alternative to glass-reinforced composites for application in automotive components. Natural fiber components are capable of delivering high-performance and non-brittle fractures. Owing to low weight, design flexibility, resistance to corrosion and cheaper production, the use of nanofiber-reinforced plastic components/composites has gained considerable significance (Hill et al. 2012). The most common application of composites in the automobile industry is in the exterior body panel and the passenger compartment (Bartus et al. 2005). Natural fibers like hemp, jute, abaca, banana, kenaf, flax and sisal have been serving as successful reinforcing fibers in polymer composites from polypropylene (PP), polyethylene, nylon or thermoplastic polymers to be used in the automobile industry. Natural fiber composite thermoset matrices are now being used by automakers for automotive components (seat backs, package trays, door panels, dashboards, headliners and interior parts). Cellulose fibers have also been used in the automobile products of many companies. Studies showed the use of coconut fiber base rubber latex composites in the seat parts of Mercedes Benz. It has also been reported that the epoxy reinforced with flax/sisal fiber mat is used in the manufacture of door panels (Shinoj et al. 2011).

4.5 Other Applications

Several studies showed the application of nanobiocomposites as a sporting good, as rubber for industrial applications, aerospace components, etc. Still, glass and wooden materials are being used for the manufacture of sporting goods. Efforts have been done to produce and develop biocomposites from agricultural waste. Hemp, flax, coir and jute type of cellulose fibers have been effectively used for manufacturing sporting equipment. Wood, glass and metal used in fishing rods, ski equipment, tennis racquets, golf clubs, spars/shafts for kayak paddles, windsurfing masts and bicycle handlebars, etc., type of sporting goods have been successfully replaced by carbon-reinforced composites. Studies are being done on the utilization of carbon black, activated charcoal and silica produced out of agricultural biomass and left out residues after harvesting. Nanoscale additives can be used as natural and synthetic reinforcements (Zhou et al. 2015). Widely used silica-reinforced natural rubber is prepared via a sol–gel process (Watcharakul et al. 2011). Elastomer composites reinforced with carbon black can be used for the production of automotive components including tires (Sasthiryar et al. 2014). Aerospace industry also makes the use of polymer composites. Many aerospace components can be manufactured with the introduction of agricultural waste-reinforced composites. Nanoparticles of carbon reinforced in epoxy matrix have found application in nose cones of the space shuttle, rocket nozzles and aircraft brake disks (Wu and Radovic 2006).

5 Conclusion

Agricultural biomass and left out residues are of great cultural and economic significance all around the globe. These agriculture-based residues are being used in numerous applications including building materials, as a decorative product and as a versatile raw product especially in the geographical areas including South Asia, Southeast Asia and East Asia. Due to high strength, environment-friendly nature, low cost, availability and sustainability, the agricultural biomass is of considerable significance in composite formation. Bionanocomposites serve as a high-strength component in composite production. CNF-reinforced nanocomposites can be employed in packaging, automotive components, medical, aircraft components and other applications. Performance of any composite is influenced by the properties of natural fiber; hence, it is necessary to understand the physical, mechanical, chemical and morphological properties of any natural fiber. Agricultural biomass is a fascinating material that brings the possibility to gain plenty of interest applications in the multidisciplinary field. However, the potential applicability of these raw materials is unlimited and rapidly expanding as an alternative reinforcement agent or filler in functional composites for various engineering applications.

Acknowledgements The financial support granted to Ms. Manpreet Kaur in the form of JRF from the Council of Scientific and Industrial Research (CSIR) is thankfully acknowledged. The financial assistance from the Council of Scientific and Industrial Research (CSIR) in the form of SRF to Mr. Kamal Kumar Bhardwaj is also thankfully acknowledged.

Conflicts of Interests The author(s) declare(s) that there is no conflict of interests regarding the publication of this article.

References

Abdul Khalil HPS, Fizree HM, Jawaid M, Alattas OS (2011) Preparation and characterization of nano-structured materials from oil palm ash: a bio-agricultural waste from oil palm mill. BioResources 64:4537–4546

Abdul Khalil HPS, Bhat AH, Ireana Yusra AF (2012) Green composites from sustainable cellulose nanofibrils: a review. Carbohydr Polym 87:963–979

Abdul Khalil HPS, Aprilia NAS, Bhat AH, Jawaid M, Paridah MT, Dungani R (2013) A Jatropha biomass as renewable materials for biocomposites and its applications. Renew Sustain Energy Rev 2013:667–685

Ahmed S, Saiqa Ikram S (2015) Synthesis of gold nanoparticles using plant extract: an overview. Nano Res Appl 1:1–5

Akbari B, Tavandashti MP, Zandrahimi M (2011) Particle size characterization of nanoparticles—a practical approach. Iran J Mater Sci Eng 8:48–56

Akil HM, Omar MF, Mazuki AAM, Safiee S, Ishak ZAM, Bakar AA (2011) Kenaf fiber reinforced composites: a review. Mater Design 32:4107–4121

Alemdar A, Sain M (2008) Isolation and characterization of nanofibers from agricultural residues-wheat straw and soy hulls. Bioresour Technol 99:1664–1671

Andersen TR, Yan Q, Larsen-Olsen TT, Søndergaard R, Li Q, Andreasen B et al (2012) A nanoparticle approach towards morphology controlled organic photovoltaics (OPV). Polymers 4:1242–1258

Aprilia NS, Davoudpour Y, Zulqarnain W, Khalil HA, Hazwan CC, Hossain MS et al (2016) Physicochemical characterization of microcrystalline cellulose extracted from kenaf bast. BioResources 11:3875–3889

Ayrilmis N, Jarusombuti S, Fueangvivat V, Bauchongkol P, White R (2011) Coir fiber reinforced polypropylene composite panel for automotive interior applications. Fibers Polym 12:919–926

Azeredo HMC (2013) Antimicrobial nanostructures in food packaging. Trends Food Sci Technol 30:56–69

Azwa Z, Yousif B, Manalo A, Karunasena W (2013) A review on the degradability of polymeric composites based on natural fibers. Mater Des 47:424–442

Bajwa SG, Bajwa DS, Holt G, Coffelt T, Nakayama F (2011) Properties of thermoplastic composites with cotton and guayule biomass residues as fiber fillers. Ind Crops Prod 33:747–755

Bartus SD, Vaidya UK, Ulven CA (2005) Design and development of a long fiber thermoplastic bus seat. Compos Struct 67:263–277

Bi Y, Luo R, Li J, Feng Z, Jin Z (2008) The effects of the hydraulic oil on mechanical and tribological properties of C/C composites. Mater Sci Eng, A 484:274–276

Chaker A, Mutjé P, Vilar MR, Boufi S (2014) Agriculture crop residues as a source for the production of nanofibrillated cellulose with low energy demand. Cellulose 21:4247–4259

Chand N, Prajapati SC, Singh RK (2012) Development and characterization of sisal nano fiber reinforced polyolefin composites. J Sci Res Rev 1:26–32

Chen H (2014) Chemical composition and structure of natural lignocellulose. In: Chen H (ed) Biotechnology of lignocellulose. Springer, Dordrecht, pp 25–71

Chen WS, Yu HP, Liu YX, Chen P, Zhang MX, Hai YF (2011) Individualization of cellulose nanofibers from wood using high-intensity ultrasonication combined with chemical pretreatments. Carbohydr Polym 83:1804–1811

Chen H, Wang W, Martin JC, Oliphant AJ, Doerr PA, Xu JF et al (2013) Extraction of lignocellulose and synthesis of porous silica nanoparticles from rice husks: a comprehensive utilization of rice husk biomass. ACS Sustain Chem Eng 1:254–259

Cherian BM, Leão AL, de Souza SF, Thomas S, Pothan LA, Kottaisamy M (2010) Isolation of nanocellulose from pineapple leaf fibers by steam explosion. Carbohydr Polym 81:720–725

Dungani R, Islam N, Abdul Khalil HPS, Hartati S, Abdullah CK, Dewi M et al (2013) Termite resistance study of oil palm trunk lumber (OPTL) impregnated with oil palm shell meal and phenol-formaldehyde resin. BioResources 8:4937–4950

Durán N, Lemes AP, Seabra AB (2012) Review of cellulose nanocrystals patents: preparation, composites and general applications. Recent Pat Nanotechnol 6:16–28

Fatah IYA, Abdul Khalil HPS, Hossain MdS, Aziz AA, Davoudpour Y, Dungani R et al (2014) Exploration of a chemo-mechanical technique for the isolation of nanofibrillated cellulosic fiber from oil palm empty fruit bunch as a reinforcing agent in composites materials. Polymers 6:2611–2624

Ghaedi M, Jahromi MN, Sajedi M, Mousavi H (2015) Comparison of modified palladium nanoparticles and homemade activated carbon derived from medlar for solid phase extraction of metal ions prior to their flame atomic absorption spectrometry determination. Environ Stud Persian Gulf 2:1–14

Ghorbani F, Sanati AM, Maleki M (2015) Production of silica nanoparticles from rice husk as agricultural waste by environmental friendly technique. Environ Stud Persian Gulf 2:56–65

Hariharan V, Sivakumar G (2013) Studies on synthesized nanosilica obtained from bagasse ash. Int J ChemTech Res 5:1263–1266

Hill K, Swiecki B, Cregger J (2012) The bio-based materials automotive value chain. Center Automot Res 2012:23–27

Hua K, Carlsson DO, Ålander E, Lindström T, Strømme M, Mihranyan A et al (2014) Translational study between structure and biological response of nanocellulose from wood and green algae. RSC Adv 4:2892–2903

Huang JL, Gray DG, Li CJ (2013) A3-Coupling catalyzed by robust Au nanoparticles covalently bonded to HS-functionalized cellulose nanocrystalline films. J Org Chem 9:1388–1396

Jawaid M, Abdul Khalil HPS (2011) Cellulosic/synthetic fiber reinforced polymer hybrid composites: a review. Carbohydr Polym 86:1–18

Jiang J, Oberdorster G, Biswas P (2009) Characterization of size, surface charge and agglomeration state of nanoparticle dispersions for toxicological studies. Nanopart Res 11:77–89

Johar N, Ahmad I, Dufresne A (2012) Extraction, preparation and characterization of cellulose fibers and nanocrystals from rice husk. Ind Crop Prod 37:93–99

John MJ, Thomas S (2008) Biofibers and biocomposites. Carbohydr Polym 71:343–364

Kalia S, Dufresne A, Cherian BM, Kaith BS, Averous L, Njuguna J et al (2011) Cellulose-based bio- and nanocomposites: a review. Inter J Polym Sci 2011:1–35

Kaur M, Mehta A, Gupta R (2018) Biomedical applications of synthetic and natural biodegradable polymers. In: Shakeel Ahmed et al (eds) Green and sustainable advanced materials, vol 2. Scrivener Publishing LLC, pp 281–310

Kiziltas A, Nazari B, Gardner DJ, Bousfield DW (2013) Polyamide 6-cellulose composites: effect of cellulose composition on melt rheology and crystallization behaviour. Polym Eng Sci 54:739–746

Lavoine N, Desloges I, Dufresne A, Bras J (2012) Microfibrillated cellulose–its barrier properties and applications in cellulosic materials: a review. Carbohydr Polym 90:735–764

Leblanc JL (2002) Rubber–filler interactions and rheological properties in filled compounds. Prog Polym Sci 27:627–687

Li Z, Wang L, Wang X (2007) Cement composites reinforced with surface modified coir fibers. J Compos Mater 41:1445–1457

Lin N, Dufresne A (2014) Nanocellulose in biomedicine: current status and future prospect. Eur Polym J 59:302–325

Liu L, Ding X, Li J, Luo Z, Hu Y, Liu J et al (2015) Enzyme responsive drug delivery system based on mesoporous silica nanoparticles for tumor therapy in vivo. Nanotechnology 26:15–24

Masoodi R, El-Hajjar RF, Pillai KM, Sabo R (2012) Mechanical characterization of cellulose nanofiber and bio-based epoxy composite. Mater Des 36:570–576

Miao C, Hamad WY (2013) Cellulose reinforced polymer composites and nanocomposites: a critical review. Cellulose 20:2221–2262

Millon E, Wan WK (2006) The polyvinyl alcohol-bacterial cellulose system as a new nanocomposite for biomedical applications. J Biomed Mater Res B Appl Biomater 79:245–253

Mollick MdMR, Rana D, Dash SK, Chattopadhyay S, Bhowmick B, Maity D et al (2016) Studies on green synthesized silver nanoparticles using *Abelmoschus esculentus* (L.) pulp extract having anticancer (in vitro) and antimicrobial applications. Arabian J Chem. https://doi.org/10.1016/j.arabjc.2015.04.033

Moon RJ, Martini A, Nairn J, Simonsen J, Youngblood J (2011) Cellulose nanomaterials review: structure, properties and nanocomposites. Chem Soc Rev 40:3941–3994

Namvar F, Jawaid M, Tanir PM, Mohamad R, Azizi S, Khodavandi A et al (2014) Potential use of plant fibers and their composites for biomedical applications. BioResources 9:5688–5706

Nisha SN, Aysha OS, Rahaman JSN, Kumar PV, Valli S, Nirmala P et al (2014) Lemon peels mediated synthesis of silver nanoparticles and its antidermatophytic activity. Spectrochim Acta Part A Mol Biomol Spectros 124:194–198

Ooi ZX, Ismail H, Bakar AA (2013) Optimisation of oil palm ash as reinforcement in natural rubber vulcanisation: a comparison between silica and carbon black fillers. Polym Test 32:625–630

Pickering K (2008) Properties and performance of natural-fiber composites. Woodhead Publishing, Cambridge

Purkait BS, Ray D, Sengupta S, Kar T, Mohanty A, Misra M (2011) Isolation of cellulose nanoparticles from sesame husk. Ind Eng Chem Res 50:871–876

Regan BC, Aloni S, Jensen K, Ritchie RO, Zettl A (2005) Nanocrystal-powered nanomotor. Nano Lett 5:1730–1733

Robles E, Urruzola I, Labidi J, Serrano L (2015) Surface modified nano-cellulose as reinforcement 390 in poly(lactic acid) to conform new composites. Ind Crops Prod 71:44–53

Rowell RM, Han JS, Rowell JS (2000) Characterization and factors effecting fiber properties. In: Frollini E, Leao AL, Mattoso LHC (eds) Natural polymers and agrofibers composites. Embrapa Agricultural Instrumentation, San Carlos, Brazil, pp 115–134

Saritha S, Nair SM, Kumar NC (2013) Nano-ordered cellulose containing Iα crystalline domains derived from the algae *Chaetomorpha antennina*. BioNano Sci 3:423–427

Sasthiryar S, Abdul Khalil HPS, Bhat AH, Ahmad ZA, Islam MdZ, Zaidon A et al (2014) Nanobioceramic composites: a study of mechanical, morphological and thermal properties. BioResources 9:861–871

Saurabh CK, Dungani R, Owolabi AF, Atiqah NS, Zaidon A, Aprilia NAS et al (2016) Effect of hydrolysis treatment on cellulose nanowhiskers from oil palm (*Elaeis guineesis*) fronds: morphology, chemical, crystallinity, and thermal characteristics. BioResources 11:6742–6755

Seo HJ, Kim S, Huh W, Park KW, Lee DR, Son DW et al (2015) Enhancing the flame-retardant performance of wood-based materials using carbon-based materials. J Therm Anal Calorim 123:1935–1942

Sheltami RM, Abdullah I, Ahmad I, Dufresne A, Kargarzadeh H (2012) Extraction of cellulose nanocrystals from mengkuang leaves (*Pandanus tectorius*). Carbohyd Polym 88:772–779

Shinoj S, Visvanathan R, Panigrahi S, Kochubabu M (2011) Oil palm fiber (OPF) and its composites: a review. Ind Crops Prod 33:7–22

Siqueira G, Bras J, Dufresne A (2010) Cellulosic bionanocomposites: a review of preparation, properties and applications. Polymers 2:728–765

Siro I, Placket D (2010) Microfibrillated cellulose and new nanocomposite materials: a review. Cellulose 17:459–494

Stenstad P, Andresen M, Tanem BS, Stenius P (2008) Chemical surface modifications of microfibrillated cellulose. Cellulose 15:35–45

Sulaeman A, Dungani R, Islam MN, Abdul Khalil HPS, Sumardi I, Hermawan D et al (2016) Preliminary study of characterization of nanoparticles from coconut shell as filler agent in composites materials. MAYFEB J Mater Sci 1:1–9

Tabatabaei S, Shukohfar A, Aghababazadeh R, Mirhabibi A (2006) Experimental study of the synthesis and characterisation of silica nanoparticles via the sol-gel method. J Phys: Conf Ser 26:371–374

Tang Y, He Z, Mosseler JA, Ni Y (2014) Production of highly electro-conductive cellulosic paper via surface coating of carbon nanotube/graphene oxide nanocomposites using nanocrystalline cellulose as a binder. Cellulose 21:4569–4581

Thomas S, Paul SA, Pothan LA, Deepa B (2011) Cellulose fibers: bio-and nano-polymer composites. In: Kalia S, Kaith BS, Kaur I (eds) Natural fibers: structure, properties and applications. Springer, Berlin, pp 3–42
Vaccaro L, Spallino L, Agnello S, Buscarino G, Cannas M (2013) Defect-related visible luminescence of silica nanoparticles. Phys Status Solidi C 10:658–661
Vaibhav V, Vijayalakshmi U, Mohana Roopan S (2015) Agricultural waste as a source for the production of silica nanoparticles. Spectrochim Acta, Part A 139:515–520
Wambua P, Ivens J, Verpoest I (2003) Natural fibers: can they replace glass in fiber reinforced plastics? Compos Sci Technol 63:1259–1264
Wang S, Zhang Y, Xing C (2007) Effects of strand drying methods on wood surface wettability. Holz Roh Werkst 65:437–442
Watcharakul N, Poompradub S, Prasassarakich P (2011) In situ silica reinforcement of methyl methacrylate grafted natural rubber by sol–gel process. J Sol-Gel Sci Technol 58:407–418
Wu X, Radovic LR (2006) Inhibition of catalytic oxidation of carbon/carbon composites by phosphorus. Carbon 44:141–151
Yahaya S, Giwa SO, Ibrahim M, Giwa A (2016) Extraction of oil from jatropha seed kernels: optimization and characterization. Int J ChemTech Res 9:758–770
Zhou Y, Fan M, Chen L, Zhuang J (2015) Lignocellulosic fiber mediated rubber composites: an overview. Compos Part B 76:180–191
Zhu H, Fang Z, Preston C, Li Y, Hu L (2014) Transparent paper: fabrications, properties, and device applications. Energy Environ Sci 7:269–287

Surface Modification of Bio-polymeric Nanoparticles and Its Applications

K. S. Yoha, S. R. Priyadarshini, J. A. Moses and C. Anandharamakrishnan

Abstract Nanoparticle research is a vibrant area of interest due to its unique functionality and composition. It finds its application in wide variety of fields including biomedical, food processing and drug delivery systems. Polymers are used as a widespread medium for carriers of bioactive components. Biopolymers play a major role because of its biodegradable and biocompatible nature. These biopolymers-based nanoparticles have certain limitations because of their restricted behavior in different solvents. In order to fine-tune these nanoparticles and to make them fit into different real-time applications, surface modification becomes essential. Some of the approaches applicable to modify the polymeric surfaces of these nanoparticles are physical immobilization, adsorption coating or grafting, chemical modifications on the functional groups such as acrylation, amination and acetylation, radiation-induced grafting, UV-ray treatment and plasma surface modification. Such surface modifications on the functional groups provide improved stability and enhanced surface activity of polymeric nanoparticles which are helpful in reducing agglomeration, preventing oxidation and protecting labile compounds. Consequences of which enables specific targeting of active compounds to the site of action. In this view, the chapter explains about the various surface modifications and its applications in food preservation, packaging and controlled release of bioactives.

Keywords Biopolymers · Surface modification · Functional groups · Nanoparticles · Bioactive targeting

K. S. Yoha · S. R. Priyadarshini · J. A. Moses · C. Anandharamakrishnan (✉)
Computational Modeling and Nanoscale Processing Unit, Indian Institute of Food Processing Technology, Pudukkottai Road, Thanjavur, Tamil Nadu 613005, India
e-mail: anadharamakrishnan@iifpt.edu.in

S. Ahmed and W. Ali (eds.), *Green Nanomaterials*,
Advanced Structured Materials 126,
https://doi.org/10.1007/978-981-15-3560-4_11

1 Introduction

The life originates from a single strand of nucleic acid, clearly depicting that nanostructures have been in existence in the nature long before since the scientists began studying those in laboratories. The nanoscale engineering is an intense field of scientific research due to its multifaceted application, especially in the fields of medical science, biology and food processing. Currently, apart from the manufacture of basic nanostructures, nanoengineering strategies focus on tailoring the surface structure of nanoparticles for unique and specific properties. The polymer matrix and most of the inorganic nanoparticles have interdependent effects. Biopolymers are renewable polymers, which play major roles in the field of green technology. The need for developing biopolymeric nanoparticles as an effective carrier provides enormous applications due to its benignant behavior in living tissues. The surface modification with these biopolymers improves the properties like stability, surface tension, surface charge and tensile strength (Platt 2006). It can be a potential replacement for synthetic polymers. The primary units of these biopolymers are natural polysaccharides, proteins and lipids which render the biodegradable nature (Aravamudhan et al. 2014). Biopolymers, namely starch, cellulose, guar gum and other polysaccharides, are from plant origin; alginates, galactans and carrageenan are from algal origin; collagen, gelatin, albumin and some animal proteins are from animal origin. Similarly, dextran, gellan gum, pullulan and polyhydroxybutyrate are from microbial origin. Some polymers synthesized chemically from naturally originated monomers are also considered as biopolymers like polylactic acid (Chandra 1998). Even though biopolymers are biocompatible and biodegradable in nature, it has limited industrial applications at large scale because of its poor thermal, mechanical and barrier properties. Surface modifications of such biopolymers and tailoring at nanoscale level improve their properties which widens their scope of application. The challenge in obtaining the desired properties of polymeric nanoparticles can be overcome by surface modifications.

2 Concept of Surface Modification

The characteristic behavior of a nanoparticle is dependent on the surface layer of the bioactive component, and survival of the component can be manipulated by modifying its surface. Surface modification is carried out by altering the surface of a material from its original state using physical, chemical or biological methods. Such modifications can be done through size reduction, introducing functional groups, immobilization, grafting, etc. By altering the surface, the specific properties of polymers can be improved for targeted applications. Biopolymer-based nanoparticles offer the green nanomaterials. Different varieties of biopolymers used in biomedical, food packaging and cosmetic field depend upon their nature and its compositions (Anirudhan and Sandeep 2011; Sen et al. 2010, 2011). Surface modifications in nanoparticles are performed by recasting the functional groups physically by grafting, adsorption and immobilization, chemically by cross-linking, alkali treatment, hydrolysis, acidification, oxidation and surfactant interactions and other techniques such as gamma irradiation, ozone and plasma treatment. Metal–polymer nanocomposites also modified form of polymeric material to improve its mechanical properties.

3 Methods of Surface Modification

Natural biopolymers possess inferior mechanical properties due to its structural incompatibility. Surface modification requires for improving the reaction sites by introducing new functional groups or modifying/removing the existing functional groups on surface. These modifications enhance the better mechanical properties of biopolymeric materials. The biopolymeric material has to be carefully chosen depending on its biocompatibility and biodegradability. It should exist in such a way that it could improve or repair a defect in living tissues. Surface property suitable for cellular adhesion is also an important factor because a tough adhesion to the surface may cause immobilization which is also not desirable. Another significant property is the pore size with tangled network for better attachment (Wang and Hon 2002; Wilson et al. 2005; Jiao and Cui 2007).

3.1 Physical Surface Modifications

Physical surface modification method ranges from simple covalent and non-covalent surface functionalization, physical adsorption to more complicated radiation and plasma-induced graft copolymerization process. Advantage of physical methods is the ability of achieving more specific surface modifications without addition of chemical agents which preventing the contamination by chemical residues, especially in food packaging applications. Some chemicals are not allowed by FDA (Food and Drug Administration) for food packaging applications because residues of those chemicals may interact with foods and it may lead to toxic effects.

3.1.1 Covalent and Non-covalent Functionalization

Covalent functionalization depends on bond between the biopolymeric nanoparticles and any functional groups of interest to enhance the biopolymer property (Kuila et al. 2012). Functional groups may be organic or inorganic such as alkyl (–R), alkoxy (–OR), amino (–NH_2), amine (–NHR) immobilized with biopolymeric nanoparticles (Ribeiro et al. 2015). Covalent attachment of biotin-binding proteins in gelatin nanoparticles targets the biotinylated drugs (Wartlick et al. 2004). Non-covalent functionalization depends on physical interaction of functional groups with biopolymeric nanoparticles which is initiated by some intermolecular forces like van der Waals, H-bonding and π–π interactions (Sharma et al. 2013). Dopamine monomers are linked together with quinhydrones by macrostructural assembly through non-covalent interactions such as hydrogen bonding, cationic–anionic interactions, polar charge transfer and π-stacking (Della Vecchia et al. 2013; Dreyer et al. 2013). Nanocarriers for biological applications generally uses polyethylene glycol (PEG) as encapsulating agent. However, PEG limits the diffusive behavior of the encapsulates. Wen et al. (2012) have studied redox-responsive PEG detachment mechanism for the delivery of doxorubicin (DOX) in the circulation system and reported that PEGylated nanographene oxide enhances the targeted delivery of DOX in tumor treatments. In their study, DOX loaded onto the disulfide-linked methoxy PEGylated nanographene oxide through π–π stacking and hydrophobic interactions. These encapsulates enable selective detachment of PEG from nanographene oxide based on the detection of glutothione levels related to tumors and release DOX at the target sites. Xiong et al. (2018) discussed the interfacial interactions of carbon nanomaterials with biopolymer matrices. Graphene and carbon nanotube components interacted with proteins via hydrophobic–hydrophobic interactions which improved the performance of bio-nanocomposites. Possible enabling covalent/non-covalent interactions between carbon nanomaterials and biopolymers were shown in Fig. 1.

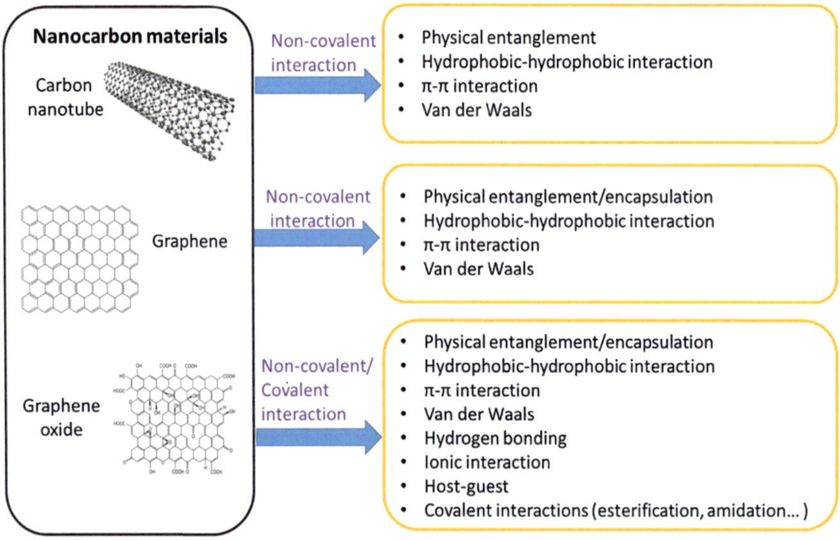

Fig. 1 Enabling interactions between carbon nanomaterials and biopolymers (*Source* Reproduced with the permission from Xiong et al. (2018). Copyright © 2018, Elsevier)

3.1.2 Physical Adsorption

The physical surface modification through adsorption is simple technique when compared to the chemical grafting. Li et al. (2001) reported that the half-life of bovine serum albumin (BSA)–poly-lactic-co-glycolic acid (PLGA) nanoparticles increased by steric PEG coating. Apart from hydrophilic PEG coating, adjusted surface charges of the nanocarriers may play a major role in the body. Due to these changes in surface charge, the unspecified cellular level internalization and the adsorption of protein can be affected during the circulation (Harush-Frenkel et al. 2008). Polyoxyethylene (PEO) physical adsorption on polystyrene matrix improved the physical properties such as transparency and thermal stability as well as mechanical performance for melt processing (Lin and Dufresne 2013). Azouz et al. (2012) demonstrated the physical adsorption study on cellulose nanocrystals (CNCs) and observed the improved thermal stability of nanocrystals due to PEO adsorption on the surface of CNCs. This adsorption hides the sulfate groups of CNCs and clearly explicits large number of hydroxyl groups at the surface which improves the thermal stability and compatibility of PEO-adsorbed cellulose CNCs as well as improves dispersion in hydrophobic matrices. Nanocrystals having such adsorption property are used as nanofillers. Figure 2 shows SEM (Scanning Electron Microscope) images of LDPE (Low density polyethylene)-based extruded films. (A) shows the smooth and uniform surface of the LDPE film, (B–D) shows the fractured rough surfaces of nanocomposites; because of their interaction with liquid nitrogen, spotted arrows denote the presence of nanocrystal aggregates.

3.1.3 Radiation-Induced Graft Copolymerization

Gamma irradiation is a nonthermal process. When the polymeric structures subjected to γ-radiation, it induces the ions, free radicals and electrons which are responsible for modifying the surface of the particles. Such modifications creates the reactive sites on the surface of the polymer through which the monomers can be grafted with the backbone of the polymer (Lv et al. 2013). Kong et al. (2016) revealed that gamma-ray activation is the energy-dependent process; high-energy gamma irradiation can degrade the glycoside bonds which converts the starch molecules into their fragments that greatly affects the physicochemical properties of starches. Bhat and Karim (2009) had already suggested the gamma irradiation treatment for commercial starch modifications which targets to obtain uniform particle size. They have studied the changes in crystalline nature of the starch and observed that due to irradiation process, there was a slight decrease in crystallinity of corn and fermented cassava starch, whereas increase in crystallinity of potato starch at 15 kGy energy level. Thus, they have concluded that the irradiation process affects the texture and shape of the particles in addition to size reduction effect. Shawrang et al. (2013) also reported that 15 kGy of gamma irradiation decreased the starch crystallinity index of the barley upto 6%, but in potato and corn starches, the crystallinity index gets increased at the same conditions. Therefore, these results indicates that the effect

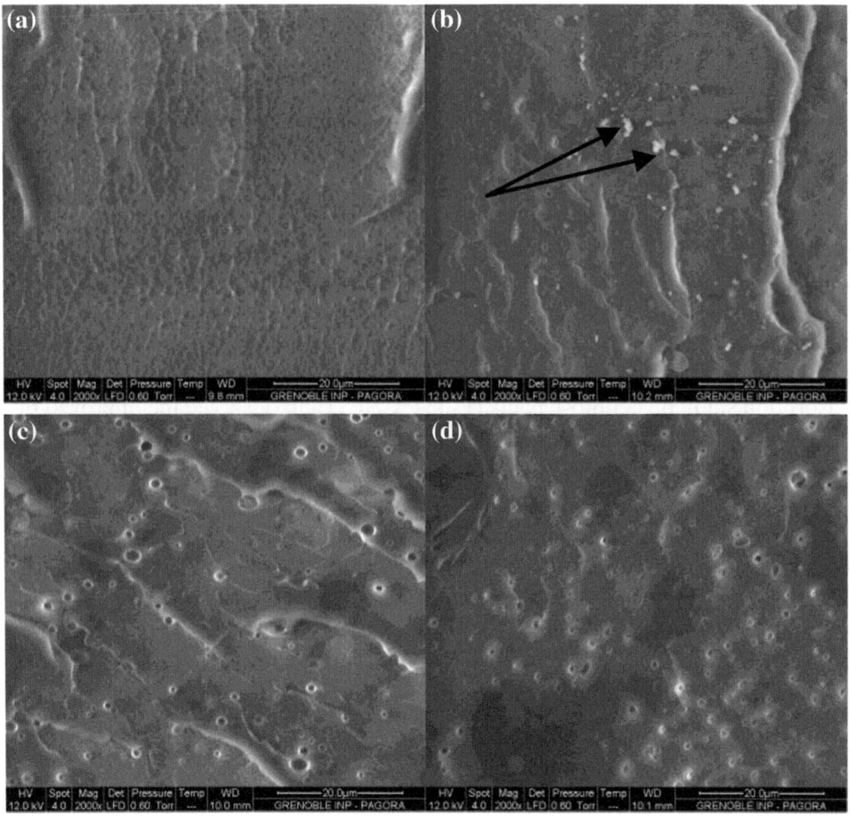

Fig. 2 SEM images of **a** neat LDPE and LDPE reinforced with **b** 9 wt% CNC, **c** 6 wt% PEO-adsorbed CNC (3.5×104 g mol^{-1}) and **d** 9 wt% PEO-adsorbed CNC (5×106 g mol^{-1}) (*Source* Reproduced with permission from (Azouz et al. 2012). Copyright © 2012, American Chemical Society)

of gamma irradiation treatment also depends on the nature of starch material. Cieśla et al. (2010) studied the radiation effects on film property and reported that the induction of radiation improved the hydrophobicity of potato and wheat starch films. Strength and flexibility were compared for both potato and wheat starch films; the results revealed that in potato starch films, strength as well as flexibility improved, whereas in wheat starch films, strength increased, but according to that, flexibility decreased. Improvement on functional and structural properties of films leads to improved resistance to breakage (Bashir et al. 2017). Color of films is an important sensory attribute because consumers usually preferred better color and clarity of the film. Teixeira et al. (2017) studied the effects on color of the films according to the irradiation dose and reported that in corn and sour cassava starch films color parameter b^* (yellow color) was slightly increased because of gamma irradiation at 15 kGy level, but there is no significant effect on parameter L^* (lightness).

Ultraviolet radiation is an another inexpensive method for surface modifications of polymers (Zhang and Henthorn 2009). Polymer surface functionalization improves the specific characteristics of the polymer, such as increased adhesion properties and surface wettability. Ultraviolet irradiation of polymer surfaces can be done simultaneously with chemical reaction directly or pre-irradiation before chemical grafting. Pre-irradiation process creates the reactive radical sites or peroxides on the polymer surface. Such created active site on the substrate polymer induces further chemical grafting.

3.1.4 Plasma-Induced Graft Copolymerization

Plasma is ionized gas containing a mixture of electrons, ions, neutrals and photons which improves the surface characteristics such as porosity and polarity of polymers and proteins without altering the integral properties. The plasma treatment for surface modification of materials is an inexpensive, rapid and green process (Reyna-martínez et al. 2018). Discharge of reactive gases such as O_2, NH_3 and CO_2 directly employed the desired functional groups on the surface (Cheng et al. 2014). Plasma treatment can be applied for addition of functional groups to polymeric surfaces through creating the nanopores, and also, it can be applied for symmetric nanostructural alteration of the polymeric membranes (Liston et al. 1993). Hot plasmas are characterized by a thermodynamic equilibrium between all particles when electrons, ions and neutrons have the same temperature. Cold plasmas are characterized by a thermodynamic nonequilibrium between the heavy particles and electrons (Khelifa et al. 2016). Among them, cold plasma is a safe, efficient and eco-friendly process. Applications have seen in many fields, especially in textile, food packaging and biomedicine. In textile and paper industries, plasma treatment of cellulose has been used intensively for active cleaning and sterilization.

In plasma treatment, high-energy electrons induced the excitation, ionization and dissociation processes on the substrate surface (Wenzel et al. 2000). This excited plasma electron stimulates cleavages in chemical bond of polymeric structure and creates the macromolecule radicals which lead to initiation of graft polymerization. The intensity of plasma treatments at the surface is too strong which leads to rapid treatment at the surface with an enriched functional layer. Plasma treatment ionizes the biomaterial surface in the presence of oxygen which improves the ionic binding of hydrogen thereby enhancing the hydrophilic nature of the material.

Plasma treatment creates a promising platform for the further surface-initiated polymerization or functional group grafting or immobilization of bioactive compounds. Oehr et al. (1999) modified the polymeric surface with primary amines by plasma activation method. Such plasma functionalized surface amine groups involved in subsequent biotin immobilization which creates the complete biotin monomolecular layer on the polymer surface.

3.2 Chemical Surface Modifications

The chemical modification of polymeric materials is one of the extensive methods in recent years. Large number of reactive groups are existing (i.e., hydroxyl, amino, thiol, alkyl and carboxyl) to modify the polymers chemically to obtain the polymers with specific thermal and mechanical properties based on applications.

3.2.1 Chemical Oxidation

Periodate oxidation is the most commonly used chemical oxidation technique which modifies the carbon-carbon (C-C) bond present in the compounds. The polysaccharide structures with aldehyde functional groups involved in this kind of modifications (Balakrishnan et al. 2005). Park and Lee (2008) studied the effect on the ion-binding properties of the alginate polymer and they have observed the cleavage of C-C bond present in the cis-diol group of uronate residues during the periodate oxidation of alginate. Such conformational changes in the structure, reduces the molecular weight of alginate which depends on both pH and temperature of the medium. Periodic acid used for oxidation mediates the aldehyde formation within the polysaccharide molecule. Reactive cellulose derivatives are prepared by the oxidation process, and further, the oxidized cellulose can be functionalized. Maekawa and Koshijima (1984) studied the modification of cellulose molecules by the sequential chemical oxidation of cellulose with periodate and less acidified sodium chlorite, respectively and reported that such modifications help in complete solubilization of cellulose. Oxidation by oxoammonium salts [(2,2,6,6-Tetramethylpiperidin-1-yl)oxyl or (2,2,6,6-tetramethylpiperidin-1-yl)oxidanyl] is known as TEMPO oxidation, which oxidizes the hydroxyl groups in biopolymers and converts them to aldehydes and ketones. The oxidation is typically performed under a catalytic process which generates oxoammonium cation by oxidation. Isogai et al. (2011) reported that TEMPO oxidation introduces carboxylates and aldehyde groups into the amorphous regions of cellulose, to maintain crystalline nature. In cellulose nanofiber, one of the two C_6 groups in the repeating unit can be oxidized; due to this phenomenon, only one of the C_6 hydroxyl group is oriented toward outside from the surface of the fiber.

3.2.2 Anionic/Cationic Surfactant Interactions

Surfactants act as wetting agent which reduces the surface tension and allows easier spreading. The interaction of biopolymers with the cationic and anionic surfactants forms the polymer-surfctant complex thereby it improves the mechanical strength of polymer. The mechanisms behind the polymer–surfactant interactions are polyelectrolyte absorption, hydrophobic and ionic interactions (Richert et al. 2004). Cationic CTAB (cetyltrimethylammonium bromide), anionic SDS (sodium dodecyl sulfate) and non-ionic TritonX-100 are widely using surfactants in industrial applications. Polar nature of biopolymer surface also takes importance in surfactant surface modification, which forms either micelles or bilayers. On nonpolar polymers, the nonpolar hydrophobic tails of the surfactant interact with the polymer surface; hence, the polar hydrophilic moieties stay outwards. Thus, the adsorption free energy of surfactants at the hydrophobic surfaces forms micelles. Whereas on polar polymeric surfaces, the surfactants adsorb with their polar hydrophilic moieties; hence, the non-polar hydrophobic tails stay outwards; thus, it forms bilayer structures (Holmberg et al. 2002). Bonnaud et al. (2010) studied the interaction of cationic lauric arginate (LAE) with various food-grade biopolymers and reported that LAE act as an antimicrobial agent, and the amount of LAE bound and the solubility of the surfactant-biopolymer aggregates greatly depend on concentration of surfactant and biopolymer type. Thongngam and McClements (2004) reported that the interaction of anionic surfactant SDS with cationic chitosan biopolymer formed the complex structure which is used for the encapsulation or controlled release of certain drug or food components, and also, he revealed that such surface modification is useful for the designing the functionalized chitosan-based food ingredients with specific characteristics.

3.2.3 Chemical Cross-Linking

Cross-linking improves the mechanical strength and the chemical resistivity of natural biopolymer to stabilize the biopolymers through intra or intermolecular attraction forces. Mostly, in ionic cross-linking, $CaCl_2$ is used as cross-linking agent, in which by increasing the concentration of $CaCl_2$, tensile strength and swelling ratio of biopolymers increased. Costa et al. (2018) reported that tensile strength of alginate edible film increases with decreased mannuronic and guluronic acid residues ratio (M/G ratio) and increased $CaCl_2$ concentration. Covalent cross-linking of alginate can be done using various chemicals such as glutaraldehyde (Chan et al. 2008), epichlorohydrin (Pawar and Edgar 2011), furfurylamine and bismaleimide (García-Astrain and Avérous 2018). Lee and Lim (2016) studied the effect of chemical cross-linking of fucoidans (algal polymer), in which the negatively charged sulfate groups present in fucoidans undergoes ionic cross-linking with the positively charged groups of chemical cross-linkers. Reys et al. (2016) explained the photo-cross-linking modification of fucoidan with methacrylic groups and reported that through cross-linking, according to concentration of the methacrylic anhydride, the particle size can be controlled. Zhou et al. (2016) modified the starch nanocrystals by cross-linking using citric acid as a reactant, and they observed such modification increases the surface roughness of the starch nanocrystals, and the size gets reduced. The process is clearly explained in Fig. 3, in which longer heating time increases the number of hydroxyl group that leads to higher hydrophobicity of the cross-linked nanocrystals.

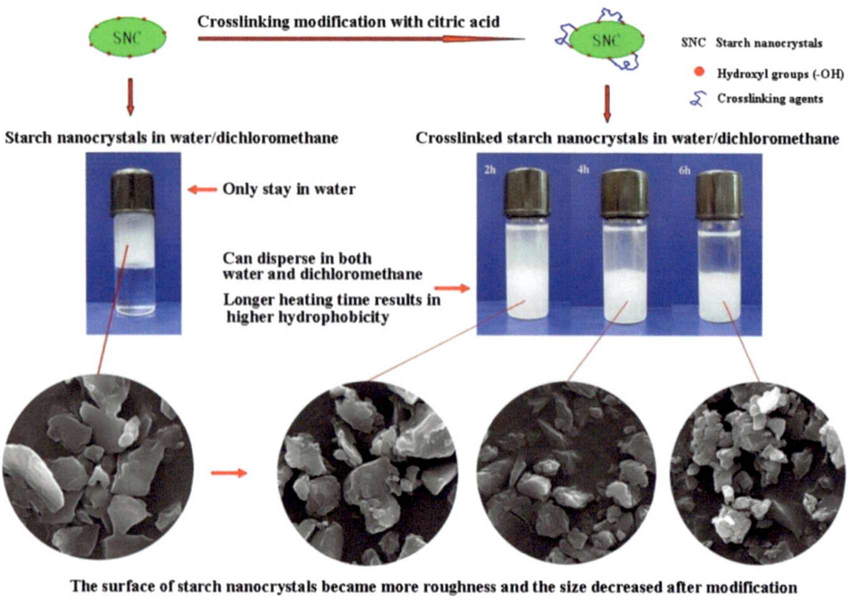

Fig. 3 Surface modification of starch nanocrystals by chemical cross-linking (*Source* Reproduced with the permission from Zhou et al. (2016). Copyright © 2018, Elsevier)

3.2.4 Chemical Grafting

Among the chemical surface modifications of polymers, grafting is the most versatile technique to impart various functional groups into a polymeric structure. It is useful for synthesizing the polymer with more vital properties. In grafting process, the monomers are covalently bonded onto the backbone of the polymer which does not affect the backbone polymer, but properties can be altered according to our requirements. Grafting reactions are mostly free-radical addition copolymerization reactions. Poly (N-Vinyl-2-Pyrrolidone) grafting of sodium alginate is widely studied for drug delivery applications (Işıklan et al. 2008). Xu et al. (2017) reported that deferoxamine grafted alginate was synthesized through amidation of sodium alginate by deferoxamine mesylate.

3.3 Advanced Surface Modifications for Bio-nanocomposites

Biopolymers interacted with nanomaterials and formed bio-nanocomposites through the structural modifications of natural polymers. Sources of naturally derived biopolymer are shown in Fig. 4. In the past decade, surface modifications were done through various physical and chemical methods as discussed earlier such as cross-linking, grafting, functional group activation like oxidation, amination, irradiation,

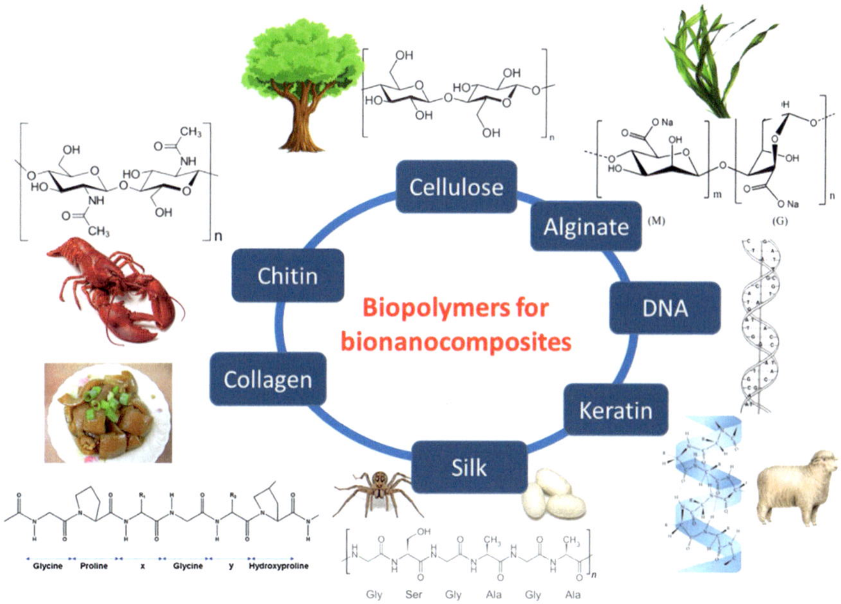

Fig. 4 Molecular structures and sources of naturally derived biopolymer components for bio-nanocomposites (*Source* Reproduced with the permission from Xiong et al. (2018). Copyright © 2018, Elsevier)

etc. Emerging techniques for the creation of micro or nanopatterned structural modifications are 2D- or 3D-organized printing technique which improves the physical properties and localized distributions of bio-nanocomposites. Techniques like soft lithography, inkjet printing and 3D printing techniques provide adapting properties such as self-folding, self-rolling and actuation by mediating stresses, deformations and electrical conductivities.

3.3.1 Lithographic Techniques

Lithographic technique is a prominent surface modification method for the development of smart materials. Photolithography or soft lithography creates pores on polymer surfaces which helps in fabricating the porous micro or nanoparticles on polymer surface for drug delivery (Kim et al. 2009). Kang et al. (2010) investigated the approach for superhydrophobic surface modification, in which superhydrophobic–hydrophilic alternating surface was created by soft lithography because of the hydrophilic conversion of the superhydrophobic surface. Due to this surface modification, the modified superhydrophobic surface has a good water-capturing capacity. Guyomard-Lack et al. (2011) applied the trypsin lithographic patterning method to create a pectin/PLL (poly-L-lysine) layer, in which alternating deposition of polymers produced the PLL/pectin/PLL thin film through spin-coating method for the creation of polyelectrolyte thin films; these selective modifications on the surface created various chemical functions on the surface of the film, and it can be applied in various fields such as biomedical field for selective capturing or releasing of cells, food industries for coating the smart packaging materials.

3.3.2 Inkjet Printing

Inkjet printing process mainly involves printing of polymeric microparticles with functional surface which finds applications in coating of transdermal patches/films and stents, drug loading, printing images on food materials and edible film coating. Functional biopolymers, biocompatible, biodegradable materials and drug coatings can be processed through inkjet printing. This process can print the droplet size of the polymer drug, in which multiple analytes can be simultaneously patterned using separate independent cartridges. This renders the coating material with uniformity and promotes release profiles of drug (Perkins et al. 2014). Mau et al. (2016) used the drop-on-demand inkjet print head for the deposition of drugs into microsized reservoirs and reported that by using this technique, surface wettability improved which plays a significant role for deposition of drug into the microsized reservoirs. Suntivich et al. (2014) created the nest-shaped multilayer structured silk arrays for cell hosting applications using inkjet printing technology. The silk dot arrays were printed onto the hydrophobic substrate which consist up to 400 silk dots with a dimension of 100 μm diameter and 100–600 nm average thickness. The nest shapes contain depleted central region and elevated rims. This "locked-in" silk nest succeeded in providing a well-organized and biocompatible platform for cell anchorage to the substrate during incubation. In food industries, drop-on-demand inkjet printing is used for creating the images by dispensing the edible liquids onto food surfaces (De Grood and De Grood 2011). Low-viscous edible inks can be easily ejected through the orifices on the print head, and printing at low temperature leads to lower the surface energy which reduces the ink spreading across the surface (Shastry et al. 2006).

3.3.3 3D Printing

3D printing is an advanced technique for the fabrication of 3D nanostructures that uses layer-wise dispersion printing and extrusion of melts. In 3D bioprinter, bio-ink is printed over deposited platform under computational control with chemical cross-linkers, thermal transitions and light (Ozbolat and Yu 2013). Various proteins, polysaccharides and biopolymers are used as bio-inks to fabricate 3D nanostructures. Biomaterials' 3D printing is a well-known approach used in tissue engineering and regenerative medicines. Lozano et al. (2015) used peptide-modified biopolymers and gellan gum as a bio-ink to print brain-like 3D structures which encapsulate the discrete layers of primary neural cells. It has also been used in skin replacement, by replacing damaged skin with 3D-printed skin. One of the major complications is to achieve the mechanical strength to tolerate the external stress and maintenance of shape after implantation. In recent years, 3D food printing is being widely applied in food sector to customize the food designs and to improve nutrient personalization. This suitable printing method improves the binding mechanisms, rheological properties and thermal properties during pretreatment and post-processing of foods. 3D printing has been applied for nutrient personalization of military and space foods (Liu et al. 2017). 3D printing can personalize the nutrition profile and the data for energy requirements according to the individuals' physical and nutritional status (Severini and Derossi 2016).

4 Safety of Surface-Modified Polymeric Nanoparticles

Surface modifications of nanoparticles posses a promising scope for medical applications like drug delivery and drug targeting in terms of transcellular transport, adhesion and invasion on cells. Generally, surface modification of the nanoparticles reduces the definite toxic effect of the nanoparticle. The unmodified silica nanoparticles have a toxic effect on the biological tissues of lungs due to the formation of reactive oxidative species. Surface-modified chitosan nanoparticles have been consider as one of the compatible and degradable nanomaterials. However, these materials have their limitations due to high molecular weight of the chitosan. A low molecular weight chitosan polymer is used for surface modification due to their wide application in biosafety of delivery systems. Iron oxide nanoparticles coated with surface-modified low molecular weight chitosan offered almost no cytotoxic effect in the human alveolar region (Dyawanapelly et al. 2017).

Balan and Verestiuc (2014) reported that the blood compatibility of chitosan has been improved by surface modifications. Figure 5 shows the modification steps applied to improve the blood compatibility of chitosan. Liposome is an extensively used carrier material for drug delivery due to the simplicity in modification of its surface. In pulmonary delivery system of elcatonin (a derivative of calcitonin), chitosan oligosaccharides and polyviniyl alcohol with hydrophobic anchor have been deployed for the surface modification of liposomes. These surface-modified liposomes have biocompatible phosopholipids which render a safe medium for delivery systems (Murata et al. 2012). In a cytotoxicity study of silica nanoparticles, the ability of silver nanoparticles surface modified with carboxyl or amino groups exhibited a greater viability when compared to unmodified silver nanoparticles on parenchymal heptocytes of a target group with liver disorder (Nagano et al. 2017).

The coating of superparamagnetic iron oxide nanoparticles with pullulan evaded its in vitro cytotoxicity (Gupta et al. 2004). A reduction in in vitro cytotoxicity was observed in dextran and albumin reducted iron oxide nanoparticles (Berry et al. 2003). Surface modification of bioactives with polyethyleneglycol (PEG) created the protective coatings and which indicated an extended occurrence of bioactives in the blood circulation (Bazile et al. 1995). One of the reported surfactants used in solid lipid nanoparticles (SLN) induced cytotoxicity during the in vitro study.

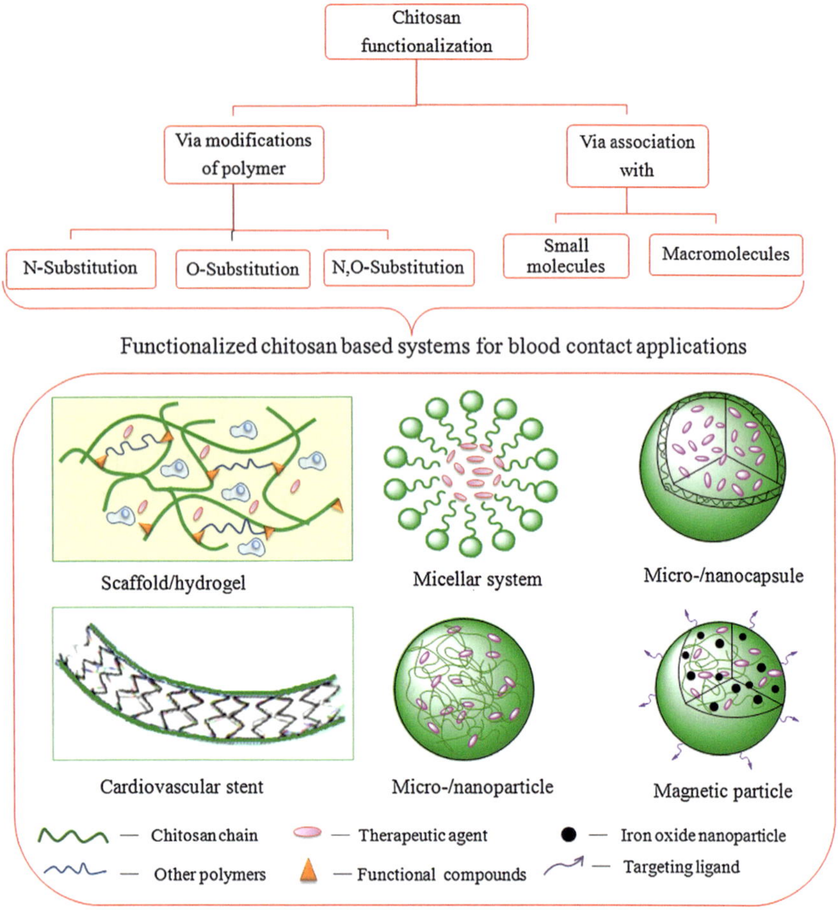

Fig. 5 Strategies used to improve the blood compatibility of the chitosan (*Source* Reproduced with permission from (Balan and Verestiuc 2014). Copyright © 2014, Elsevier)

This reliance of nanoparticles on the surfactant was used for material stabilization (Müller et al. 1997; Olbrich et al. 2004). SLNs showed reduced toxicity due to stabilizing surfactants when binding with the nanoparticles (Maia et al. 2002). Increased adherence leads to an improved activity of drugs. Weissenböck et al. (2004) studied the surface engineering of drug loaded bio-polymer (PLGA) with wheat germ agglutinin (WGA) - carbohydrate binding ligand. Covalent binding of WGA-biorecognitive ligands to the surface of PLGA nanospheres was activated by carbodiimide /N-hydroxysuccinimide. WGA grafted PLGA nanospheres have improved surface roughness due to increased ligand density, which enhances the cytoadhesion of drug loaded-PLGA nanospheres. The composition of lipid matrix also has been reported to have effect on the cell toxicity of SLN (Schöler et al. 2002). A surface-modified nanoparticle was formulated that specifically deliver selective

RNA (Ribonucleic acid) with high delivery efficiency in case of lung metastasis, and this formulation has a promising prospective for metastasis treatment as it showed a immunotoxic effect (Li et al. 2008).

5 Applications in Food Preservation, Packaging and Controlled Release of Food Bioactives

The modification of biopolymers possess a numerous perspective such as preparation of bio and nanocomposites specifically for packaging applications, chemical linkages with specific inorganic and organic components and compatible linkages with any type of polymers. Studies have shown that the release mechanism of drugs from the nanoparticles is by diffusion. Surface modification has also been exploited in the cytotoxicity of cancer cells. In a typical example, the surface-modified vinorelbine nanoparticles are used for the killing of cancer cells. Vinorelbine bitartrate is a partially synthetic alkaloid component used for the treatment of breast and lung cancer. The cell cytotoxicity of vinorelbine nanoparticles increased with increase in exposure time and concentration. Higher concentration release could be achieved by unmodified vinorelbine bitartrate nanoparticles. In order to achieve a sustained release and to prolong the release time, the surface of the vinorelbine bitartrate is modified in way such that the steric properties of vinorelbine bitartrate are changed with poly(ε-caprolactone) nanoparticles using hydrophilic surface coatings (Sathyamoorthy et al. 2017).

The surface modification also changes the particle mobility related to particle size and solution viscosity. Surface-modified nanoparticles are able to foam which are steady even under pressure but have a limited lifetime than usual surfactant stabilized foams (Baran and Cabrera 2006). The surface of the nanoparticles modified with polymer grafting prevents aggregation and increase the affinity of the modified surface in polymer matrix or solvent. This sequentially affects the dispersion stability of the extremely fine inorganic particles in polymer matrix or in organic solvents (Shirai et al. 1999). The dispersion stability of unmodified and modified iron oxide and titanium dioxide nanoparticles was compared, and the results of surface-modified nanoparticles showed an enhanced dispersion (Kobayashi et al. 2006). Similarly, the surface-modified alumina nanoparticles showed an improved dispersibility in a solvent (Rong et al. 2002).

Surface-modified TiO_2 and SiO_2 with PEG promoted the increase in pore size, hydrophilic nature, band gap and surface area. Acidity of TiO_2 (catalyst) was increased by incorporating the silica; hence, the hydrophilic properties of catalyst still retained and are active even at less irradiation. This was used in the application of antifogging and self-cleaning packaging material (Tristantini and Mustikasari 2011). Gold nanoparticles are well-known for their resistance toward oxidation. The surface modification with proper ligands is well acceded by living tissues. These modified particles are easy to synthesis and are efficient vehicles as tracers into the biological system in case of tumor detection (Warner and Hutchison 2004). Niikura et al. (2012) described a simple one-pot ligand modification process which involved in production of hollow drug carrier for encapsulating drug molecules. The ligand exchange from citric acid to semifluorinated ligands (SFLs) in tetrahydrofuran produced the SFL-AuNPs, and the subsequent processes are explained in Fig. 6.

Heavy metals present as contaminants in food and water are toxic when consumed for a long term. These heavy metals are mixed in water from industrial effluents. Numerous separation processes like adsorption and filtration have been employed to remove these heavy metals. However, these processes are cost effective at a large scale. For this, grafted copolymers are used in selective absorption of metals. These

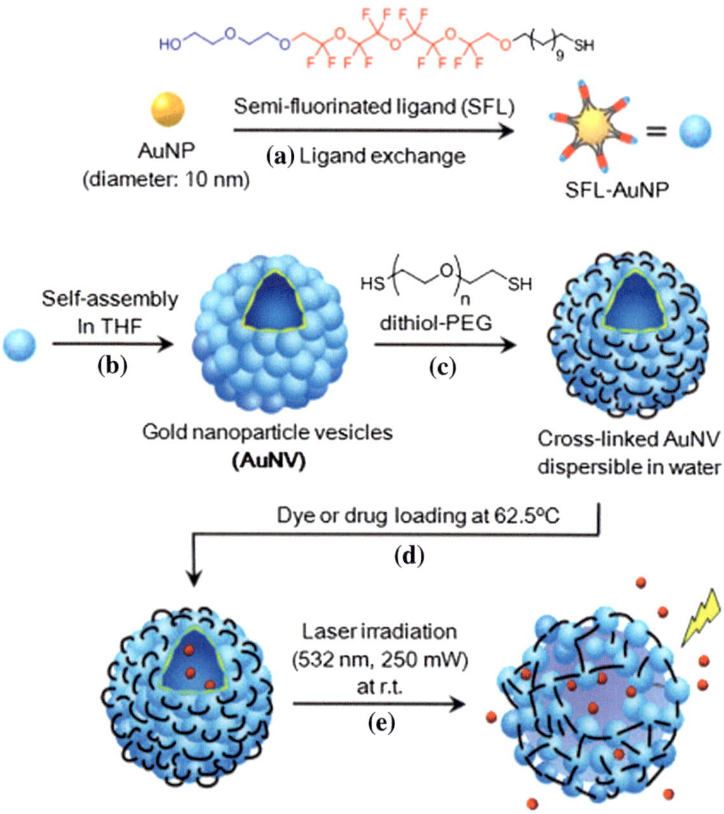

Fig. 6 Modification of gold NPs to produce a hollow drug carrier for encapsulating drug molecules and their subsequent light-triggered release process: **a** surface modification of gold NPs with semi-fluorinated ligands; **b** self-assembly of semifluorinated-coated gold NPs; **c** cross-linking of gold NPs of the nanocarrier with PEGylated dithiol to produce water-dispersible NPs; **d** encapsulation of DOX at 62.5 °C, and subsequent cooling to room temperature; **e** laser irradiation induces the release of DOX molecules (*Source* Reproduced with the permission from Niikura et al. (2012) Copyright © 2012, American Chemical Society)

copolymers are basically chelating polymers possessing a backbone of polysaccharides that can bind metal ions by coordination or ionic interactions (Bhattacharya et al. 2009). The introduction of bio-organoclays to food packaging has improved the barrier properties of packaging materials, for example incorporation of palygorskite and sepiolite fibrous clays with zein protein enhances the hydrophobic nature of material. These nanoclays are known for their bio-hybrid properties in packaging materials (Alcântara et al. 2012).

6 Conclusion

Surface modifications improve the functional properties of biopolymers and polymeric nanomaterials by different methods such as physical-, chemical- and radiation-induced surface modifications are described in this chapter. And also, surface modifications-dependent factors are discussed. The surface-modified polymers exhibited better performance than the unmodified form of polymers; such surface modifications particularly improve the mechanical strength, protection, controlled release, resistance to breakage, porosity, flexibility and polarity of the polymers. By selecting the suitable surface modification method, we can achieve the desired functionality of the polymers. The best suited application of surface-modified nanoparticles is in the targeted drug delivery that includes the treatment of neurological disorder because these surface-modified particles help in repairing the misfolding of specific proteins. Bio-nanocomposites are widely used in many fields; hence, the surface modification technique is a boon to scientific community to design the bio and nanocomponents for specific uses. The food packaging also gets a wide application from the surface-modified nanoclays and biohybrid polymers. Thus, the potential surface modification and further optimization can develop the novel functional green bio-nanomaterial. Safety issues of surface-modified polymers and its applications are also discussed in various fields such as in food industries, pharma industries, biomedical and medical fields.

References

Alcântara AC et al (2012). Zein–fibrous clays biohybrid materials. European Journal of Inorganic Chemistry 2012(32):5216-5224. https://doi.org/10.1002/ejic.201200582

Anirudhan TS, Sandeep S (2011) Synthesis and characterization of molecularly imprinted polymer of N-maleoylchitosan-grafted-2-acrylamido-2-methylpropanesulfonic acid and its controlled delivery and recognition of bovine serum albumin. Polym Chem 2(9):2052–2061. https://doi.org/10.1039/c1py00059d

Aravamudhan A et al (2014) Natural polymers: polysaccharides and their derivatives for biomedical applications. In Natural and synthetic biomedical polymers. Elsevier Inc. pp 67–89. https://doi.org/10.1016/B978-0-12-396983-5.00004-1

Azouz KB et al (2012) Simple method for the melt extrusion of a cellulose nanocrystal reinforced hydrophobic polymer. ACS Macro Lett 1(1):236–240. https://doi.org/10.1021/mz2001737

Balakrishnan B et al (2005) Periodate oxidation of sodium alginate in water and in ethanol-water mixture: a comparative study. Carbohyd Res 340(7):1425–1429. https://doi.org/10.1016/j.carres.2005.02.028

Balan V, Verestiuc L (2014) Strategies to improve chitosan hemocompatibility: A review. Eur Polym J 53:171–188. https://doi.org/10.1016/j.eurpolymj.2014.01.033

Baran JR, Cabrera OJ (2006) Use of surface modified nanoparticles for oil recovery. Washington, DC: U.S. Patent and Trademark Office., U.S. Paten. https://doi.org/10.1016/j.(73)

Bashir K, Jan K, Aggarwal M (2017) Thermo-rheological and functional properties of gamma-irradiated wholewheat flour. Int J Food Sci Technol 52(4):927–935. https://doi.org/10.1111/ijfs.13356

Bazile D et al (1995) Stealth me. PEG-PLA nanoparticles avoid uptake by the mononuclear phagocytes system. J Pharm Sci 84(4):493–498. https://doi.org/10.1002/jps.2600840420

Berry CC et al (2003) Dextran and albumin derivatised iron oxide nanoparticles: Influence on fibroblasts in vitro. Biomaterials 24(25):4551–4557. https://doi.org/10.1016/S0142-9612(03)00237-0

Bhat R, Karim AA (2009) Impact of radiation processing on starch. Compr Rev Food Sci Food Saf 8(2):44–58. https://doi.org/10.1111/j.1541-4337.2008.00066.x

Bhattacharya A, Rawlins J, Ray P (2009) Polymer grafting and crosslinking. J Am Chem Soc Am Chem Soc 131(20):7204. https://doi.org/10.1021/ja902814m

Bonnaud M, Weiss J, McClements DJ (2010) Interaction of a food-grade cationic surfactant (Lauric Arginate) with food-grade biopolymers (pectin, carrageenan, xanthan, alginate, dextran, and chitosan). J Agric Food Chem 58(17):9770–9777. https://doi.org/10.1021/jf101309h

Chan OCM, So KF, Chan BP (2008) Fabrication of nano-fibrous collagen microspheres for protein delivery and effects of photochemical crosslinking on release kinetics. J Controlled Release 129(2):135–143. https://doi.org/10.1016/j.jconrel.2008.04.011

Chandra R (1998) Biodegradable polymers. Prog Polym Sci 23(7):1273–1335. https://doi.org/10.1016/S0079-6700(97)00039-7

Cheng YL et al (2014) Non-thermal plasma assisted polymer surface modification and synthesis: a review. Int J Agric Biol Eng 7:1–9. https://doi.org/10.3965/j.ijabe.20140702.001

Cieśla KA, Nowicki A, Buczkowski MJ (2010) Preliminary studies of the influence of starch irradiation on physicochemical properties of films prepared using starch and starch-surfactant systems. Nukleonika 55:233–242

Costa MJ et al (2018) Physicochemical properties of alginate-based films: Effect of ionic crosslinking and mannuronic and guluronic acid ratio. Food Hydrocoll 81:442–448. https://doi.org/10.1016/j.foodhyd.2018.03.014

De Grood, JPW, De Grood PJ (2011) Method and device for dispensing a liquid. Patent Application Publication (10) Pub. No.: US 2011/0121016 A1 1(19)

Della Vecchia NF et al (2013) Building-block diversity in polydopamine underpins a multifunctional eumelanin-type platform tunable through a quinone control point. Adv Func Mater 23(10):1331–1340. https://doi.org/10.1002/adfm.201202127

Dreyer DR et al (2013) Perspectives on poly(dopamine). Chem Sci 4(10):3796–3802. https://doi.org/10.1039/c3sc51501j

Dyawanapelly S et al (2017) Assessing safety and protein interactions of surface-modified iron oxide nanoparticles for potential use in biomedical areas. Colloids Surf *B*: *Biointerfaces*. 154:408–420. https://doi.org/10.1016/j.colsurfb.2017.03.050

García-Astrain C, Avérous L (2018) Synthesis and evaluation of functional alginate hydrogels based on click chemistry for drug delivery applications. Carbohydr Polym 190:271–280. https://doi.org/10.1016/j.carbpol.2018.02.086

Gupta M et al (2004) In vitro cytotoxicity studies of hydrogel pullulan nanoparticles prepared by AOT/N-hexane micellar system. Scanning Electron Microsc 7(1):38–46

Guyomard-Lack A et al (2011) Site-selective surface modification using enzymatic soft lithography. Langmuir 27(12):7629–7634. https://doi.org/10.1021/la2007843

Harush-Frenkel O et al (2008) Surface charge of nanoparticles determines their endocytic and transcytotic pathway in polarized MDCK cells. Biomacromol 9(2):435–443. https://doi.org/10.1021/bm700535p
Holmberg K, Bo J, Kronberg B (2002) Surfactants and polymers in aqueous solution. John Wiley & Sons Ltd., Chichester.
Işıklan N, Inal M, Yiğitoğlu M (2008) Synthesis and characterization of poly(N-Vinyl-2-pyrrolidone) grafted sodium alginate hydrogel beads for the controlled release of indomethacin. J Appl Polym Sci 110:481–493. https://doi.org/10.1002/app.28577
Isogai A, Saito T, Fukuzumi H (2011) TEMPO-oxidized cellulose nanofibers. Nanoscale 3(1):71–85. https://doi.org/10.1039/c0nr00583e
Jiao Y, Cui F (2007) Surface modification of polyester biomaterials for tissue engineering. Biomedical Materials 2(4): 24-37. https://doi.org/10.1088/1748-6041/2/4/r02
Kang SM et al (2010) One-step modification of superhydrophobic surfaces by a mussel- inspired polymer coating. Angewandte Chemie International Edition 49(49): 9401–9404. https://doi.org/10.1002/anie.201004693
Khelifa F et al (2016) Free-radical-induced grafting from plasma polymer surfaces. Chemical Reviews 116(6):3975–4005. https://doi.org/10.1021/acs.chemrev.5b00634
Kim S et al (2009) Engineered polymers for advanced drug delivery. Eur J Pharm Biopharm 71(3):420–430. https://doi.org/10.1016/j.ejpb.2008.09.021
Kobayashi M et al (2006) Precise surface structure control of inorganic solid and metal oxide nanoparticles through surface-initiated radical polymerization. Sci Technol Adv Mater 7(7):617–628. https://doi.org/10.1016/j.stam.2006.07.008
Kong X et al (2016) Effects of gamma irradiation on physicochemical properties of native and acetylated wheat starches. Int J Biol Macromol 91:1141–1150. https://doi.org/10.1016/j.ijbiomac.2016.06.072
Kuila T et al (2012) Chemical functionalization of graphene and its applications. Prog Mater Sci 57(7):1061–1105. https://doi.org/10.1016/j.pmatsci.2012.03.002
Lee EJ, Lim KH (2016) Formation of chitosan-fucoidan nanoparticles and their electrostatic interactions: quantitative analysis. J Biosci Bioeng 121(1):73–83. https://doi.org/10.1016/j.jbiosc.2015.05.006
Li YP et al (2001) PEGylated PLGA nanoparticles as protein carriers: synthesis, preparation and biodistribution in rats. J Control Release 71(2):203–211. https://doi.org/10.1016/S0168-3659(01)00218-8
Li SD, Chono S, Huang L (2008) Efficient gene silencing in metastatic tumor by siRNA formulated in surface-modified nanoparticles. J Controlled Release 126(1):77–84. https://doi.org/10.1016/j.jconrel.2007.11.002
Lin N, Dufresne A (2013) Physical and/or chemical compatibilization of extruded cellulose nanocrystal reinforced polystyrene nanocomposites. Macromolecules 46(14):5570–5583. https://doi.org/10.1021/ma4010154
Liston EM, Martinu L, Wertheimer MR (1993) Plasma surface modification of polymers for improved adhesion: a critical review. J Adhes Sci Technol 7(10):1091–1127. https://doi.org/10.1163/156856193X00600
Liu Z et al (2017) 3D printing: Printing precision and application in food sector. Trends Food Sci Technol. 69:83–94. https://doi.org/10.1016/j.tifs.2017.08.018
Lozano R et al (2015) Biomaterials 3D printing of layered brain-like structures using peptide modified gellan gum substrates. Biomaterials 67:264–273. https://doi.org/10.1016/j.biomaterials.2015.07.022
Lv X et al (2013) Gamma radiation-induced grafting of acrylamide and dimethyl diallyl ammonium chloride onto starch. Carbohydr Polym 92(1):388–393. https://doi.org/10.1016/j.carbpol.2012.10.002
Maekawa E, Koshijima T (1984) Properties of 2,3-dicarboxy cellulose combined with various metallic ions. J Appl Polym Sci 29(7):2289–2297. https://doi.org/10.1002/app.1984.070290705

Maia CS et al (2002) Drug targeting by solid lipid nanoparticles for dermal use. J Drug Target 10(6):489–495. https://doi.org/10.1080/1061186021000038364

Mau R et al (2016) Adjusting inkjet printhead parameters to deposit drugs into micro-sized reservoirs 2(1):387–390. https://doi.org/10.1515/cdbme-2016-0086

Müller RH et al (1997) Cytotoxicity of solid lipid nanoparticles as a function of the lipid matrix and the surfactant. Pharm Res 14(4):458–462. https://doi.org/10.1023/a:1012043315093

Murata M et al (2012) Pulmonary delivery of elcatonin using surface-modified liposomes to improve systemic absorption: polyvinyl alcohol with a hydrophobic anchor and chitosan oligosaccharide as effective surface modifiers. Eur J Pharm Biopharm 80(2):340–346. https://doi.org/10.1016/j.ejpb.2011.10.011

Nagano T et al (2017) Modifying the surface of silica nanoparticles with amino or carboxyl groups decreases their cytotoxicity to parenchymal hepatocytes. Biol Pharm Bull 40(5):726–728. https://doi.org/10.1248/bpb.b16-00917

Niikura K et al (2012) Gold nanoparticles coated with semi-fluorinated oligo(ethylene glycol) produce sub-100 nm nanoparticle vesicles without Templates. Journal of the American Chemical Society 134(18):7632–7635. https://doi.org/10.1021/ja302122w

Oehr C et al (1999) Plasma grafting—a method to obtain monofunctional surfaces. Surf Coat Technol 116–119:25–35. https://doi.org/10.1016/S0257-8972(99)00201-7

Olbrich C et al (2004) Cytotoxicity studies of Dynasan 114 solid lipid nanoparticles (SLN) on RAW 264.7 macrophages-impact of phagocytosis on viability and cytokine production. J Pharm Pharmacol 56(7):883–891. https://doi.org/10.1211/0022357023754

Ozbolat I, Yu Y (2013) Bioprinting towards organ fabrication: challenges and future trends. IEEE Transactions on Biomedical Engineering 60(3):691–699. https://doi.org/10.1109/tbme.2013.2243912

Park H, Lee KY (2008) Alginate hydrogels as matrices for tissue engineering. In Natural-based polymers for biomedical applications. Wood head Pub, 515–532. https://doi.org/10.1533/9781845694814.4.515

Pawar SN, Edgar KJ (2011) Chemical modification of alginates in organic solvent systems. Biomacromol 12:4095–4103. https://doi.org/10.1021/bm201152a

Perkins J et al (2014) Direct writing of bio-functional coatings for cardiovascular applications. Journal of Biomedical Materials Research Part A 4290–4300. https://doi.org/10.1002/jbm.a.35105

Platt D (2006) Biodegradable polymers: market report. iSmithers Rapra Publishing

Reynamartínez R et al (2018) Use of cold plasma technology in biomaterials and their potential utilization in controlled administration of active substances 4(5). https://doi.org/10.19080/jojms.2018.04.555649

Reys LL et al (2016) Fucoidan hydrogels photo-cross-linked with visible radiation as matrices for cell culture. ACS Biomater Sci Eng 2(7):1151–1161. https://doi.org/10.1021/acsbiomaterials.6b00180

Ribeiro H et al (2015) Multifunctional nanocomposites based on tetraethylenepentamine-modified graphene oxide/epoxy. Polym Testing 43:182–192. https://doi.org/10.1016/j.polymertesting.2015.03.010

Richert L et al (2004) Improvement of stability and cell adhesion properties of polyelectrolyte multilayer films by chemical cross-linking. Biomacromolecules 5:284–294. https://doi.org/10.1021/BM0342281

Rong MZ et al (2002) Graft polymerization of vinyl monomers onto nanosized alumina particles. Eur Polymer J 38(8):1573–1582. https://doi.org/10.1016/S0014-3057(02)00037-X

Sathyamoorthy N, Magharla DD, Vankayalu SD (2017) Effect of surface modification on the in vitro protein adsorption and cell cytotoxicity of vinorelbine nanoparticles. J Pharm & Bioallied Sci 9(2):135–143. https://doi.org/10.4103/jpbs.JPBS_258_16

Schöler N et al (2002) Effect of lipid matrix and size of solid lipid nanoparticles (SLN) on the viability and cytokine production of macrophages. Int J Pharm 231:167–176. https://doi.org/10.1016/s0378-5173(01)00882-1

Sen G et al (2010) Microwave initiated synthesis of polyacrylamide grafted guar gum (GG-g-PAM)-Characterizations and application as matrix for controlled release of 5-amino salicylic acid. Int J Biol Macromol 47(2):164–170. https://doi.org/10.1016/j.ijbiomac.2010.05.004

Sen G et al (2011) Microwave initiated synthesis of polyacrylamide grafted Psyllium and its application as a flocculant. Int J Biol Macromol 50(2):369–375. https://doi.org/10.1016/j.ijbiomac.2011.12.014

Severini C, Derossi A (2016) Could the 3D printing technology be a useful strategy to obtain customized nutrition? Journal of Clinical Gastroenterology 50:175–178. https://doi.org/10.1097/mcg.0000000000000705

Sharma P et al (2013) Bio-functionalized graphene-graphene oxide nanocomposite based electrochemical immunosensing. Biosens Bioelectron 39(1):99–105. https://doi.org/10.1016/j.bios.2012.06.061

Shastry A, Ben-Joseph E, Collins TM (2006) Ink-jet printing on surface modified edibles and products made. United States Patent. No.: US 7,500,744 B2 2(12)

Shawrang P, Sadeghi AA, Ghorbani B (2013) The effect of electron beam irradiation on ß-glucan content, x-ray diffraction of starch, protein subunit patterns, and in vivo digestibility of barley grain in cockerels. Turkish J Vet Anim Sci 37(4):443–448. https://doi.org/10.3906/vet-1202-25

Shirai Y, Kawatsura K, Tsubokawa N (1999) Graft polymerization of vinyl monomers from initiating groups introduced onto polymethylsiloxane-coated titanium dioxide modified with alcoholic hydroxyl groups. Prog Org Coat 36(4):217–224. https://doi.org/10.1016/S0300-9440(99)00046-6

Suntivich R, Drachuk I, Calabrese R, Kaplan DL, Tsukruk VV (2014) Inkjet printing of silk nest arrays for cell hosting. Biomacromolecules 15(4):1428–1435. https://doi.org/10.1021/bm500027c

Teixeira BS et al (2017) Effects on natural corn and potato starches and modified cassava starch. Radiat Phys Chem 142:44–49. https://doi.org/10.1016/j.radphyschem.2017.09.001

Thongngam M, McClements DJ (2004) Characterization of Interactions between chitosan and an anionic surfactant. J Agric Food Chem 52(4):987–991. https://doi.org/10.1021/jf034429w

Tristantini D, Mustikasari R (2011) Modification of TiO_2 Nanoparticle with PEG Application Modification of TiO_2 Nanoparticle with PEG and SiO_2 For anti-fogging and self-cleaning application. Int J Eng Technol 11(2):73–78

Wang J W, Hon MH (2002) Sugar-mediated chitosan/poly (ethylene glycol)-β-dicalcium pyrophosphate composite: Mechanical and microstructural properties. J Biomed Mater Res A 64(2): 262-272. https://doi.org/10.1002/jbm.a.10358

Warner M, Hutchison JE (2004) Synthesis and assembly of functionalized gold nanoparticles. ChemInform 35(1). https://doi.org/10.1002/chin.200401219

Wartlick H et al (2004) Highly specific HER2-mediated cellular uptake of antibody-modified nanoparticles in tumour cells. J Drug Target 12(7):461–471. https://doi.org/10.1080/10611860400010697

Weissenböck A, Wirth M, Gabor F (2004) WGA-grafted PLGA-nanospheres: preparation and association with Caco-2 single cells. J Controlled Release 99(3):383–392. https://doi.org/10.1016/j.jconrel.2004.07.025

Wen H et al (2012) Engineered redox-responsive PEG detachment mechanism in PEGylated nano-graphene oxide for intracellular drug delivery. Small 8(5):760–769. https://doi.org/10.1002/smll.201101613

Wenzel A et al (2000) Effects of preparation condition of photoinduced graft filling-polymerized membranes on pervaporation performance. J Membr Sci 179(1):69–77. https://doi.org/10.1016/S0376-7388(00)00502-0

Wilson CJ et al (2005) Mediation of biomaterial—cell interactions by adsorbed proteins: a review. Tissue Engineering 11(1). https://doi.org/10.1089/ten.2005.11.1

Xiong R et al (2018) Naturally-derived biopolymer nanocomposites : Interfacial design, properties and emerging applications. Mater Sci & Eng R 125:1–41. https://doi.org/10.1016/j.mser.2018.01.002

Xu G et al (2017) Synthesis and rheological investigation of self-healable deferoxamine grafted alginate hydrogel. J Polym Sci, Part B: Polym Phys 55(11):856–865. https://doi.org/10.1002/polb.24334

Zhang P, Henthorn DB (2009) Synthesis of PEGylated single wall carbon nanotubes by a photoinitiated graft from polymerization. AIChE J 56(6):1610–1615. https://doi.org/10.1002/aic.12108

Zhou J et al (2016) Hydrophobic starch nanocrystals preparations through crosslinking modification using citric acid. Int J Biol Macromol 91:1186–1193.https://doi.org/10.1016/j.ijbiomac.2016.06.082

Biopolymer Nanocomposites and Its Application in Food Processing

R. Preethi, M. Maria Leena, J. A. Moses and C. Anandharamakrishnan

Abstract At present, nanotechnology is recognized as one of the promising and challenging areas for technology development in the food processing sector. Development of biopolymer nanocomposites is an evolving field in the food preservation research area, due to their added functionalities and enhanced structural and physical properties in food packaging industry, and delivery of nutraceutical products. Furthermore, the biopolymer and nanocomposite are deriving from the natural sources, which are abundant and available in nature, biodegradable and eco-friendly as well as economical. The conventional nanocomposites are synthetic non-degradable based on nylon commonly, considered as a solid waste, which is depleting our mother nature. Nowadays, consumer demand and expectation increase for fresh, nutraceutical-rich and good-quality foods, which enforce the industrial needs of biodegradable or edible materials extracted from renewable resources. Advanced bio-derived polymer nanocomposites will create a promising path to encounter the current and future needs and to overcome the environmental risk attributed by the conventional polymer.

Keywords Biopolymer · Nanocomposites · Application · Food processing

1 Introduction

A promising application of nanocomposites nowadays expands the use of edible and biodegradable films in the food processing and preservation industry to enhance the excellence and shelf life of food products. Biopolymer nanocomposites are fabricated with biopolymers and bio-fillers, which possess improved physical, mechanical and thermal properties with biocompatibility and biodegradability functionalities. Due to the environmental safety concern and to replace non-biodegradable polymers, recent research is trending on the production of environmentally friendly

R. Preethi · M. Maria Leena · J. A. Moses · C. Anandharamakrishnan (✉)
Computational Modeling and Nanoscale Processing Unit, Indian Institute of Food Processing Technology, Pudukkottai Road, Thanjavur, Tamil Nadu 613005, India
e-mail: anandharamakrishnan@iifpt.edu.in

S. Ahmed and W. Ali (eds.), *Green Nanomaterials*,
Advanced Structured Materials 126,
https://doi.org/10.1007/978-981-15-3560-4_12

polymers. Biopolymers are produced from the renewable resources, such as starch, carrageenan, collagen, gum arabic, chitosan, pectin and polylactic acid, which added more advantages in the field of food preservation and agriculture nowadays (Al et al. 2018).

2 Biopolymer

For past few years, researches are focusing on the replacement of hazardous petroleum-based polymer to overcome and control the pollution and toxic level generated by the synthetic polymers. The "green" materials (biopolymer) are made up of nontoxic components which can be easily degraded by microorganism, which reduce the fossil fuels. Based on the status of global warming, we are in need of recyclable environmental-friendly materials which protect both "mother nature" and living creatures. Biopolymers are biodegradable under suitable conditions with applicable temperature, oxygen, moisture and humidity; the biopolymers degrade into fragments or disintegrate, which is further degraded by microorganisms, with no toxic and environmental-friendly residue. Biopolymers are broadly classified under three categories based on the origin (Fig. 1). The types of polymers will be explained as follows. These biopolymers are biodegradable under bioactive environment, with the help of enzymes produce by the microorganism convert polymer into non-toxic residues, as biomass, humic matter act as compost for the plant growth. Various plant-based polysaccharides are used as the biopolymer, which can degrade through fermentative routes based on the cyclic process as explained in Fig. 2.

2.1 *Polysaccharides*

It is available in various resources such as vegetables, microorganism and animal, which is predominantly used as an encapsulation agent, thickener, gelling agent and stabilizer in the food industry. Polysaccharides are hydrophilic polymers and called

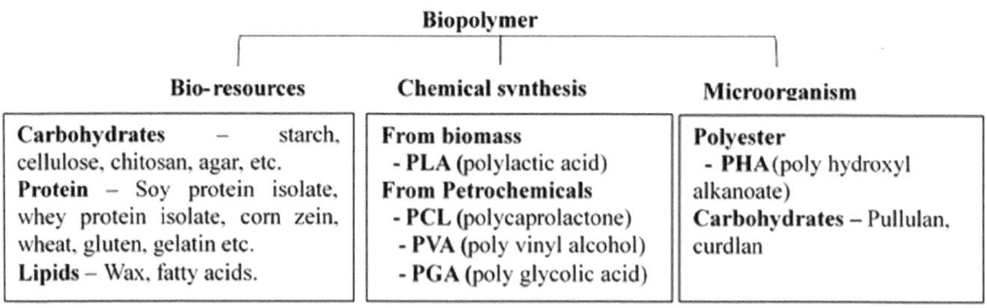

Fig. 1 Classification of biopolymer (Rhim et al. 2013)

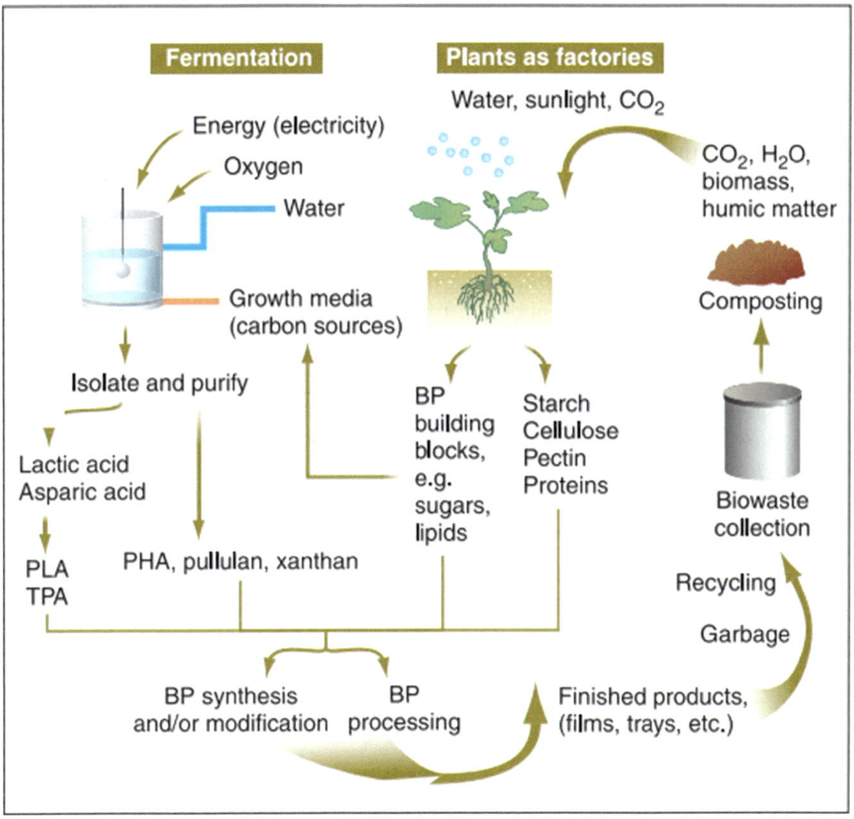

Fig. 2 Cyclic process of biopolymer production and degradation (Gross and Kalra 2002)

as hydrocolloids; examples are starch, amylose, xanthan gum, carrageenan, gum arabic and carboxymethylcellulose briefly explained below.

(a) **Cellulose**

Cellulose is the major building blocks of higher plants, also present in algae, fungi, bacteria and marine-living creatures. They are hygroscopic material, insoluble but swell in water, soluble in dilute acid and solvents. Cellulose is modified chemically into microfibrillated cellulose, microcrystalline cellulose, cellulose ethers, sodium carboxymethylcellulose (CMC), methylcellulose (MC), hydroxypropyl methylcellulose (HPMC) and hydroxypropyl cellulose (HPC) to enhance the indigenous property. Cellulose base product in the food industry act as a low-calorie food. It also increase the flowability of the products and, controlled water interaction and textural attributes in the food products (Klemm et al. 2005). The cellulose nano-reinforcements are attractive materials for the preparation of low weight and low cost with high mechanical strength. Various concentrations of cellulose nanofiber were incorporated with mango puree to decrease water vapor permeability, where the nanofiller cellulose increases the tensile strength and elongation of the film (Azeredo et al. 2011). Protein-based foods frequently need a stabilizer to improve the shelf life; methylcellulose interacts with proteins by the reaction of hydrophobic–hydrophilic mechanism, which provides stable thermal property while frying the foodstuffs. Cellulose products act as the binder in food pattie and increase the viscosity of sauces, gravies, soups and syrups. They also serve as an excellent emulsifier, which can be used as nondairy whipped toppings, desserts and mousses and increase the volume in the baked foods (Klemm et al. 2005; Putri et al. 2017).

Cellulose derivatives are generally with medium strength, transparent, colorless, tasteless, water-soluble and moderate oxygen barriers. The incorporation of fatty acid with the HPMC enhanced the moisture barrier and utilised as a packing material, which can biodegrade by the microbial degradation mechanism. Integration of filler materials like nanoclay or inorganic/organic nanoparticles and food additives improves the properties of the cellulose derivatives which creates a demand in industry for food packing sector (Shit and Shah 2014).

(b) **Gum Arabic (GA)**

Gum arabic (GA) is the exudate extracted from *Acacia senegal* and *Acacia seyal* tree. The gum has a desirable low viscosity, binding, emulsifying, stabilizing and shelf-life-enhancing properties, which derived attention in the food industry. Due to extensive property, the gum powder gains more attention in the food processing and preservation sector and it is incorporated in food system like ice creams, candies, soft drinks, syrups, fruit juices and confectionery coatings. The various literature showed that GA possesses antimicrobial, anti-inflammatory and anticoagulant properties (Anandharamakrishnan and Ishwarya 2015). Maqbool et al. demonstrated the impact of GA with chitosan as an edible coating to extend the storage time for 33 days. At lower temperature and higher humidity conditions, the fruits can store for a long time without fruit firmness and no biochemical changes (Maqbool et al. 2011; Goyal 2014).

(c) **Chitin, Chitosan and Chitooligosaccharides**

Chitin is a cellulose-like polysaccharide, found abundant in marine crustaceans, shrimp, crabs and marine sponges with a (1,4)-b-*N*-acetyl glycosaminoglycan-repeating structure. Based on the strong intermolecular hydrogen bond, it is insoluble in water and organic solvents and soluble in hexafluoroisopropanol, hexafluoroacetone and chloroalcohols. Chitin whiskers are prepared from the chitin by acid hydrolysis, which also used as the nanofillers due to the low-cost, odorless, tasteless, excellent biocompatibility and biodegradability property. The mechanical property and gelling ability increased in the combination of gelatin and chitin whiskers as a reinforcing agent (Wan and Tai 2013; Ge et al. 2018).

Chitosan is the *N*-deacetylated derivative of chitin, and the major source of chitin is marine crustacean shells. It is the only sophisticated cationic polymer with positive charges on its amino group, nontoxic and biocompatible; hence, it is widely used in various fields like pharmaceuticals, food, agriculture and food engineering. Recent studies evidence the use of chitosan as an edible packing due to its antioxidant and antimicrobial property which also possesses the property to enhance the sensory and nutritional quality of the food (Thadathil and Velappan 2014; Hamed et al. 2016). Chitooligosaccharides are the depolymerized form of chitosan and contain low molecular weight with low viscosity, which attract various food industries. It also has cholesterol-lowering property and immune-enhancing effects. Hence, chitosan and chitooligosaccharides act as thickening agent and as a source of dietary fiber to stimulate the beneficial bacteria in the gastrointestinal tract (Lodhi et al. 2014; Kumar et al. 2018).

(d) **Starch and derivatives**

Starch contains two types of polymeric molecules, amylose and amylopectin. Amylose is a linear chain with (1→4)-α-D-glucopyranosyl units and amylopectin, which possesses a backbone of amylose and is highly branched with side units of D-glucopyranosyl linked by α-1,6-glycosidic bonds. Amylose is isotropic, tasteless, odorless and colorless which has the ability to form a film. Hence, tapioca starch and sweet potato starch have incorporated along with chitosan or potassium sorbate to create an edible film, which acts as an antimicrobial packing. Food additives were also added with the cornstarch to improve the water vapor permeability property; similarly, cornstarch and cellulose acetate were used to reduce the ability of water sorption (Campos et al. 2011).

(e) **Seaweeds**

Alginate, carrageenan and agar are three essential polysaccharides obtained from the seaweeds. These three polymers are extensively used for the film formation, where they act as a precursor. Seaweeds have high rigidity with low deformability, nontoxic, degradable and biocompatible, which exploited in the field of food industry. The film fabricated from the seaweed has poor water vapor barrier properties due to its hydrophilic nature, but this disadvantage improved by the incorporation of other nanocomposites and polymers. Few seaweed matrices and its advantages are given in Table 1 (Khalil et al. 2017).

(f) **Plant seed mucilage**

Plant seed mucilage also acts as polysaccharide hydrocolloid source, which has more health benefits and is used in food and nutraceutical industry applications. Due to unique properties of mucilages, like texturizing, emulsification and thickening along with the dietary aspects such as low glycemic index, potential to increase the beneficial microflora and bioavailability, they gained attention towards the food industry. Polysaccharides of plant seed are classified into three groups: (a) galactomannans–endosperm non-starch components, (b) mucilaginous constituents of the seed coat (e.g., flax, chia, psyllium, yellow mustard) and (c) cell wall material of the endosperm (e.g., xyloglucans and soybean hemicelluloses). Mucilage around the seed hydrated and forms gelatin like an envelope, which is highly hydrophilic in nature. Mucilage of flax, yellow mustard basil, chia and quince hold two types of

Table 1 Combination of seaweed polysaccharide with other components and its advantages (Khalil et al. 2017)

Seaweed	Other components	Film characteristics and improvements
Algnate	Apple puree and essential oil	Exhibits antibacterial activity. No adverse effect found
	Sago starch and lemongrass oil/glycerol	The film exhibits antibacterial activity. Decrease mechanical properties and increase water vapor permeability
	Montmorillonite (MMT)	Film with low water solubility and water vapor permeability and high mechanical properties
	Silver nanoparticles/glycerol	Silver particles improve the mechanical strength and water vapor barrier properties. This film exhibits a UV screening effect and strong antimicrobial activity
Carrageenan	Grapefruit seed extract (GSE)/glycerol	Yellowish tint and great antibacterial activity film. The addition of GSE increases the moisture content, water vapor permeability and surface hydrophilicity, but decreases tensile strength and elastic modulus of film
	Clay mineral and silver particles/glycerol	The nanocomposite film improves the mechanical and water vapor barrier properties as well as antimicrobial activity
	Chitin nanofibrils (CNFs)	The film shows transparent, strong antibacterial activity and improved mechanical properties.
Agar	Starch/glycerol	The addition of starch degrades surface resistance to water wetting and mechanical properties of film
	Silver (Ag) nanoparticles	The film exhibits good mechanical stability, water vapor and gas barrier as well as strong antimicrobial activity
	Nanoclay/glycerin	Incorporation of clay (up to 10%) increases the tensile strength and decreases the water vapor permeability
	Grapefruit seed extract (GSE)	The addition of GSE increases the color, UV barrier, moisture content, water solubility and water vapor permeability, but decreases the surface hydrophobicity, tensile strength and elastic modulus of film. The film exhibits distinctive antimicrobial activity

(continued)

Table 1 (continued)

Seaweed	Other components	Film characteristics and improvements
	Banana powder and silver (Ag) nanoparticles/glycerol	The addition of banana powder increases the UV light absorption, water vapor barrier property and antioxidant activity, but decreases the mechanical properties of bilayer film. The composite film exhibits distinctive antimicrobial activity and mechanical properties
	Fish gelatin and TiO_2 nanoparticles	The addition of TiO_2 decreases water vapor permeability and increases tensile strength, UV light barrier property, swelling ratio and moisture content of film

polysaccharides, rhamnogalacturonan and hemicellulose. These polysaccharides are extracted through chemical and physical method, and the content of mucilage varies according to the species (Francoz et al. 2015; Tsai et al. 2017; Soukoulis et al. 2018). Lim et al. studied the feasibility of biopolymer extracted from garden cress seeds as coagulant aid for turbidity removal in agricultural wastewater. The results showed that the pollutant is removed through the coagulation–flocculation process, where the turbid matters settle down to remove (Lim et al. 2018).

2.2 *Proteins*

Protein polymers are the strong candidates for the formation of bio-nanocomposites. The chemical and physical characteristic of the protein depends on the presence of amino acid residues and its arrangement along with the polypeptide chain. Blends of protein with other polymer improve the film strength, flexibility and water vapor permeability, for example keratin–cellulose–polypropylene, keratin–chitosan, gluten–methylcellulose and keratin–polypropylene. Protein-based film acts as carrier materials for flavor, antioxidant and bacteriostats, which improve the food quality as well as health benefits to end users. Source of protein is soybean, corn, milk, peanut, sunflower seeds, corn, etc., which employed in the edible film showed tremendous gas barrier properties other than carbohydrates and lipids. The various films like corn zein film, gelatin film, milk protein film and silk protein film have been generated to extend the state of food products (Gupta and Nayak 2014).

Liu and his co-workers developed gluten-based film; in this film, flaxseed gum (FG), oligomeric procyanidins (OPCs) and lauric acid (LA) were incorporated to increase the mechanical and barrier properties, which are used as seasoning packaging for 75 days (Liu et al. 2018). To extend the shelf life of fresh-cut apples, phenolic antioxidant was incorporated in the soy protein-based coating, and the results demonstrated the extension of shelf life at 10 °C (Alves et al. 2017).

2.3 Lipids

Lipids are extracted from either plants or animal sources. Due to the hydrophobicity nature of the lipid, it is widely used in the food industry especially in edible film or packing material, where it gives water-resistant properties. Few disadvantages of lipids are low mechanical properties, melt in high temperature and less soluble. These limitations overcome by applying polymer materials along with the fat. Starch has a brittle and hard nature; by incorporating the fatty acid, the hardness of the polymer improves. Wax extracted from bee, candelilla, carnauba and jojoba liquid, phospholipids, mono- and diglycerides of fatty acid (oleate and linoleate), fatty acid, mineral oil, soybean oil, olive oil, thyme essential oil, palm oil, lavender oil, peppermint oil, rosemary oil, cinnamon oil, tea oil, sunflower oil, eugenol, linseed oil, vegetable oil from peanut and castor oil is being studied in various researches along with suitable polymers to deliver the nutraceutical products and as edible coatings/films in the field of food processing and preservation. Proteins and polysaccharides have poor moisture barrier; hence, the combination of lipids along with them enhances the barrier properties of the biopolymers agingist moisture, oxygen, carbon dioxide and aromas (Ali et al. 2016; Castro-rosas et al. 2016).

Certain disadvantages of biopolymer restrict their uses in industry in bulk production due to its limited mechanical and barrier properties, brittleness, low heat distortion temperature and processing cost. The reach of the biopolymers to end user has strongly limited their applications, especially in the food industry. This issue can overcome through the incorporation of nanotechnology into nanocomposites. The nano-sized composites improve all barriers of the biopolymer to withstand in the food industry market. The following chapter detailed the advantage of nanocomposites (Rhim et al. 2013).

3 Filler Components

The bio-nanocomposites with potent thermal, mechanical property, gas and moisture barrier create an interest among food industrialist. The use of nanocomposites in food packing materials protects the food and increases the shelf life in an environmentally friendly concept. Most of the conventional packing materials are fabricated with non-degradable materials, which create the environmental pollution and rise the waste issues. The property of packing materials can be enhanced by the addition of inorganic/organic particles with multiple functionalities into a biopolymer matrix

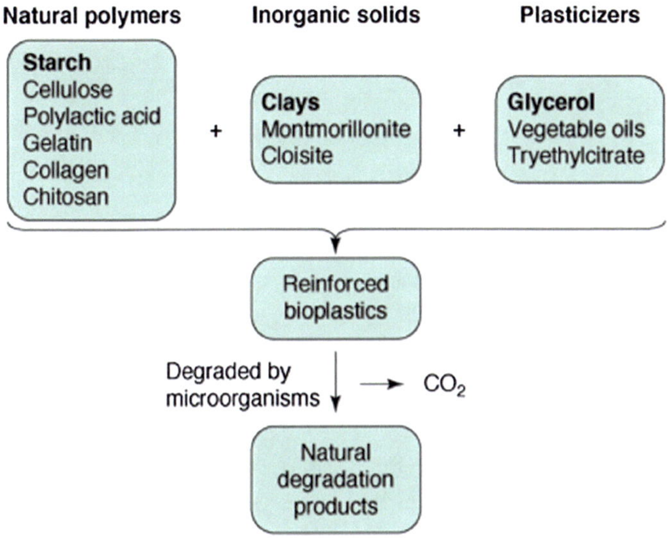

Fig. 3 Formation of biopolymer nanocomposites (Sozer and Kokini 2009)

to improve the biodegradability (Fig. 3). The fabrication of biopolymer nanocomposites depends on the physicochemical properties, concentration and the viscosity of biomaterial to develop a different form of composites like fibers, aerogels and patterned microstructure through the method of solution assembly technique. But nowadays various wet chemistry methods like layer-by-layer assembly, cast drying, electrospinning, freeze drying, ink-jet printing, mask-based micropatterns and 3D printing are used to construct polymer nanocomposites.

3.1 Nanoclays

Nanoclays have been the most used filler materials for the construction of biopolymer composites. Due to the high availability, low cost with high performance and easy processing, the incorporation of nanoclays in biopolymer research started in the 1990s. Nanoclays are also called as layered silicates. The blending of silicates and polymer results in different types of nanoscale composites, as phase-separated and intercalated or exfoliated nanocomposite (Fig. 4). Exfoliated nanocomposites improve the tensile strength of the film in food packaging.

Different shapes of mineral nanoparticles, for instance 0D nanoparticles, and 1D and 2D nanoplatelets, define reinforcing structures and are easy to construct. These mineral materials such as montmorillonite (MMT), $CaCO_3$, hydroxylapatite and boron nitride were utilized to build biopolymer nanocomposites, which mimic natural structures. The functional groups of biopolymer facilitate hydrophobic–hydrophobic

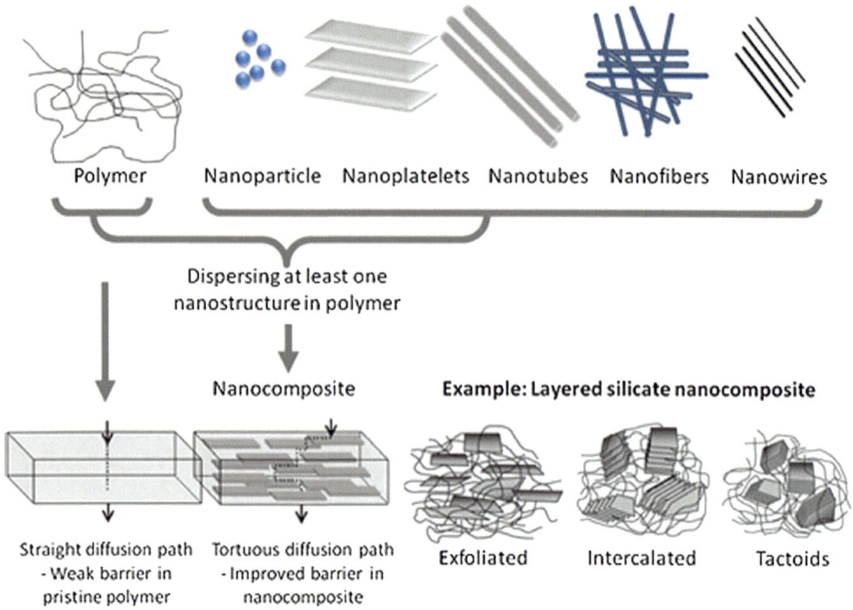

Fig. 4 Formation of composites from polymer–clay interaction (Mihindukulasuriya and Lim 2014)

pairings through multiple hydrogen bonding with clay nanoparticles. These clays can be chemically modified to develop the interfacial strength by addition of covalent bonding. Dopamine was loaded with the cellulose nanofiber surface to intensify the interaction of montmorillonite and cellulose (Fig. 5), which generate a robust mechanical property to withstand at a high humid environment and exhibit high tensile strength even at high moisture environment (Yao et al. 2017). Dispersion of organic matrices is challenging due to the hydrophilic property of the clay, which has overcome by organoclays, for example, organomontmorillonite (OMMT); the inorganic cations of MMT are replaced with the ammonium ions to enhance the interaction of MMT and polymer. The primary function of clay to incorporate with polymers is to better the mechanical strength (Azeredo et al. 2011).

Other minerals such as $CaCO_3$ and calcium phosphate infused in the polymer matrix. Protein-based polymer acts as a good template with these minerals for the formation of superior bio-nanocomposite. Calcium phosphate composites are promising source for remineralization. A study has proved the significant antibacterial effect and mechanical properties of silver nanoparticles and quaternary ammonium dimethacrylate with calcium phosphate against *Streptococcus* spp. than the conventional film (Cheng et al. 2012). Similarly, another research also proved that organically modified MMT (Cloisite 30B) with chitosan exhibits antimicrobial activity against *Listeria monocytogenes* and *Staphylococcus aureus* than standard MMT with chitosan (Fig. 6). The significant advantages of thermal, mechanical and flammability of

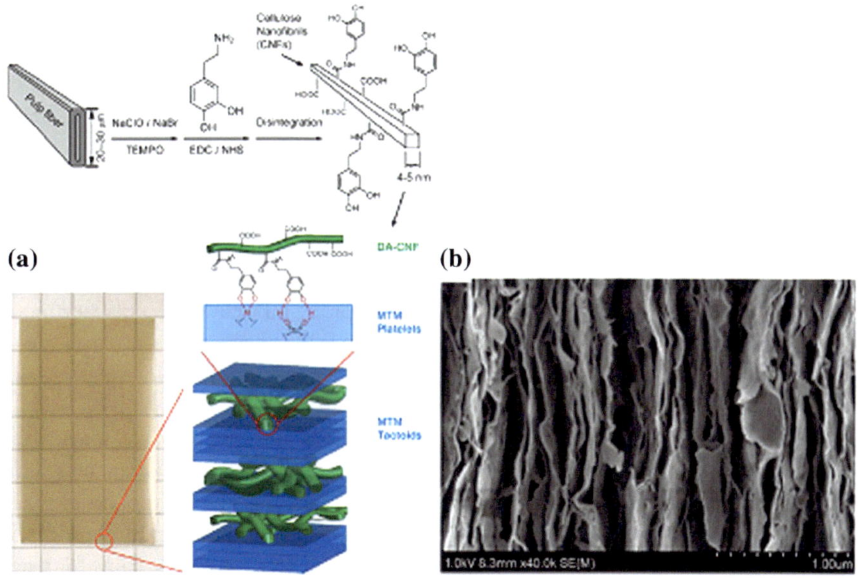

Fig. 5 **a** Dopamine-conjugated MMT and cellulose nanofibers and **b** SEM images of the conjugation. Reproduce from Yao et al. (2017) with permission. Copyright from American Chemical Society

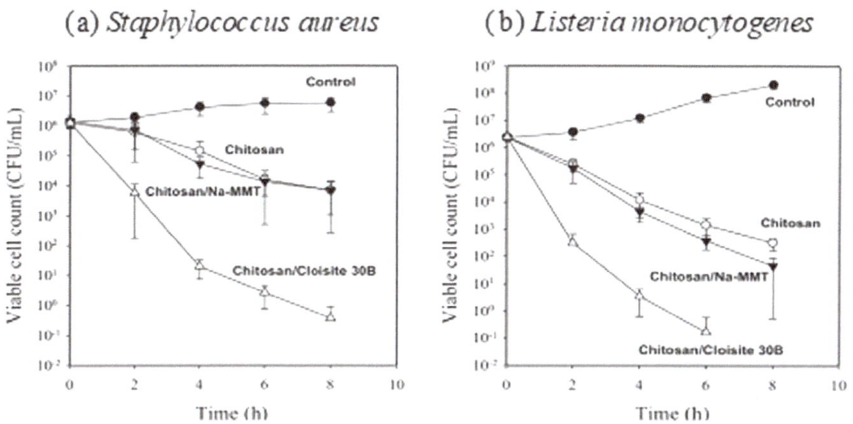

Fig. 6 Antimicrobial activity of MMT (Cloisite 30B) with chitosan against *S. aureus* and *L. monocytogenes* (Rhim et al. 2013)

kaolinite make researchers to disperse in the polymeric material. Due to the sophistication of kaolinite, which is very cheap compared to montmorillonite, it is slowly implemented in the field of food industry. The kaolinite is modified with the application of dimethyl sulfoxide, methanol and octadecylamine and intercalated with

PVC, to improve the thermal resistance, barrier properties, flame retardancy and glass transition temperature (Rehim et al. 2010; Müller et al. 2017).

3.2 Metal and Metal Oxides

Various metal nanoparticles have been infused into the polymer matrix due to their unique properties such as optical, catalytic, electrical, thermal and antimicrobial properties. Two general methods have been used for the fabrication of polymer/metal nanocomposites, (a) in situ—polymer is used as the reaction medium and (b) ex situ—particles are synthesized and incorporated in the polymer matrix, with the existence of various macromolecules along with functional groups (Fig. 7).

Metal nanoparticles like silver play a dynamic role in the innovative packaging materials along with biopolymer, which possess a potent antimicrobial agent against a broad spectrum of bacteria, fungi and viruses. Hence, silver nanoparticles are incorporated into the biopolymer film like chitosan, starch, zein to exhibit profound antimicrobial activity and extension of shelf life of the food products. The efficiency of silver-lined biopolymer depends on numerous factors like particle size, shape, particle dispersion and interaction of silver with a polymer, concentration of silver ions and particle agglomeration of silver nanoparticles (Kim et al. 2007).

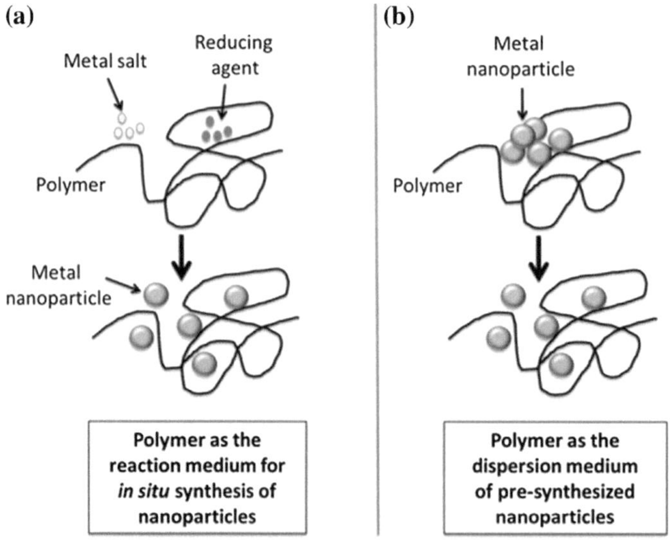

Fig. 7 Routes of polymer/metal nanocomposites: **a** in-situ route and **b** ex-situ route (Zhou et al. 2009)

High stable silver nanoparticles were synthesized in the hydrogel networks through an *in situ* method in the presence of silver nitrate along with sodium borohydride, which showed significant antimicrobial activity against both gram-positive and gram-negative microorganisms. Similarly, silver nanocomposites are synthesized in carbohydrate polymer reaction medium like starch, gum acacia and carboxymethylcellulose. All the polymer-based silver nanocomposites showed a antimicrobial trend; based on the size of the silver nanoparticles, the functional mechanism of biopolymer silver nanocomposites varies (Murthy et al. 2008; Murali Mohan et al. 2010; Palem et al. 2018).

Other metal ions like copper, TiO_2, ZnO and MgO have been used in the preparation of biopolymer nanocomposites due to their high stability and antimicrobial property. Copper is considered as toxic material due to its oxidation property which creates an obnoxious biochemical reaction with food, hence, it was not used in the food industry in the packing system. (Karlsson et al. 2013; Settanni et al. 2016). Due to specific limitation like a need for ultraviolet excitation sources, modified TiO_2 has been applied in the polymer with enhanced photocatalytic and antimicrobial actives. The multiple functionalities of modified TiO_2 along with whey protein provide protection not only against food-borne pathogens but also against staining, odor, deterioration, which have signification biodegradability and water vapor barrier. Combination of cellulose nanofibers, polylactic acid and TiO_2 acts as excellent active packaging materials against bacterial infestation (Zhou et al. 2009; El-gendy et al. 2017). Few biopolymers with inorganic nanocomposites used as an edible coating in various foods are given in Table 2.

3.3 *Natural Fibers*

Natural fibers extracted from plant and animal origin are used as the filler material. Other than this fiber, various naturally derived carbohydrate and protein derivatives also act as reinforcement residues in a biopolymer, which are described in biopolymer segment (2). Fibers are extracted from oil palm, pineapple, kenaf, jute, flax, cotton, banana, rice husk, sisal and coir. This fiber consists of cellulose, hemicellulose and lignin with minimal sugars, starch and protein, which act as reinforcing elements in the polymer matrix. Due to low cost, low density, availability, recyclability, degradability and compatibility, it is interesting to incorporate in polymer nanocomposites. The incorporated polymer showed high stiffness increasing the tensile strength of composite and improved thermal, mechanical and barrier (Valdés et al. 2014).

4 Application of Biopolymer Nanocomposites

The global food industry is now focusing on meeting the consumer's demand regarding healthy, fresh and safe food. Based on the food safety regulations, it is a great

Table 2 Biopolymeric inorganic nanocomposites in edible coating (Zambrano-Zaragoza et al. 2018)

Food/product	Nano-inorganic component	Biopolymer matrix	Functionality
Apples	Nano-SiO_x	Soy protein isolate (SPI)	Quality preservation and shelf life extender
Fresh-cut papaya and pear	Montmorillonite (MMT)	Whitemouth croaker/oregano essential oil	Antimicrobial psychotropic microorganism (fungi and yeasts) and shelf life extender
Strawberries	Montmorillonite (MMT)	Whey protein isolate (WPI)/calcium caseinate	Antifungal effect, increase storage time
Lamb meat	TiO_2	Cellulose nanofibers, WPI and rosemary essential oil	*E. coli., L. monocytogenes, S. aureus*
Ready-to-eat pomegranate	Nano-ZnO_2	Carboxymethylcellulose (CMC)	Quality preservation and shelf life extender
Citrus fruit	Silver nanoparticles (AgNPs)	*Fantasia japônica* leaf extract	*E. coli, S. aureus, Penicillium italicum*
Kinnow (*Citrus reticulata*)	Silver nanoparticles (AgNPs)	CMC/guar gum	Retention of volatile compound

challenge for the industrialist to meet all the environmental-friendly requirements. In this regard, naturally derived polymers act as the protective agent, which possess antioxidant and antimicrobial activity along with the hands of nanotechnology by adding nanoclays and nanoparticles in the polymer complex. The application of nanoscience in the polymer creates enormous attention toward food researchers. Comparatively to the macroscale and nanoscale, biopolymer nanocomposites hold a high surface-to-volume ratio and surface activity with improved biodegradability and biocompatibility. Various applications of biopolymer nanocomposites in the food industry are explained below.

4.1 Active Packaging

The concept of active packaging is to incorporate the active compounds into the packing material to enhance the protection function. The active compounds act as an antimicrobials, oxygen absorbers, water vapor absorbers and ethylene removers (Fig. 8). Hence, active components are assembled together using nanotechnology to

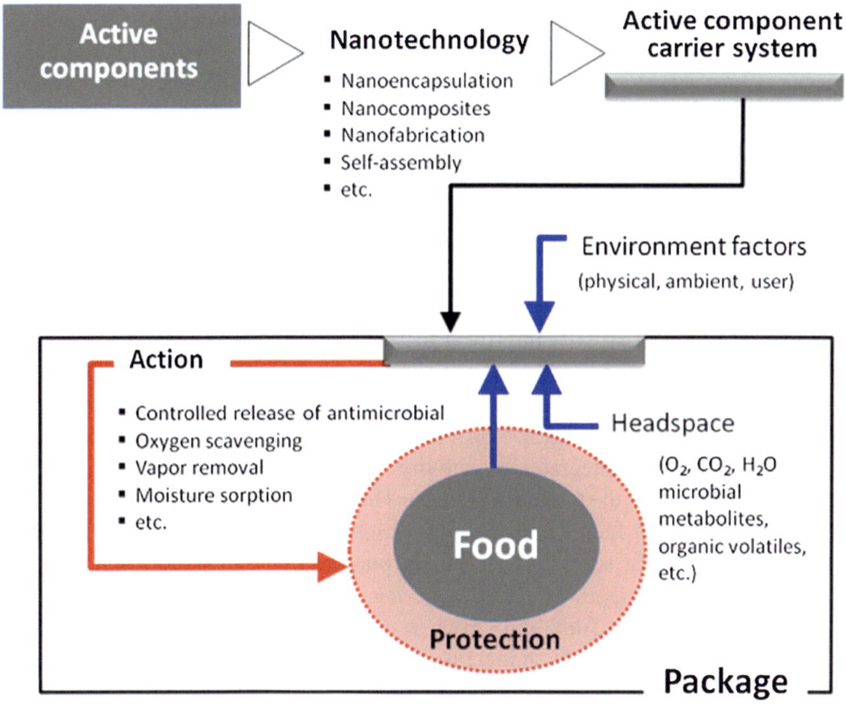

Fig. 8 Concept of active packaging (Mihindukulasuriya and Lim 2014)

create an active food packing material.

Metal nanoparticles such as gold (Au), silver (Ag) and zinc (Zn) have been widely used in the active packaging. Nano-sized particles with higher surface in the packaging materials prevent the proliferation of microorganism on the surface of the food materials. Among the metal nanoparticles, silver has been commonly used for active packaging, due to its bactericidal effect against both gram-positive and gram-negative microorganisms. The silver ions penetrate into inner membranes and disturb the polysaccharides and protein component present in the membrane; it also inhibits the respiratory chain enzymes to reduce the ARP levels, which are required for the production of nucleic acid (Bumbudsanpharoke et al. 2015). The use of nanoclays (montmorillonite) and nanosilver as additives for polymer matrix (polypropylene) film showed signification antibacterial properties against *Escherichia coli* and *S. aureus* bacteria (Oliani et al. 2016).

Different concentrations of Ag nanoparticles were added with low-density polyethylene and manufactured via extrusion method. The microbial quality of the film was evaluated for chicken breast fillets, the results showed the significant microbial control and lengthen the shelf life of the fillets, and it also slows down the oxidation process of chicken (Azlin-Hasim et al. 2015). Wheat gluten bio-nanocomposite embedded with 7.5% of nano-cellulose crystal and 0.6% TiO_2

enhanced both mechanical and water barrier properties, and also reduced about 98.5% of microbial count (*Saccharomyces cerevisiae*, *E. coli* and *S. aureus*) (El-Wakil et al. 2015).

Bacterial cellulose is impregnated with nisin to develop a packaging film against *L. monocytogenes* and total aerobic bacteria. The microbial count was under control when compared with standard cellulose film. Antibacterial sausage casings were further enhanced with addition of polylysine and cellulose, against *S. aureus*, with significant improvement in its heat resistance, tensile and barrier properties (Azeredo et al. 2017). Methylcellulose films with Ag nanoparticles of *Lippia alba* extract showed excellent antioxidant and antimicrobial property against *S. aureus* and *E. coli*. The incorporation of AgNPs improved the MC film elasticity and hydrophobicity (Nunes et al. 2018).

Tomatoes are packaged in chitosan film containing nano-sized titanium dioxide to extend the quality and storage period. The coated tomatoes are exposed to UV light for ethylene photodegradation activity which delays the ripening process and changes the quality of the tomatoes (Fig. 9). The large surface area of the nano-sized TiO_2 is an important factor in achieving a high efficiency by the higher hydroxyl groups adsorbed on the surface of TiO_2 in the photocatalytic oxidation of the ethylene (Kaewklin et al. 2018).

In the chitosan-based film, lipids are used as a plasticizer to improve the mechanical properties of the film which increase the flexibility and decrease the brittleness. Hence, various natural sources of lipid-based oil like olive, grape, sunflower and soybean are incorporated to enhance the biological properties of the film. Improved

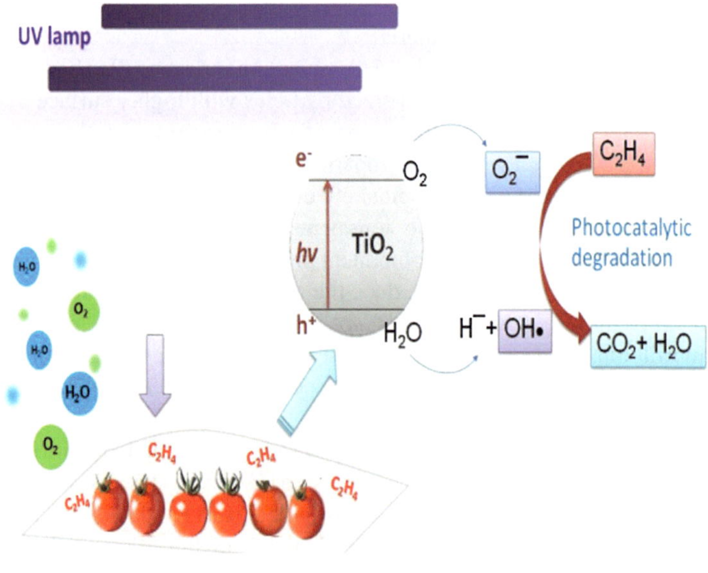

Fig. 9 Chitosan–TiO_2 tomatoes (Kaewklin et al. 2018)

antioxidant property and tensile strength found in grape seed oil integrated chitosan film (Nunes et al. 2018). Biological molecules are currently being evaluated for their anti-biofilm activity. Chitosan–streptomycin conjugates/gold nanoparticles were evaluated for their anti-biofilm properties against gram negative (*P. aeruginosa* and *S. typhimurium*) and gram positive (*L. monocytogenes* and *S. aureus*). The chitosan–streptomycin gold nanoparticles showed a disruption effect on biofilms formed by gram-negative or gram-positive bacteria, and an inhibition effect on the biofilm formation of gram-negative bacteria. Incorporation of streptomycin to chitosan and gold nanoparticles facilitated its penetration into the biofilm matrix and improved the contact with the bacterial surface, enhancing its bactericidal effect (Mu et al. 2016). Oregano essential oil due its high antioxidant and antimicrobial activity was integrated with the fish gelatin/chitosan nanoparticles, which act as an antimicrobial active packaging against *S. aureus, L. monocytogenes, S.enteritidis* and *E. coli* (Perinelli et al. 2018).

Kombucha tea is a beverage prepared by the fermentation of sugared tea with a symbiosis of *Acetobacter*, including *Acetobacter xylinum*, *Acetobacter xylinoides*, *Bacterium gluconicum*, *Acetobacter aceti* and *Acetobacter pasteurianus* and various yeasts, such as the *Schizosaccharomyces pombe*, *Saccharomycodes ludwigii*, *Kloeckera apiculata*, *S. cerevisiae*, *Zygosaccharomyces bailii*, *Brettanomyces bruxellensis*, *Brettanomyces lambicus*, *Brettanomyces custersii*, *Candida* and *Pichia species*. The presence of various phenolic compounds in tea such as catechin, epicatechin, epicatechin gallate, gallocatechin, epigallocatechin and epigallocatechin gallate and groups of vitamin B, makes them superior antimicrobial, antioxidant, anti-carcinogenic and antidiabetic agent. For eaxmple, incorporation of Kombucha tea in chitosan film enhanced the antimicrobial activity of the biopolymer film against *E. coli* and *S. aureus* in the wraped minced beef (Ashrafi et al. 2018). Gold nanoparticles were synthesized and integrated into lignocellulose composite papers used for the production of active packaging to preserve the foods. The nanoparticles are directly synthesized from soft unbleached wool, which showed an exceptional antioxidant performance of about 98% activity (Bumbudsanpharoke and Ko 2018).

Silicon oxide (SiO_2) has high mechanical and barrier properties, which is added as fillers in food packing materials to help retard the penetration of O_2, other gases and loss of moisture. The combination of silica nanoparticles and polydimethylsiloxane showed strong water repellent characteristics. Other metal nanoparticles like zinc oxide have good antibacterial activity. Therefore, polybutylene succinate (PBS) and zinc oxide are used to develop composite film against *E. coli* and *S. aureus*. Similarly, essential oils were incorporated along with polybutylene succinate (PBS) and zinc oxide to improve odor and bacterial activity (Sharma et al. 2017).

4.2 Smart Packing/Intelligent Packaging

Nanodevices or nanosensors in a combination of polymers are implemented to monitor food-borne pathogens and chemical reaction during the transport and storage

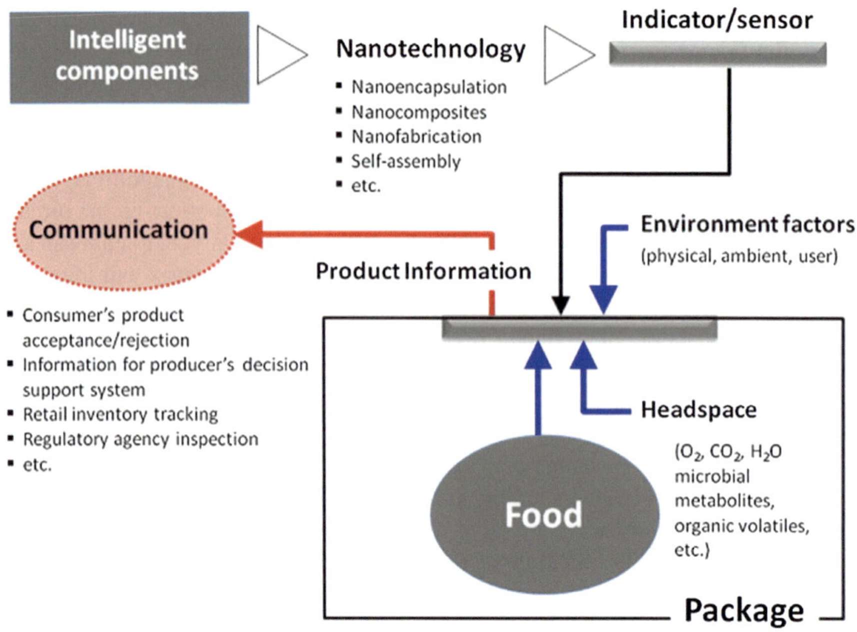

Fig. 10 Concept of intelligent packaging (Mihindukulasuriya and Lim 2014)

period. This method comes with smart packaging, to ensure the integrity and authenticity of the food product. The nanosensor device can be able to track the whole storage history of the product including temperature, packed and expiry period to monitor and prevent any food-borne diseases (Fig. 10). Various researches have proven that nanosensors can be able to detect toxins produced by food-borne pathogens, food composition and flavor analysis. Molecularly imprinted polymers with nanoparticles have been developed to identify small molecules, a large protein and macromolecules and to determine the butylhydroquinone and trypsin, glucose, catechol and ascorbic acid in food products (Pathakoti et al. 2017).

(a) **pH indicators**

Fish and meat products are highly perishable when compared with other food products. pH plays a essential role in the food spoilage. Various chemical seasons have been developed to find the changes of pH, which act as the indicator for microbial growth. In this regard, Mustapha et al. produced a hydroxyethylcellulose/polyaniline (HEC/PANI) film as a pH indicator and sensor, which indicate the color changes in the film as the pH varies. Similarly, the release of volatile amines from food stuffs was observed by changes in the pH, where the color changes from green to blue which can be observed through the naked eyes (Fig. 11) (Mohebi and Marquez 2015; Zambrano-Zaragoza et al. 2018). Several conjugated polymer nanocomposites are used as sensors in food spoilage detection though colorimetric technique, which is a simple and low-cost technique applicable in the food industry to meet customers'

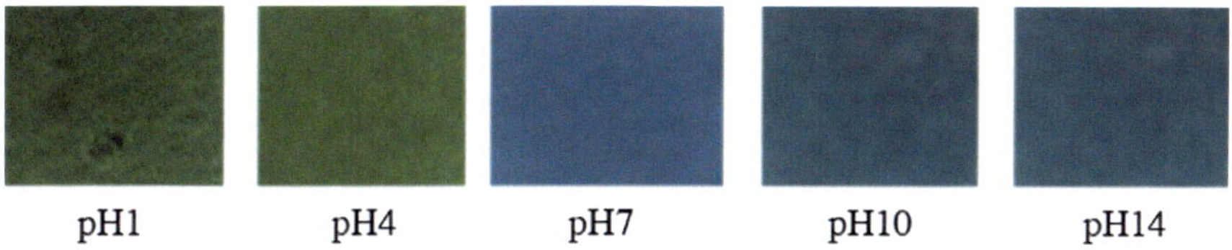

Fig. 11 Color changes of HEC/PANI film as the pH varies (Zambrano-Zaragoza et al. 2018)

satisfaction, and the wide range of applications are brief in Fig. 12 (Pavase et al. 2018).

(b) **Detection of spoilage and pathogenic microorganism**

Polypyrrole and polyaniline are the excellent sources of conducting polymer, which is utilized to detect the food-borne microorganism (*L. monocytogenes, Bacillus cereus* and *E. coli*) with the biomolecules of antibody and DNA. These polymer-based

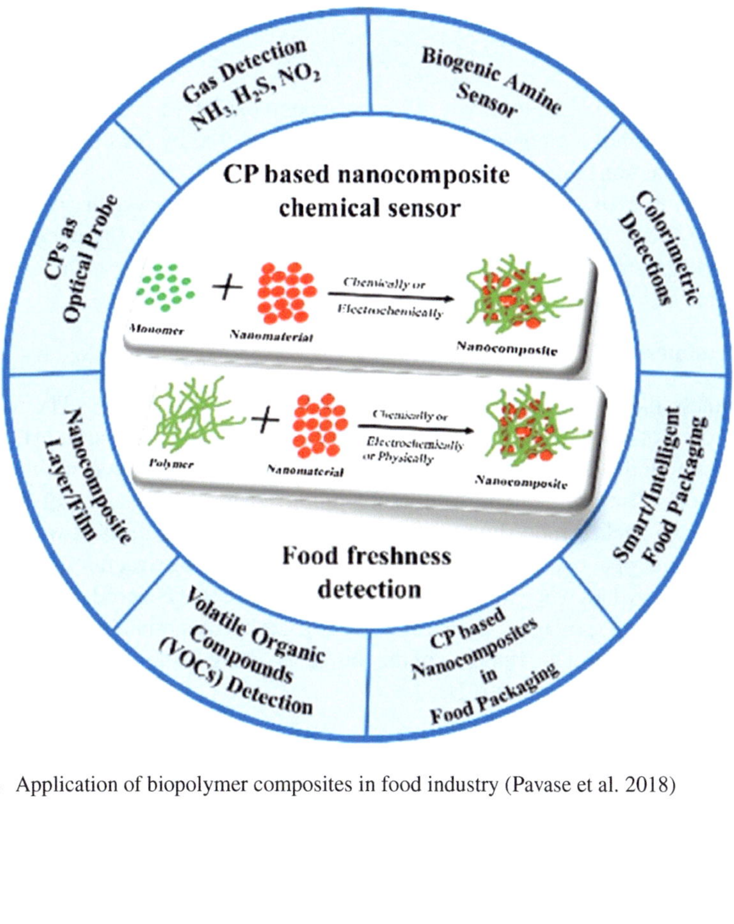

Fig. 12 Application of biopolymer composites in food industry (Pavase et al. 2018)

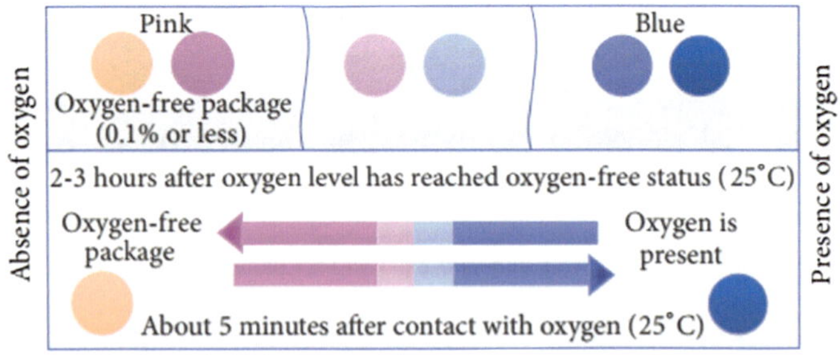

Fig. 13 Gas indicators (Fuertes et al. 2016)

biosensors are rapid and sensitive, which is making a promising effect in the identification of pathogens in the food chain supply (Arshak et al. 2009). Grain quality monitoring nanosensors were developed using the conducting polymer by the researches of Canadian Wheat Board Centre for Grain Storage Research, University of Manitoba, Canada. The sensor accurately detected the presence of insects and fungus inside the stored grains (Neethirajan et al. 2007).

(c) **Gas indicators**

Gas indicators are useful to monitor the toxic composition of gases generated from food decomposing in a container, which can be measured by the color through the enzymatic reaction (Fig. 13).

UV-activated oxygen was developed to determine the oxygen in the modified atmospheric packing. TiO_2 nanoparticles, glycerol and methylene blue were dispersed and spun to develop a membrane to detect the presence of oxygen (Mihindukulasuriya and Lim 2013).

(d) **Freshness indicators**

A freshness indicator helps to determine the quality of the product. The indicator works based on the changes in the composition of chemical substances, pH, gas and other compounds like volatile nitrogen compounds, amines, organic acids, ethanol, glucose, or sulfur compounds, which are produced by the presence of microbial growth in the packed food (Fuertes et al. 2016). The spoilage of meat was detected by a silver or copper coating plastic film or as paper packaging structures. The coating on the film turns dark when it reacts with sulfide volatiles that are produced from fresh meats undergoing spoilage (Sharma et al. 2017). Carbon nanotube-based sensors are implemented to detect the ripening of the tomatoes owing to the emission of ethylene from tomatoes (Esser et al. 2012).

(e) **Pesticide and adulterant detectors**

The present world requires rapid, accurate, simple, low-cost material to detect the food adulterants like benzoic acid and pesticides which cause the severe illness and create major life-killing disease like cancer (Table 3) (Silva et al. 2013). Biopolymer

Table 3 Nanocomposites as sensors (Silva et al. 2013)

Biological component	Analyte
Oxalate oxidase immobilized on chitosan	Oxalate (anti-nutrients)
Acetylcholinesterase immobilized on multiwalled carbon nanotubes	Carbaryl (pesticides)
Acetylcholinesterase–choline oxidase on a gold–platinum bimetallic nanoparticles	Paraoxon ethyl, aldicarb and sarin (pesticides)
Acetylcholinesterase immobilized on polyaniline and multiwalled carbon nanotubes	Carbaryl and methomyl (pesticides)
Mushroom tissue homogenate/enzyme inhibition	Benzoate/benzoic acid
PPO gel made of titanium modified with carbon nanotubes/enzyme inhibition	Benzoate/benzoic acid

nanocomposites act as a sensor to detect the culprits hidden in the foodstuff. Every day, researches are focusing on developing a technique to monitor or identify the contaminators in food to protect both human and environmental health. Malathion is an organophosphorous-type pesticide, which is highly toxic to human. Gold nanoparticles–chitosan–ionic liquid hybrid nanocomposite senses the malathion by voltammetry measurement in apples and tomatoes (Bolat and Abaci 2018). Cost-effective plant esterase–chitosan/gold nanoparticles–graphene nanosheet (PLaE-CS/AuNPs-GNs) biosensor is developed for the sensitive detection of methyl parathion and malathion in carrot and apples (Bao et al. 2015).

4.3 *Edible Film/Coatings*

Edible film and coating are also called primary packaging, composed of consumable components. The food material is coated or wraps with a film to protect the ingredients or food materials without changing the physical or chemical property of food. The edible film has various improved properties than the conventional polymer film, like potent moisture and gas barriers, prevention of microbial infestation, improved mechanical strength and sensory perceptions and increases the shelf life of the food materials for a more extended period. Recent studies showed the incorporation of bioactive component, nutritional supplements such as vitamins, in the film which increases the health benefits.

Edible films are used on the food product with about 50–250-μm thickness as pouches or wrap. Three main ingredients such as polysaccharides, proteins and lipids are applied to produce edible films. Polysaccharides like carrageenan or chitosan possess strong oxygen barrier and hydrophilic nature, which provides strong bonding with food additives like color, flavor and minerals. Protein edible film with good mechanical strength decreases the physical injuries happened during handling and transportation. Due to the hydrophilic properties, it has a low moisture barrier, which can be overcome by a combination of lipid. The hydrophobic nature of lipid acts as an excellent moisture barrier. The blend of natural polymers in preparation for edible film will overcome the disadvantages of the individual negative impact and increase the moisture, oxygen, gases, fats and oils barriers (Al et al. 2018).

The edible films (Table 4) have been commercially applied to protect meat and poultry product to prevent and protect the product from microbial invasion (Quintavalla and Vicini 2002). This edible coating holds antimicrobial agents and acts as an antioxidant, which reduces moisture loss, rancidity, discoloration occurred by myoglobin oxidation and heat resistance. The microorganism on the surface of the meat is controlled by the edible coating, which can be consumed along the meat, creating the attention for safety and edibility to incorporate antimicrobial agents.

Conventionally, edible films are flavorless, colorless and transparent, but the trend is slowly changing, and several researchers have been carried out to prepare an edible film from fruits and vegetable purees, in combination with film-forming polymers. Consumption of edible film along with nutrient, bioactive component and antioxidant enriched to improve the health benefits as well as the shelf life of the product. The biomass generated in the food industry is rich in dietary fibers and phytochemicals, which also subjected to edible film preparation along with natural plasticizers. Mono-, di- and oligosaccharides present in vegetables and fruits act as plasticizers other than natural resources like gelatin, corn starch, glycerol, fructose–glucose syrup and honey to increase the flexibility, stretchability and toughness of edible film. Functional additives like secondary metabolites like essential oil of garlic, cinnamon, lemongrass and clove used are implemented to increase the antioxidant and antimicrobial property (Pathakoti et al. 2017).

The edible coating is a thin layer coating of selected formulations of edible materials on the surface of the food materials applied by various techniques like dipping, spraying and spreading. FDA approved food grade ingredients which are generally recognized as safe (GRAS) to use in the food industry are used to form thin layer coating of the films with the help of plasticizers. The presence of high water activity in the fresh fruits and vegetables promotes the fast degradation like enzymatic browning, off-flavor development and microbial contamination. Undesirable properties of food products are disliked by the consumers or end users, which overcome through the application of edible coating research. Various researches have focused on using edible coating on a wide range of fruits and vegetables to retain the natural flavor, taste, color and nutrient and to prevent microbial infestation (Table 4; Fig. 14) (Brychcy-rajska 2017; Valdés et al. 2017; Zambrano-Zaragoza et al. 2018). Inorganic compounds also have a vital role in the edible coating, in minimal proportion along with the polysaccharides, protein and lipid polymers, which have excellent antimicrobial properties and retain flavor and nutrient, shown in Table 5.

Probiotics are live microorganisms that confer a health benefit to the host when administered in permissible amount. Recently, probiotics have been incorporated into biopolymeric matrix to establish an active food packaging as an alternative method for controlling non-degradable polymer, food-borne microorganisms, improving and enhancing health benefits. The organism is encapsulated by various techniques, electrospinning/electrospray, impinging aerosol, micro-nozzle array and extraction, and

Table 4 Applications of edible food packaging with antimicrobial agents (Quintavalla and Vicini 2002; Valdés et al. 2017)

Food	Product	Matrix	Antimicrobial agent
Fish	Sliced fresh Channa argus	Chitosan and polyethyleneimine	Thyme essential oil
	Fish sausages	Chitosan and warm water fish gelatin	Shrimp concentrate from Litopenaeus vannamei cooking juice
	Trout fillets	Chitosan	Lactoperoxidase enzyme
		Whey protein	Lactoperoxidase enzyme
	Fresh Indian salmon	Gelatin from waste of *Nemipterus japonicus*	Garlic and lime
	Silvery pomfret	Chitosan	Gallic acid
	Gilthead seabream fillets	Methylcellulose	*Satureja thymbra* (L.) essential oil
	Fresh silver carp fillets	Methylcellulose	*Pimpinella affinis* essential oil
	Surimi	Zein and alginate	Iron chelator and glucose oxidase
Meat	Cooked cured chicken breasts, fresh chicken breasts	k-carrageenan and chitosan	Mustard extract
	Pork	Sodium alginate	Thyme and propionic acid
		Oleic acid as part of starch	Lactic acid, nisin and lauric arginate
		Chitosan	Clove oil and/or ethylenediaminetetraacetate
	Dry-cured ham	Propylene glycol, xanthan gum and carrageenan with propylene glycol alginate	–
	Ham	Soybean meal and xanthan	Lactoperoxidase
	Frankfurters and ham	Sodium alginate	Ethanol
	Roast beef	Chitosan	Lauric arginate ester, lactic and levulinic acids
	Rucola	Modified chitosan	Lemon, mandarin, oregano or clove essential oils
	Bologna, cooked ham, pastrami	Chitosan	Acetic acid
	Lean beef muscle	Alginate	Lactic acid
	Bologna, cooked ham, pastrami	Chitosan	Propionic acid

(continued)

Table 4 (continued)

Food	Product	Matrix	Antimicrobial agent
	Chicken breast	Starch/glycerol	Potassium sorbate
	Beef tissue	Silicon coating	Nisin
	Cooked meats	Cellulose	Pediocin

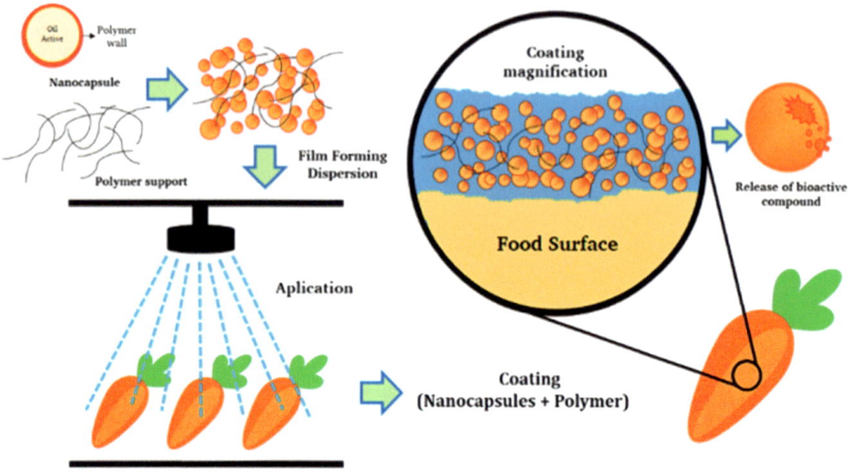

Fig. 14 Polymer nanoparticles in edible coating (Zambrano-Zaragoza et al. 2018)

combined in the active packaging to replace the antibiotics or additives (Espitia et al. 2016). Some probiotic organism used in the edible film is given in Table 6.

4.4 Biopolymer Nanocomposites in Nutraceutical Delivery

Nutraceuticals are nutritional substances isolated from natural food sources which are not just used for the nutritional purpose but also used as medicine. It may include one or more of the following dietary ingredients: a mineral, a vitamin, an amino acid, antioxidants, plant metabolites and combination of these ingredients (Nasri et al. 2014). Some popular nutraceutical components are β-carotene, curcumin, resveratrol, ginseng, *Echinacea*, green tea, glucosamine, omega-3, lutein, folic acid, cod liver oil, etc. Generally, most of the nutraceutical products possess multiple therapeutic properties including anticancer, antithyroid, antioxidant, anti-inflammatory, antimicrobial, cardio protective and many more. However, poor stability, lack of water solubility and poor bioavailability hinder its incorporation and supplementation through processed foods and its use as nutraceutical ingredients. So, encapsulation of nutraceutical molecules in micro- and nano-size is in practice to offer

Table 5 Antimicrobial coatings developed for fruits and vegetables (Valdés et al. 2017)

Food applicability	Product	Matrix	Antimicrobial agent
Fruits	Watermelon	Sodium alginate, pectin and calcium lactate	Trans-cinnamaldehyde
	Cantaloupe melon	Chitosan and pectin	Trans-cinnamaldehyde
	Strawberries	Pectin, pullulan and chitosan	Sodium benzoate and potassium sorbate
	Raspberries	Pectin and calcium chloride	Eugenol, citral, ascorbic acid
	Pineapple	Sodium alginate	Lemongrass essential oil
	Fuji apples	Sodium alginate	Lemongrass essential oil
	Blueberry	Chitosan	Carvacrol, cinnamaldehyde, trans-cinnamaldehyde
	Avocado	Gum arabic, aloe vera and chitosan	Thyme oil
Vegetables	Pepper	Chitosan	Lemongrass essential oil
		Pullulan	Leather bergenia leaves ethanolic extracts
	Pumpkin	Xanthan gum, guar and chitosan	–
		Starch	Carvacrol
		Zein	Benzoic acid
	Cherry tomatoes	Zein	Cinnamon, mustard essential oil, commercial wax
	Fresh-cut broccoli	Chitosan	Bioactive compounds and essential oils
	Cauliflower florets	Maltodextrins and methylcellulose	Lactic acid, citrus extract, lemongrass essential oil
	Green beans	Modified chitosan	Mandarin essential oil
	Rucola	Modified chitosan	Lemon, mandarin, oregano or clove essential oils

the following advantages: (i) to protect the nutraceutical compound from undesired interactions with the environment and thus protect its degradation during processing or storage; (ii) mask undesirable tastes; (iii) improve its solubility and dispersibility and thus facilitate its processability; (iv) control and/or prolong its release; and (v) improve its oral bioavailability (Penalva et al. 2014). For this purpose, wide ranges of synthetic and natural polymers have been used to encapsulate nutraceutical compound in micro- and nano-size particles. Owing to its biocompatibility and biodegradability, biopolymer-based carrier systems are preferred for nutraceutical

Table 6 Probiotic incorporated edible biopolymer (Pavli et al. 2018)

Biopolymer material	Probiotics	Additives	Application
Alginate/gellan	*B. lactis Bb-12*	Glycerol	Fresh-cut apples and papayas
Sodium caseinate	*L. sakei*	Sorbitol	Fresh beef
Alginate/starch	*C. maltaromaticum*	Glycerol	Smoked salmon
Starch	*L. acidophilus*	–	Bread
Agar	*B. animalis spp. lactis* and *L. paracasei spp. paracasei*	Green tea extract, glycerol	Hake fillets

and drug delivery than synthetic polymers. Range of biopolymers including proteins from plant source like soy protein isolate, zein, gliadin, hordein and kafirin and from animal source like whey, egg albumen, collagen, gelatin, silk fibroin, and polysaccharides from plant source (like cellulose, starch, pectin, guar gum, xylan), animal source (like chitosan, chondroitin, heparin, hyaluronic acid) and microbial source (xanthan, dextran, pullulan, cyclodextrin, alginate, carrageenan) are studied for their efficacy in nutraceutical delivery. In the past decade, biopolymer-based nanocarrier systems attained high attention and proved its efficacy for nutraceutical and drug delivery applications (Table 7).

For example, curcumin one of the predominantly used nutraceutical molecules has been effectively encapsulated in various biopolymers to enhance its solubility, stability, processability and bioavailability. Zein, a water-insoluble storage protein from corn was used as potential biopolymeric matrix to encapsulate curcumin through anti-solvent method and produced as colloidal nanoparticles with average particle size of 100–150 nm. These particles were further stabilized with sodium caseinate. Figure 15 shows the pristine curcumin and nano-encapsulated zein–curcumin composite colloidal particles prepared at zein: curcumin ratio of 5:1 and stabilized with 2% w/v sodium caseinate under stimulated pH, gastro intestinal and UV irradiation. The results show that the encapsulation in biopolymer matrix increases the pH stability of curcumin in both acidic and basic conditions and UV stability (Fig. 15a, b). Further upon stimulated gastro intestinal conditions, more than 60% curcumin retained at the end of 120 min (Fig. 15c) and exhibited excellent mucoadhesion in Caco-2 cells (Patel et al. 2010).

Similarly, Pan et al. (2013) encapsulated curcumin in casein nanoparticles through spray-drying technique. Encapsulation of curcumin in casein nanoparticles through hydrophobic interaction increased its dispersibility by four times and thermal stability up to 250 °C and exhibited better antioxidant and bioactivity than pristine curcumin. Curcumin encapsulation in zein nanoparticles through electrospraying process improved its dispersibility in aqueous food matrix and semi-skimmed milk in addition to the storage stability up to three months of storage period at 23 °C and 43% relating humidity (Gomez-Estaca et al. 2012). Curcumin encapsulation in zein nanofibers through electrospinning process released curcumin at sustained rate,

Table 7 Usage of biopolymer to encapsulate nutraceutical compounds and benefits attained

Nutraceutical	Biopolymer carrier	Size and shape	Advantage/applications	References
Folic acid	Casein	150 nm	Better gastro resistance and released only in intestinal condition, thus improving the bioavailability	Penalva et al. (2014)
Resveratrol	Zein: pectin core–shell	235 nm	Higher antioxidant and anticancer activity and anti-proliferation of hepatocarcinoma Bel-7402 cells	Huang et al. (2017)
Resveratrol	Zein core–caseinate or caseinate–dextran shell	–	UV protection; shell coating provides stability under gastrointestinal condition and improved bioavailability	Davidov-Pardo et al. (2015)
β-carotene	Zein	80-130 nm	Improvement in oral bioavailability and higher cytotoxicity and apoptotic activity in MCF-7 cells	Jain et al. (2018)
β-carotene	Whey protein capsules	<100 nm 3.0	Increased stability of the WPC capsules containing β-carotene against photooxidation	López-rubio and Lagaron (2012)
β-carotene	Zein and whey protein	~ 800 nm	Higher bioaccessibility than pristine compound	Perez-masi and Laura (2017)
α-linolenic acid	Gelatin, whey protein concentrate (WPC) and a soy protein isolate (SPI)	–	Improved thermal stability even after 27 h of thermal treatment	Gómez-mascaraque and López-rubio (2016)

(continued)

Table 7 (continued)

Nutraceutical	Biopolymer carrier	Size and shape	Advantage/applications	References
Tea catechins	Chitosan–tripolyphosphate	128 nm	Controlled release of catechins	Nasri et al. (2014)
Moringa oleifera extract	Gelatin	Nanofibers	Enhanced thermal stability and antioxidant activity	Hani et al. (2017)
Alpha-tocopherol stability	Soluble dietary fiber (SDF)	Nanofibers	Increased water solubility	Nasri et al. (2014)
Alpha-tocopherol	Whey protein isolate (WPI), zein and soy protein isolate (SPI)	Nanoparticles and nanofibers	Controlled release of bioactive food packaging application	Fabra et al. (2016)
Quercetin	Zein	Nanoribbons	Sustained release	Li et al. (2014)
Folic acid	Amaranth protein isolate: pullulan	Nanofibers	Improved thermal stability and UV stability with no degradation even after 2 h of UV exposure	Aceituno-medina et al. (2015)
DHA	Zein	190-nm nanofibers	Greater oxidative stability	Moomand and Lim (2014)
Fish oil	Chitosan and low methoxyl pectin solution	Nano-emulsion	Increase the digestibility as well as permeation through skin	Jung et al. (2016)

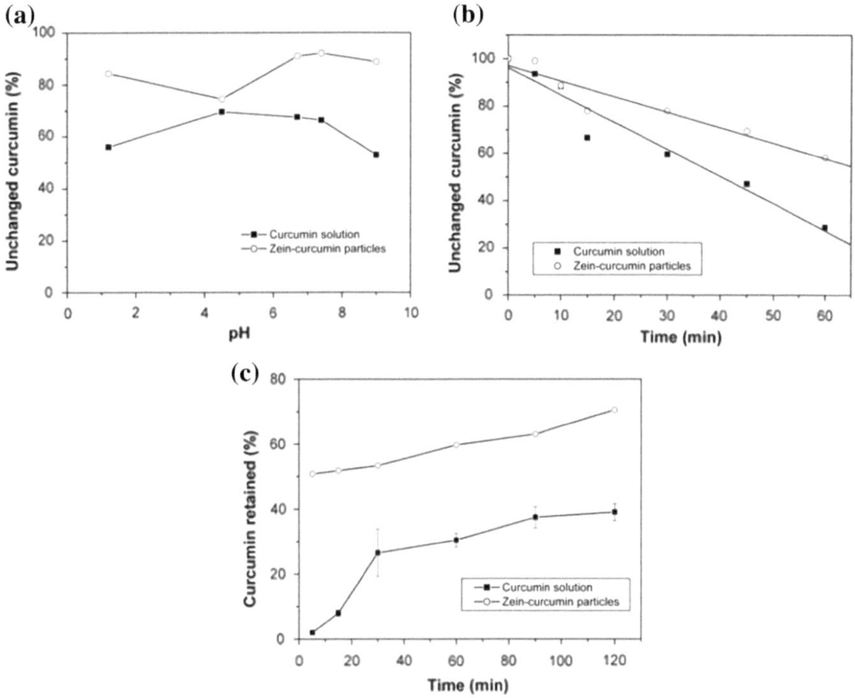

Fig. 15 Pristine curcumin and nano-encapsulated curcumin: **a** pH stability; **b** percent unchanged curcumin during UV irradiation as a function of time; and **c** detected in media during gastric and intestinal incubation (Patel et al. 2010)

supported growth of fibroblast and proved its efficacy for drug delivery and scaffold for soft tissue engineering (Brahatheeswaran et al. 2012).

Chitosan and gelatin biopolymers were also used to encapsulate curcumin, were proved to improve its solubility up to 20 mg/ml and released curcumin at sustained rate (Farnia et al. 2016). Another biopolymer isolated from Bombyx mori silkworm, the silk fibroin, is known for its biocompatibility, biodegradability, amphiphilic chemistry and excellent mechanical properties. Using physical adsorption and co-precipitation (Montalbán et al. 2018), encapsulated curcumin in silk fibroin is used as effective vehicle for curcumin delivery. The curcumin-encapsulated silk nanoparticles with mead diameter of approximately 170 nm showed higher cytotoxicity to carcinogenic cells like neuroblastoma cells than hepatocarcinoma cells, while not decreasing viability of healthy cells thus proving the compatibility of biopolymer carriers for nutraceutical molecule.

Recent studies report the co-delivery of more than one nutraceutical component through nanocarrier system and enhanced health benefits. Baspinar et al. (2018) reported co-delivery of curcumin and piperine through zein–chitosan core–shell nanoparticles with mean size of 550 nm produced by electrospraying process.

The co-administered curcumin and piperine showed higher cytotoxicity against neuroblastoma cells.

Apart from simple encapsulation matrix and carrier system, biopolymer-based nanocarriers have also been designed for targeted delivery as well as for bioimaging. For example (Aswathy et al. 2012), fabricated zein nanoparticle is conjugated with ZnS: Mn quantum dots for delivery of drug 5-fluorouracil. The fluorescent nanoparticles attached to zein nanocarriers offered simultaneous imaging of cancer cells, while zein matrix released anticancer drug at controlled rate.

5 Conclusion

Biopolymer-based biodegradable food packaging materials could potentially replace non-biodegradable plastics. Integration of nanotechnology, along with polymer, improved their mechanical and barrier properties, and created a new path toward biodegradability, which reduces the wastage and acts environmentally friendly. Only, a few bio-nanocomposite products are commercialized but most of the polymer nanocomposites are still under research stage. The commercialization of this technology needs advancement in bulk production in the processing sector. This challenge should be addressed to improve the feasibility of bio-nanocomposite in the commercial market, which is nontoxic, eco-friendly and substantially low cost. This biopolymer nanocomposite can assist in the detection, monitoring, tracking and recording throughout the supply chain, which helps to channelize the food security.

References

Aceituno-medina M et al (2015) Photoprotection of folic acid upon encapsulation in food-grade amaranth (*Amaranthus hypochondriacus* L.) protein isolate—pullulan electrospun fibers. LWT—Food Sci Technol. https://doi.org/10.1016/j.lwt.2015.02.025

Al G et al (2018) Preparation and characterization of biopolymer nanocomposites from cellulose nanofibrils and nanoclays. J Compos Mater 52(5):689–700. https://doi.org/10.1177/0021998317713589

Ali M, Rehman S, Zuber M (2016) Lipid functionalized biopolymers: a review. Int J Biol Macromol. https://doi.org/10.1016/j.ijbiomac.2016.09.071

Alves MM, Gonçalves MP, Rocha CMR (2017) Effect of ferulic acid on the performance of soy protein isolate-based edible coatings applied to fresh-cut apples. LWT—Food Sci Technol 80:409–415. https://doi.org/10.1016/j.lwt.2017.03.013

Anandharamakrishnan C, Ishwarya PI (2015) Spray drying technique for food ingredient encapsulation. Wiley

Arshak K et al (2009) Conducting polymers and their applications to biosensors: emphasizing on foodborne pathogen detection. IEEE Sens J 9(12):1942–1951

Ashrafi A, Jokar M, Mohammadi A (2018) Preparation and characterization of biocomposite film based on chitosan and kombucha tea as active food packaging. Int J Biol Macromol 108:444–454. https://doi.org/10.1016/j.ijbiomac.2017.12.028

Aswathy RG et al (2012) 'Biocompatible fluorescent zein nanoparticles for simultaneous bioimaging and drug delivery application. Adv Nat Sci: Nanosci Nanotechnol 3(2). https://doi.org/10.1088/2043-6262/3/2/025006

Azeredo H, Mattoso L, McHugh T (2011) Nanocomposites in food packaging—a review. Adv Diverse Ind Appl Nanocomposites 57–78. https://doi.org/10.5772/1931

Azeredo HMC, Rosa MF, Mattoso LHC (2017) Nanocellulose in bio-based food packaging applications. Ind Crops Prod 97:664–671. https://doi.org/10.1016/j.indcrop.2016.03.013

Azlin-Hasim S et al (2015) Effects of a combination of antimicrobial silver low density polyethylene nanocomposite films and modified atmosphere packaging on the shelf life of chicken breast fillets. Food Packag Shelf Life 4:26–35. https://doi.org/10.1016/j.fpsl.2015.03.003

Bao J et al (2015) plant esterase-chitosan/gold nanoparticles-graphene nanosheet composite-based biosensor for the ultrasensitive detection of organophosphate pesticides. J Agric Food Chem 63(47):10319–10326. https://doi.org/10.1021/acs.jafc.5b03971

Baspinar Y et al (2018) Curcumin and piperine loaded zein-chitosan nanoparticles: development and in-vitro characterisation. Saudi Pharm J 26(3):323–334. https://doi.org/10.1016/j.jsps.2018.01.010

Bolat G, Abaci S (2018) Non-enzymatic electrochemical sensing of malathion pesticide in tomato and apple samples based on gold nanoparticles-chitosan-ionic liquid hybrid nanocomposite. Sensors 18(3):773. https://doi.org/10.3390/s18030773

Brahatheeswaran D et al (2012) Hybrid fluorescent curcumin loaded zein electrospun nanofibrous scaffold for biomedical applications. Biomed Mater 7(4):45001. https://doi.org/10.1088/1748-6041/7/4/045001

Brychcy-rajska E (2017) Edible protective films and coatings in food industry. J Food Microbiol 1(1):3–4

Bumbudsanpharoke N, Ko S (2018) The green fabrication, characterization and evaluation of catalytic antioxidation of gold nanoparticle-lignocellulose composite papers for active packaging. Int J Biol Macromol 107:1782–1791. https://doi.org/10.1016/j.ijbiomac.2017.10.046

Bumbudsanpharoke N, Choi J, Ko S (2015) Applications of nanomaterials in food packaging. J Nanosci Nanotechnol 15(9):6357–6372. https://doi.org/10.1166/jnn.2015.10847

Campos CA, Gerschenson LN, Flores SK (2011) Development of edible films and coatings with antimicrobial activity. Food Bioprocess Technol 4(6):849–875. https://doi.org/10.1007/s11947-010-0434-1

Castro-rosas J et al (2016) Biopolymer films and the effects of added lipids, nanoparticles and antimicrobials on their mechanical and barrier properties : a review. Int J Food Sci Technol 1–12. https://doi.org/10.1111/ijfs.13183

Davidov-Pardo G et al (2015) Improving resveratrol bioaccessibility using biopolymer nanoparticles and complexes: impact of protein-carbohydrate maillard conjugation. J Agric Food Chem 63(15):3915–3923. https://doi.org/10.1021/acs.jafc.5b00777

El-Gendy A, Abou-Zeid RE, Salama A, Diab MA, El-Sakhawy M (2017) TEMPO-oxidized cellulose nanofibers/polylactic acid/TiO_2 as antibacterial bionanocomposite for active packaging. Egypt J Chem 60(6):1007–1014

El-Wakil NA et al (2015) Development of wheat gluten/nanocellulose/titanium dioxide nanocomposites for active food packaging. Carbohydr Polym 124:337–346. https://doi.org/10.1016/j.carbpol.2015.01.076

Espitia PJP et al (2016) Probiotics and their potential applications in active edible films and coatings. Food Res Int 90:42–52. https://doi.org/10.1016/j.foodres.2016.10.026

Esser B, Schnorr JM, Swager TM (2012) Selective detection of ethylene gas using carbon nanotube-based devices: utility in determination of fruit ripeness. Angewandte Chemie - Int Ed 51(23):5752–5756. https://doi.org/10.1002/anie.201201042

Fabra MJ, López-Rubio A, Lagaron JM (2016) Use of the electrohydrodynamic process to develop active/bioactive bilayer films for food packaging applications. Food Hydrocoll 55:11–18. https://doi.org/10.1016/j.foodhyd.2015.10.026

Farnia P et al (2016) Improvement of curcumin solubility by polyethylene glycol/ chitosan-gelatin nanoparticles (CUR-PEG/ CS-G-nps). Biomed Res 27(3):659–665

Francoz E et al (2015) Arabidopsis seed mucilage secretory cells: regulation and dynamics. Trends in Plant Sci 1–10. https://doi.org/10.1016/j.tplants.2015.04.008

Fuertes G et al (2016) Intelligent packaging systems: sensors and nanosensors to monitor food quality and safety. J Sens. https://doi.org/10.1155/2016/4046061

Ge S et al (2018) Enhanced mechanical properties and gelling ability of gelatin hydrogels reinforced with chitin whiskers. Food Hydrocoll 75:1–12. https://doi.org/10.1016/j.foodhyd.2017.09.023

Gomez-Estaca J et al (2012) Formation of zein nanoparticles by electrohydrodynamic atomization: Effect of the main processing variables and suitability for encapsulating the food coloring and active ingredient curcumin. Food Hydrocoll 28(1):82–91. https://doi.org/10.1016/j.foodhyd.2011.11.013

Gómez-mascaraque, LG, López-rubio, A (2016) Protein-based emulsion electrosprayed micro- and submicroparticles for the encapsulation and stabilization of thermosensitive hydrophobic bioactives. J Colloid Interface Sci 465:259–270. https://doi.org/10.1016/j.jcis.2015.11.061

Goyal, SP, & A (2014, November) 'ce pt cr t', International Jounral of Food Properties, pp. 37–41. https://doi.org/10.1080/10942912.2013.809541

Gross RA, Kalra B (2002) Biodegradable polymers for the environment. Science 297(5582):803–807. https://doi.org/10.1126/science.297.5582.803

Gupta, P, Nayak, KK (2014) Characteristics of protein-based biopolymer and its application. Polym Eng Sci 55(3):485–498. https://doi.org/10.1002/pen

Hamed I, Özogul F, Regenstein JM (2016) Industrial applications of crustacean by-products (chitin, chitosan, and chitooligosaccharides): a review, Trends Food Sci Technol 48:40–50. https://doi.org/10.1016/j.tifs.2015.11.007

Huang X et al (2017) Resveratrol encapsulation in core-shell biopolymer nanoparticles: impact on antioxidant and anticancer activities. Food Hydrocoll 64:157–165. https://doi.org/10.1016/j.foodhyd.2016.10.029

Jain A et al (2018) Beta carotene-loaded zein nanoparticles to improve the biopharmaceutical attributes and to abolish the toxicity of methotrexate: a preclinical study for breast cancer. Artif Cells Nanomed Biotechnol 1–11. https://doi.org/10.1080/21691401.2018.1428811

Jung EY et al (2016) Effect of layer-by-layer (LbL) encapsulation of nano-emulsified fish oil on their digestibility ex vivo and skin permeability in vitro. Prevent Nutrition Food Sci 21:85–89

Kaewklin P et al (2018) Active packaging from chitosan-titanium dioxide nanocomposite film for prolonging storage life of tomato fruit. Int J Biol Macromol 112:523–529. https://doi.org/10.1016/j.ijbiomac.2018.01.124

Karlsson HL et al (2013) Cell membrane damage and protein interaction induced by copper containing nanoparticles—importance of the metal release process. Toxicology 313(1):59–69. https://doi.org/10.1016/j.tox.2013.07.012

Khalil HPSA et al (2017) Biodegradable polymer films from seaweed polysaccharides: a review on cellulose as a reinforcement material. EXPRESS Polym Lett 11(4):244–265

Kim JS et al (2007) Antimicrobial effects of silver nanoparticles. Nanomed: Nanotechnolo Biol Med 3(1):95–101. https://doi.org/10.1016/j.nano.2006.12.001

Klemm D et al (2005) Polymer science cellulose: fascinating biopolymer and sustainable raw material. Angew Chem Int Ed 44(22):3358–3393. https://doi.org/10.1002/anie.200460587

Kumar M et al (2018) Bioconversion of chitin to bioactive chitooligosaccharides: amelioration and coastal pollution reduction by microbial resources. Mar Biotechnol 20(3):269–281. https://doi.org/10.1007/s10126-018-9812-x

Li XY et al (2014) Electrospun quercetin-loaded zein nanoribbons. Bio-Med Mater Eng 24(6):2015–2023. https://doi.org/10.3233/BME-141011

Lim B, Lim J, Ho Y (2018) Garden cress mucilage as a potential emerging biopolymer for improving turbidity removal in water treatment. Process Saf Environ 119:233–241. https://doi.org/10.1016/j.psep.2018.08.015

Liu R et al (2018) Edible gum—phenolic–lipid incorporated gluten films for food packaging. J Food Sci 83(6):1622–1630. https://doi.org/10.1111/1750-3841.14151
Lodhi G et al (2014) Chitooligosaccharide and its derivatives: preparation and biological applications. BioMed Res Int 2014:e654913. https://doi.org/10.1155/2014/654913
López-rubio A, Lagaron JM (2012) Whey protein capsules obtained through electrospraying for the encapsulation of bioactives. Innov Food Sci Emerg Technol 13:200–206. https://doi.org/10.1016/j.ifset.2011.10.012
Hani NM, Torkamani AE, Azarian MH, Mahmood KW, Ngalim SH (2017) Characterization of electrospun gelatine nanofibres encapsulated with *Moringa oleifera* bioactive extract. J Sci Food Agric 97(10):3348–3358. https://doi.org/10.1002/jsfa.8185
Maqbool M et al (2011) Effect of a novel edible composite coating based on gum arabic and chitosan on biochemical and physiological responses of banana fruits during cold storage. J Agr Food Chem 59(10):5474–5482
Mihindukulasuriya SDF, Lim LT (2013) Oxygen detection using UV-activated electrospun poly(ethylene oxide) fibers encapsulated with TiO_2 nanoparticles. J Mater Sci 48(16):5489–5498. https://doi.org/10.1007/s10853-013-7343-4
Mihindukulasuriya SDF, Lim LT (2014) Nanotechnology development in food packaging: a review. Trends Food Sci Technol 40(2):149–167. https://doi.org/10.1016/j.tifs.2014.09.009
Mohebi E, Marquez L (2015) Intelligent packaging in meat industry: an overview of existing solutions. J Food Sci Technol 52(7):3947–3964. https://doi.org/10.1007/s13197-014-1588-z
Montalbán M et al (2018) Production of curcumin-loaded silk fibroin nanoparticles for cancer therapy. Nanomaterials 8(3):126. https://doi.org/10.3390/nano8020126
Moomand K, Lim L (2014) Oxidative stability of encapsulated fish oil in electrospun zein fibres. Food Res Int 62:523–532. https://doi.org/10.1016/j.foodres.2014.03.054
Mu H et al (2016) Gold nanoparticles make chitosan-streptomycin conjugates effective towards Gram-negative bacterial biofilm. RSC Adv 6(11):8714–8721. https://doi.org/10.1039/c5ra22803d
Müller K et al (2017) Review on the processing and properties of polymer nanocomposites and nanocoatings and their applications in the packaging, automotive and solar energy fields. Nanomaterials 7(4):74. https://doi.org/10.3390/nano7040074
Murali Mohan Y et al (2010) Controlling of silver nanoparticles structure by hydrogel networks. J Colloid Inter Sci 342(1):73–82. https://doi.org/10.1016/j.jcis.2009.10.008
Murthy PSK et al (2008) First successful design of semi-IPN hydrogel-silver nanocomposites: a facile approach for antibacterial application. J Colloid Interface Sci 318(2):217–224. https://doi.org/10.1016/j.jcis.2007.10.014
Nasri H et al (2014) new concepts in nutraceuticals as alternative for pharmaceuticals. Int J Prevent Med 5(12):1487–1499
Neethirajan S, Jayas DS (2007) Sensors for grain storage. ASABE Ann Int Meet 17–20. https://doi.org/10.13031/2013.23510
Nunes C, Coimbra MA, Ferreira P (2018a) Tailoring functional chitosan-based composites for food applications. Chem Rec 18(7):1138–1149. https://doi.org/10.1002/tcr.201700112
Nunes MR et al (2018b) Antioxidant and antimicrobial methylcellulose films containing *Lippia alba* extract and silver nanoparticles. Carbohyd Polym 192:37–43. https://doi.org/10.1016/j.carbpol.2018.03.014
Oliani WL et al (2016) Processing and antimicrobial efficacy of polypropylene/montmorillonite/silver nanocomposites as food packaging films. In: AIP conference proceedings, vol 1779, pp 1–6. https://doi.org/10.1063/1.4965499
Palem RR et al (2018) Green synthesis of silver nanoparticles and biopolymer nanocomposites: a comparative study on physico-chemical, antimicrobial and anticancer activity. Bull Mater Sci 41(2):55. https://doi.org/10.1007/s12034-018-1567-5
Pan K, Zhong Q, Baek SJ (2013) Enhanced dispersibility and bioactivity of curcumin by encapsulation in casein nanocapsules. J Agric Food Chem 61(25):6036–6043. https://doi.org/10.1021/jf400752a

Patel A et al (2010) Synthesis and characterisation of zein-curcumin colloidal particles. Soft Matter 6(24):6192–6199. https://doi.org/10.1039/c0sm00800a
Pathakoti K, Manubolu M, Hwang HM (2017) Nanostructures: current uses and future applications in food science. J Food Drug Anal 25(2):245–253. https://doi.org/10.1016/j.jfda.2017.02.004
Pavase TR et al (2018) Recent advances of conjugated polymer (CP) nanocomposite-based chemical sensors and their applications in food spoilage detection: a comprehensive review. Sens Actuators B: Chem. https://doi.org/10.1016/j.snb.2018.06.118
Pavli F et al (2018) Probiotic incorporation in edible films and coatings: bioactive solution for functional foods. Int J Mol Sci 19(1). https://doi.org/10.3390/ijms19010150
Penalva R et al (2014) Casein nanoparticles as carriers for the oral delivery of folic acid. Food Hydrocoll 44(2015): 399–406. https://doi.org/10.1016/j.foodhyd.2014.10.004
Perez-masi R, Laura GG (2017) Potential of microencapsulation through emulsion-electrospraying to improve the bioaccessibility of β-carotene. Food Hydrocoll 73:1–12. https://doi.org/10.1016/j.foodhyd.2017.06.019
Perinelli DR et al (2018) Chitosan-based nanosystems and their exploited antimicrobial activity. Eur J Pharm Sci 117(2017):8–20. https://doi.org/10.1016/j.ejps.2018.01.046
Putri RDA, Setiawan A, Anggraini PD (2017) Effect of carboxymethyl cellulose (CMC) as biopolymers to the edible film sorghum starch hydrophobicity characteristics. In: AIP conference proceedings, 1818(Cmc). https://doi.org/10.1063/1.4976908
Quintavalla S, Vicini L (2002) Antimicrobial food packaing in meat industry. Meat Sci 62(3):373–380. https://doi.org/10.1016/S0309-1740(02)00121-3
Rehim MHA, Youssef AM, Essawy HA (2010) Hybridization of kaolinite by consecutive intercalation: preparation and characterization of hyperbranched poly(amidoamine)-kaolinite nanocomposites. Mater Chem Phys 119(3):546–552. https://doi.org/10.1016/j.matchemphys.2009.10.012
Rhim JW, Park HM, Ha CS (2013) Bio-nanocomposites for food packaging applications. Progr Polymer Sci 38(10–11):1629–1652. https://doi.org/10.1016/j.progpolymsci.2013.05.008
Settanni G et al (2016) Protein corona composition of PEGylated nanoparticles correlates strongly with amino acid composition of protein surface. arXiv 9(14):4739–4750
Sharma C et al (2017) Nanotechnology: an untapped resource for food packaging. Front Microbiol. https://doi.org/10.3389/fmicb.2017.01735
Shit SC, Shah PM (2014) Edible polymers: challenges and opportunities. J Ploym 13. https://doi.org/10.1155/2014/427259
Silva LMDC et al (2013) Biosensors for contaminants monitoring in food and environment for human and environmental health. In: State of the art in biosensors—environmental and medical applications, InTech, Croatia, pp 51–168. https://doi.org/10.5772/55617
Soukoulis C, Gaiani C, Hoffmann L (2018) Plant seed mucilage as emerging biopolymer in food industry applications. Curr Opin Food Sci. https://doi.org/10.1016/j.cofs.2018.01.004
Sozer N, Kokini JL (2009) Nanotechnology and its applications in the food sector. Trends Biotechnol 27(2):82–89. https://doi.org/10.1016/j.tibtech.2008.10.010
Thadathil N, Velappan SP (2014) Recent developments in chitosanase research and its biotechnological applications: a review. Food Chem 150:392–399. https://doi.org/10.1016/j.foodchem.2013.10.083
Tsai AY et al (2017) Identification and characterization of arabidopsis seed coat mucilage proteins. 173(February), pp 1059–1074. https://doi.org/10.1104/pp.16.01600
Valdés A et al (2014) Natural additives and agricultural wastes in biopolymer formulations for food packaging. Front Chem 1–10. https://doi.org/10.3389/fchem.2014.00006
Valdés A et al (2017) State of the art of antimicrobial edible coatings for food packaging applications. Coatings 7(4):56. https://doi.org/10.3390/coatings7040056
Wan ACA, Tai BCU (2013) Chitin - a promising biomaterial for tissue engineering and stem cell technologies. Biotechnol Adv 31(8):1776–1785. https://doi.org/10.1016/j.biotechadv.2013.09.007

Yao K et al (2017) Bioinspired interface engineering for moisture resistance in nacre-mimetic cellulose nanofibrils/clay nanocomposites. ACS Appl Mater Interfaces 9(23):20169–20178. https://doi.org/10.1021/acsami.7b02177

Zambrano-Zaragoza ML et al (2018) Nanosystems in edible coatings: a novel strategy for food preservation. Int J Mol Sci 19(3). https://doi.org/10.3390/ijms19030705

Zhou JJ, Wang SY, Gunasekaran S (2009) Preparation and characterization of whey protein film incorporated with TiO_2 nanoparticles. J Food Sci 74(7). https://doi.org/10.1111/j.1750-3841.2009.01270.x

Tissue Engineering Applications of Bacterial Cellulose Based Nanofibers

Semra Unal, Oguzhan Gunduz and Muhammet Uzun

Abstract Bacterial cellulose derived extracellularly by specific bacterial genera is an environmentally friendly polymeric material. The structural properties of bacterial cellulose are greater to those of herbal cellulose, as BC possesses novel features such as high purity, high crystallinity, nanostructure networks, good light transmittance, remarkable mechanical properties, stress–strain characterization and in situ formability, porosity, uniformity, inherent biocompatibility, and improvement of cell enhancing, separation, and proliferation. In recent years, bacterial cellulose has many opportunity purposes in different applications in biomedicine such as wound-dressing materials, medical membranes, biosensors, regeneration of organs, pharmaceutical industries, food, and cosmetics. Herein, the potential applications of bacterial cellulose, alone or in combination with different components, have been focused on for the use in the regenerative and tissue engineering as an implant and scaffold.

Keywords Bacterial cellulose · Natural polymer · Tissue engineering · Composites · Biomedical

S. Unal
Department of Bioengineering, Faculty of Engineering, Marmara University, Istanbul, Turkey
e-mail: unalsemra@gmail.com

O. Gunduz
Department of Metallurgical and Materials Engineering, Faculty of Technology, Marmara University, Istanbul, Turkey
e-mail: ucemogu@ucl.ac.uk

M. Uzun (✉)
Department of Textile Engineering, Faculty of Technology, Marmara University, Istanbul, Turkey
e-mail: m.uzun@marmara.edu.tr

S. Unal · O. Gunduz · M. Uzun
Center for Nanotechnology and Biomaterials Applied and Research, Marmara University, Istanbul, Turkey

S. Ahmed and W. Ali (eds.), *Green Nanomaterials*,
Advanced Structured Materials 126,
https://doi.org/10.1007/978-981-15-3560-4_13

1 Introduction

Cellulose is the richest biopolymer in the world and is often derived from herbal sources. Special bacterial strains such as Rhizobium, Gluconacetobacter, Rhodobacter, Agrobacterium, and Sarcina can generate extracellular cellulose (Chawla et al. 2009; Petersen and Gatenholm 2011). The structural properties of BC are better than those of plant cellulose, as BC has unique features such as high purity, high crystallinity, ultrafine networks, good light transmittance, remarkable mechanical properties, stress–strain characterization and in situ formability, porosity, uniformity, inherent biocompatibility, low immunogenic potential, and capability for cell enhancing, adhesion, and proliferation (Sheykhnazari et al. 2011; Cai and Kim 2010; Ueno et al. 1999; Leitão et al. 2013; Klemm et al. 2001; Svensson et al. 2005; Zang et al. 2015). Due to the above-mentioned properties, bacterial cellulose (BC) has many opportunity purposes in different applications in biomedicine, including artificial blood vessel fabrication, wound dressing, medical membranes, and biosensors (Klemm et al. 2001; Backdahl et al. 2006; Ul-Islam et al. 2015), pharmaceutical industries, food (Shi et al. 2014) and cosmetics (Lin et al. 2013a, b).

During the last decade, BC biopolymer has been developed for a variety of biomedical engineering purposes (illustrated in Fig. 1). The review study aims to provide an understanding of the biomedical engineering applicability of BC by presenting various forms of this biopolymer.

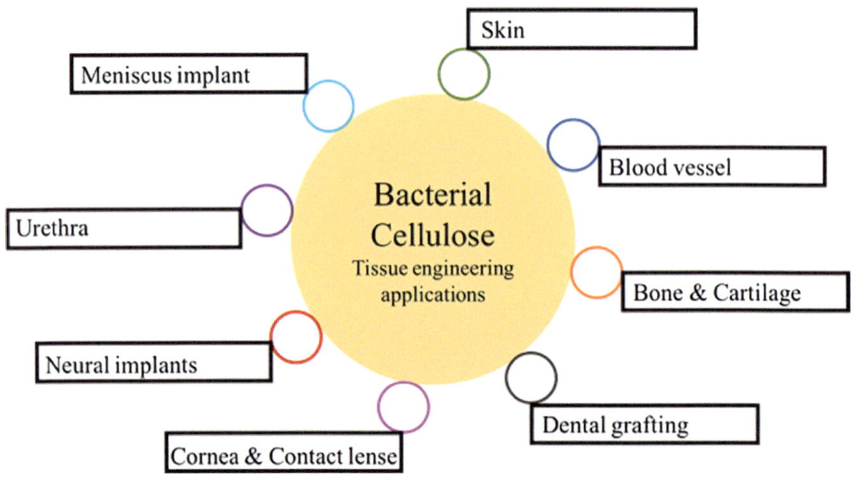

Fig. 1 Schematic illustration of biomedical engineering purpose of BC

2 Production and Synthetic Pathway of the Bacterial Cellulose (BC)

The production of bacterial cellulose is discussed and investigated by making use of several other organisms, including *Acetobacter, Agrobacterium, Zoogloea, Achromobacter, Sarcina, Aerobacter, Acanthamoeba, Rhizobium, Achromobacter, Alcaligenes,* and others. The most broadly reported bacteria strain is *Acetobacter xylinum* or currently renamed *Gluconacetobacter xylinus* due to its great yield of BC (Chawla et al. 2009; Petersen and Gatenholm 2011). It was initially described by A.J. Brown in 1986 who identified a form of an extracellular gelatinous film over the surface of a vinegar fermentation broth (Brown 1886). The first report on the production process of the BC was published by scientists, who carried on a number of experiments to grow *A. xylinum* in culture media (Tarr and Hibbery 1931) Hestrin and Schramm reported that glucose and oxygen were essential contribution to the development of the process of production of bacterial cellulose from *A. xylinum* (Hestrin and Schramm 1954). BC is produced by specified classes of bacteria by utilizing various substrate sources that mainly contain glucose or fructose as a carbon source. The biosynthesis of BC is a large number of precisely and specially regulated multistage network, including both specific enzymes and regulatory proteins to check various cellulose syntheses like phosphoglucomutase, isomerase, UDPG pyrophosphorylase glucokinase, and cellulose synthase, involving the divided hexose monophosphate pathway (HMP) and tricarboxylic acid cycle (Yu et al. 2012; Lin et al. 2013a, b).

Some of the critical mechanism and system of the glucose polymerization in the cellulose chains are still not entirely known, and on the other hand, the process and mechanism of the synthesis of uridine diphosphoglucose (UDPGlc) have been relatively well clarified. Cellulose biosynthesis includes the UDPGlc production which is the cellulose precursor, tracked by polymerization of glucose units forming a β-1 $\rightarrow$ 4 glucan chain. The single cellulose chains were transferred outside of the cell, causing in a ribbon-like structure consisting of hundreds and thousands of individual chains and thus the self-assembly of these ribbons caused the formation of fibrils (Lin et al. 2013a, b). This biosynthesis of bacterial cellulose is related to the catabolic processes and does not interact with other anabolic processes, including protein synthesis. The bacterial cellulose produced by *G. xylinus* is obtained following either the pentose phosphate cycle or the Krebs cycle coupled with gluconeogenesis (Ross et al. 1991; Tonouchi et al. 1996).

3 Chemical Structure of BC Cellulose

Cellulose is also a natural homopolymer composed of monosaccharide molecules bonded through β-(1,4)-glycosidic linkages which are connected by intermolecular and intramolecular hydrogen bonding. The repeating unit consists of two glucose molecules bonded to each other in such a way that a molecule is joined by 180° to

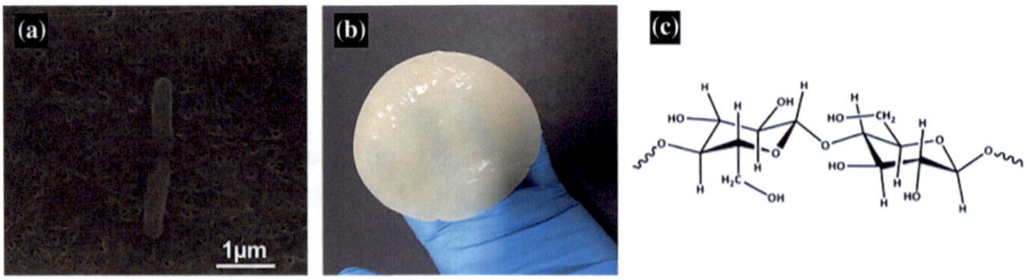

Fig. 2 **a** SEM image of BC membrane of bacterial cellulose network; **b** photograph of BC membrane after purification; **c** chemical structure of cellulose

adapt the bonding angles of the acetal oxygen bridges (Fig. 2). It is well established that chemical properties of cellulose are characterized by its hydrogen bonding network. While repeating glucose units in the BC generate a long polymer chain, the polymer chains compose strong intermolecular hydrogen bonds. The BC sheet is formed by the inter- and intramolecular hydrogen bonds and is responsible for the stabilization, crystalline, and hydrophilic structure of cellulose (Ross et al. 1991).

Bacterial cellulose produced by *A. xylinum* has two distinct physical forms: (i) the ribbon-like cellulose I and (ii) cellulose II that is the thermodynamically stable form of cellulose (Yu and Atalla 1996). The β-1 $\rightarrow$ 4 glucan chains in cellulose I lay in a parallel configuration and arranged uniaxially (Kuga and Malcolm Brown 1988), while those of cellulose II are randomly regulated (Kolpak and Blackwell 1976).

4 Bacterial Cellulose Properties

The chemical properties of BC are not different from well-known herbal cellulose, but the degree of polymerization differs in which it is 13,000–14,000 for plant cellulose and 2000–6000 for BC (Jonas and Farah 1998). Its nanofibrils' structure exhibits similarities with extracellular matrix components and more collagen (Bielecki et al. 2005; Khajavi et al. 2011). BC exhibits non-toxic nature, higher purity, greater water-storage capacity, flexibility, porosity, crystallinity, excellent mechanical properties, and hydrophilicity (Sheykhnazari et al. 2011).

Bacterial cellulose consists of microfibril ribbons, in which the dimension of ribbons is 3–4 nm thick and 70–80 nm wide. It has crystallinity of the order of 65–79% which varies with environmental conditions (Park et al. 2010). BC has better mechanical (such as stress–strain at failure and Young's modulus), purity, and biocompatibility properties than cellulose derived from other conventional sources. The mechanical properties of BC, pure cellulose I, are estimated by its high crystallinity and elastic modulus (EM) (Czaja et al. 2004). For single bacterial cellulose nanoribbons, the elastic modulus of BC has been reported as 78 $\pm$ 17 GPa by atomic force microscope (Guhados et al. 2005).

5 Tissue Engineering Applications of BC and BC-Based Composites

BC and its composites have been used for the development of biomedical applications. Some of their applications include their use as wound healing, artificial skin and skin tissue repair, blood vessels, artificial cornea and contact lenses, bone and cartilage regeneration, meniscus implant, dental root canal treatment, urinary conduits, neural implants, and dura mater (Table 1). BC has already been approved by the FDA as few commercial products used for various clinical indications that are listed in Table 2.

5.1 Wound Healing

The skin plays a role in many vital functions such as control of body temperature, keeping of water and electrolyte balance, and sensory and primary protective barrier of the body (Uzun et al. 2013). Therefore, when these functions are lost in skin injuries such as burns, ulcers, or trauma, immediate coverage is required to help repair and restore its main functions. Wound healing is a highly regulated biological process involving a number of complex interactions between different cell types, surrounding extracellular matrix (ECM) structures, and various combinations (Eming et al. 2002). It needs the coordination of a range of processes such as homeostasis, inflammation, new connective tissue formation, and remodeling which are not clearly understood yet (Balasubramani et al. 2001). Recent research in wound healing and wound dressings have been concentrated on improving and optimizing the functional elements that can quick healing so can reduce scratching, developing new and smart material or organ replacement, and identifying signals that accelerating of the healing process by regeneration and not fixing. For instance, the healing of skin wounds involves the renewal and regeneration of various tissue types and forms, cell types, and network components (Balasubramani et al. 2001).

BC offers as a potentially candidate biomaterial for skin tissue repair and wound dressing due to its attractive features including nanofibril structure, water absorption capacity, high crystallinity, mechanical strength, good permeability, non-toxicity, structural stability, low solubility, highly nanoporous structure, and moist environment for injured area (Backdahl et al. 2006). These properties provide many advantages, such as offering a moist healing and a physical prevention against infections and also allowing the transfer of antibiotics or medicines (Ul-Islam et al. 2015).

On the other hand, BC alone does not have an antimicrobial potential to avoid wound infection and has no important effect on the biological status of chronic wounds. For this reason, a variety of BC composites were prepared with different materials that can provide their antimicrobial properties against various microorganisms and thus being applicable as a skin tissue repair material.

Table 1 BC/BC-composite for biomedical applications

BC and its composite	Applications	Findings	References
BC/Chitosan	Wound dressing	Highly transparent, flexible, increased liquid storage capacity and liquid release rate, more biocompatible and display better mechanical properties, enhanced biocompatibility, cell adhesion, proliferation, promoted wound healing, and faster epithelialization rates than pure BC	Czaja et al. (2006), Ul-Islam et al. (2011), Tammelin et al. (2006), Fernandes et al. (2009)
BC/Collagen	Wound dressing	Accelerated the infiltration of inflammatory cells, stimulate fibroblast cell migration and proliferation.	Ueno et al. (1999)
BC/PEG	Wound dressing	Exhibited good biocompatibility, biomimetic capability and improved cell adhesion and proliferation	Cai and Kim (2010)
BC/Fibrin	Artificial blood vessels	Mechanical and structural properties were similar to a small-diameter blood vessel	Brown et al. (2011)
BC/AgNPs	Artificial blood vessels	Strong antimicrobial activity, improved mechanical properties and flexibility	Maneerung et al. (2008)
BC-/Cellulose-binding domain	Artificial blood vessels	Enhanced endothelial cell adhesion and angiogenesis	Andrade et al. (2010)
BC/PVA	Artificial blood vessels	Low cytotoxicity, enhanced tensile strength and the suture retention	Tang et al. (2015)

(continued)

Table 1 (continued)

BC and its composite	Applications	Findings	References
BC/PVA/nHA	Corneal tissue engineering	Demonstrated structural similarities with the human cornea and an important retinal pigment epithelial cell growth	Kharaghani et al. (2015)
BC	Contact lenses	Retained its transparency and form for more than eight weeks	Levinson and Glonek (2010)
BC/PHEMA	Contact lenses	Enhanced properties such as high strength, water-soluble	Li et al. (2010)
BC/Bone morphogenic protein	Bone regeneration	The differentiation of C2Cl2 into the osteoblast-like cell	Shi et al. (2012)
BC-HA and BC-GAG	Bone regeneration	Enhanced cell proliferation, the functional activity of hATMSC and the differentiation of hATMSC into osteoblasts and chondrocytes, respectively	Kumbhar et al. (2017)
BC/PTFE	Dental implants, grafting material	Completely reconstruct bone defect	Matuda et al. (2004)
BC	Dental root canal treatment	Showed better tensile strength, higher expansion, ten-fold greater absorption, biocompatibility and a significantly greater drug release	Yoshino et al. (2013)
BC/Gelatin	Urethral tissue regeneration	Good biocompatibility with lingual keratinocytes	Huang et al. (2015)
BC	Nerve tissue regeneration	Exhibit proliferation and expression of nerve growth factor neurotrophin	Pértile et al. (2012)
BC	Dura mater	Reduced inflammatory response	Xu et al. (2014)

Table 2 Products based on bacterial cellulose available in the market

Product	Application	Effects	Company
BIOFILL®	Ulcers, burns/artificial skin	Reduced wound pain and infection, faster healing, etc.	Robin Goad
BIONEXT®	Ulcers, burns, lacerations/wound-dressing	Reduced wound pain and infection, faster healing, etc.	Bionext Produtos Biotecnol6gicos Ltda
MEMBRACEL®	Ulcers, burns/temporary skin substitute	Fast skin regeneration	Laboratorio Celina
XCELL®	Venous ulcers/wound-dressing	Relieving pain, decreased infection, faster healing, etc.	XCELL Biologix™
DERMAFILL®	Burns/wound-dressing	Faster healing	Fibrocel
BIOPROCESS®	Ulcers, burns/artificial skin	Reduced wound pain and infection, faster healing, etc.	Biofill Produtos Biotechnologicos
GENGIFLEX®	Recovery of periodontal tissues/dental implants, grafting material	Reduced inflammation and less surgical steps needed	Biofill Produtos Biotechnologicos
SECURIAN®	Tendon repair/tissue reinforcement matrix	Exhibited good cell adhesion, proliferation	Xylos corporation

BC membranes which are based on nanoparticles can act as a template for various nanomaterials synthesis, such as silver nanoparticles, silver nanoparticles (AgNPs) which have been known to obtain antimicrobial properties to BC in via ex situ modifications. In such applications, Maneerung et al. prepared a BC membrane impregnated with silver nanoparticles (AgNPs) by immersing the BC membrane in silver nitrate ($AgNO_3$) solution after that the silver ion (Ag^+)-saturated bacterial cellulose was reduced to Ag^0 particles by using of sodium borohydride ($NaBH_4$). The BC/AgNPs composite demonstrated strong antimicrobial activity against *Escherichia coli* (Gram-negative) and *Staphylococcus aureus* (Gram-positive) (Maneerung et al. 2008). Yang et al. produced that the composites were obtained by silver nanoparticles impregnated into bacterial cellulose produced using different carbon sources with similar method and indicated that the composite derived from maltose with a silver content of 1.06% realized a more than 8-log reduction and >99.99% killing ratio in the number of viable *E. coli* after contact for 18 h (Yang et al. 2012). In another study, the AgCl nanoparticle-impregnated BC composite was obtained by immersing BC in a silver nitrate solution and sodium chloride, followed by a rinsing step. BC/AgCl membrane exhibited high hydrophilicity and strong antimicrobial activity against *E. coli* (Gram-negative) and *S. aureus* (Gram-positive) (Hu et al. 2009). Similar methods were described in other studies which showed comparable antimicrobial results (Schluesener and Schluesener 2013).

The natural polymer chitosan has been established to stimulate wound healing by accelerating the infiltration of inflammatory cells, increasing fibroblast cells' migration and proliferation, and promoting type III collagen production in the wound side (Ueno et al. 1999). Chitosan is one of the important materials for smart wound-dressing applications.

It has been reported that various BC compounds synthesized with different materials replace the traditional biomaterials used in skin regeneration. BC-chitosan (BC/Ch) nanocomposites were produced at the first time by Fernandes et al. (2009). The nanocomposite is obtained via a simple and entirely green approach by casting a water-based suspension of chitosan and bacterial cellulose nanofibrils. When compared to pure BC, adding chitosan made it highly transparent, flexible, increased water retention ability and water release rate, more biocompatible and displayed better mechanical properties (Fernandes et al. 2009; Ul-Islam et al. 2011; Kim et al. 2011). Furthermore, the addition of chitosan into BC significantly increased the growth inhibition against *E. coli*, *S. aureus*, and the epithelialization and regeneration process proceeded faster that wounds covered with BC-chitosan than wounds covered with pure BC (Lin et al. 2013a, b).

Similarly, BC-collagen composites fabricated by using in situ synthesis method have high potential tissue engineering applications. BC-collagen composites demonstrated distinguishing features through high biodegradability, low antigenicity, and cell adhesion properties that most considered them a suitable material for biomedical applications (Luo et al. 2008; Zhijiang and Guang 2011). Chai reported that the collagen was not only coated on the surface of BC nanofibrils but also impregnated inside BC and therefore formed a hydrogen bond between BC and collagen (Cai and Yang 2011; Zhijiang and Guang 2011).

To treat chronic wounds, it is also necessary to diminish high protease concentrations and increased reactive oxygen species (ROS) to allow healing of wounds (Schultz and Mast 1998; Trengove et al. 1999). Wiegand et al. concluded that the composite produced by in situ type I collagen modified BC was able to significantly reduce the concentration of interleukin and amount of proteases and ROS activity (Wiegand et al. 2006). Therefore, the novel composite seems to be a promising material for wood healing application.

The similar behavior has been found to be at BC-PEG (polyethylene glycol) composites. BC-poly (ethylene glycol) (PEG) hybrid (BC-PEG) was produced by in situ and in situ modifications to obtain biocompatible wound healing material. PEG molecules were not only coated on the BC membrane surface but also penetrated into the BC fibrils. The study was concluded that the crystallinity feature of BC was reduced when the presence of PEG affected the preferential orientation of the plane during the process of drying while the thermal stability was increased up to 293 °C due to the great connection between PEG and BC. The composite exhibited good biological material, biomimetic capability, and improved cell adhesion and proliferation on 3T3 fibroblast cells (Cai and Kim 2010).

Double-network (DN) hydrogels in combination with BC and gelatin showed excellent properties such as increased mechanical strength and fracture strength, and better elastic modulus under compressive stress. The biodegradable composite also promoted the proliferation of NIH/13 cells and presented as a candidate for skin tissue regeneration template (Nakayama et al. 2004).

Khan et al. fabricated BC-gelatin (BC/G) composite scaffolds by dissolving both BC and gelatin in N-methylmorpholine-N-oxide (NMMO). The scaffold showed enhanced cell compatibility compared with pure BC scaffold, indicating gelatin-enhanced biocompatibility (Khan et al. 2016). Recently, three-dimensionally micro-porous regenerated bacterial cellulose (rBC)/gelatin (3DMP rBC/G) scaffolds were fabricated using a porogen-assisted surface modification method, where the porogen material covered the BC fibers after dissolution, thus leading to enhanced biocompatibility. The scaffolds showed a regular and highly porous structure, while a high amount of gelatin was also penetrated into BC nanofibrils. The scaffold showed good adhesion and proliferation of human keratinocytes in vitro. The in vivo result indicated that the scaffold provided 94% skin regeneration within two weeks (Khan et al. 2018). Thus, the 3DMP rBC/G scaffolds could be used as candidate materials for skin regeneration applications. BC including sodium alginate (SA) and silver sulfadiazine (AgSD) was produced by mixing BC slurry with AgSD and SA solution, and finally, the mixture was crosslinked by $CaCl_2$ (Shao et al. 2015). The composite presented excellent antibacterial performances on various bacterial infections, including *E. coli*, Staphylococcus.

5.2 Cardiovascular System

Cardiovascular diseases (CVDs) are globally the leading cause of death impairing the quality of life of individuals. Despite the recent improvements in preventing and treating CVDs, lifestyle changes (increased use of tobacco and alcohol, unhealthy diet, decreased physical activity) and the growth in aging population have elevated the incidence of CVDs (Lee and Park 2017). The World Health Organization 2017 report indicated that CVDs lead to the death of 17.7 million people every year and is expected to further increase to 23.3 million by 2030. The narrowing or blockage of blood vessels is often associated with CVDs leading to reduced blood flow and resulting in tissue damage owing to insufficient nutrient supply. The most common disorders are coronary heart disease, cerebrovascular disease, peripheral arterial disease, and deep vein thrombosis (Lee and Park 2017).

One of the best popular interventions for cardiac disease is the coronary bypass implant operation which provides blood in the heart tissue via replacing an appropriate blood vessel. The standard for bypass surgery is to replace with another blood vessel derived from patient's or donor's body, or with an artificial one. The patient's vessel as additional surgical procedure during surgery may cause other health problems including prolonged anesthesia, blood loss, and surgical complications at the donor site, whereas the donor strategies are limited in terms of the high number of tissue rejections and improvement of arteriosclerosis (Ullah et al. 2016; Tang et al. 2017). Due to all these problems mentioned above, there is an increasing demand for the development of artificial blood vessels to replace small-caliber blood vessels (<6 mm), particularly for bypass surgery (Tang et al. 2017). For this reason, novel materials convenient for small vessel production are in high demand.

As broadly discussed in the existing literature that expanded-polytetrafluoroethylene (ePTFE), polyethylene terephthalate (i.e. Dacron®), and polyurethanes are extensively used as synthetic graft materials for artificial blood vessel development. PTFE is a synthetic fluorocarbon polymer with an extensive application range in the biomedical industry which is mostly employed in cardiovascular engineering to produce vascular grafts and heart valves. ePTFE is formed by compressing PTFE in the presence of career medium and then extruding the mixture (Yashiro et al. 2012). While PTFE is used for vascular grafts in large-diameter arteries including the aorta, ePTFE was chosen for the use of smaller arteries with biochemically inert properties, lower flow rates promoting low thrombogenicity, and lower rates of restenosis and hemostasis (Yashiro et al. 2012; Saha et al. 2011; Wang et al. 2009; Barozzi et al. 2011). Another important polymer is polyethylene terephthalate (PET) for such applications which is a family of polymers containing ester functional groups, commercially known as Dacron®. This material can be produced into many textile structures such as non-woven, woven, knitting, or yarn; however, it is mainly used as woven fabric structure for vascular grafts due to its stiffness. It is mainly because of warp and warp yarns into the woven structures that decrease blood leakage and improve its efficiency as a vascular graft (Kudo et al. 2002) and also provide enough air permeability. The hydrophobic surface of PET gives possibility to artificial blood vessels with large diameters in woven fabric form (Lessim et al. 2015). Polyurethane (PU) is formed through a reaction between isocyanates and polyol. Because of its good physiochemical and mechanical properties, PU is also widely used in cardiovascular applications (Ashton et al. 2011; Theron et al. 2010). Some of the primary properties of PU are highly biocompatible, high shear strength, elasticity, and transparency (Grasl et al. 2010). Synthetic vascular grafts were demonstrated to be satisfactory for applications that require grafts greater than 6 mm in diameter in high-flow, low-resistance circulation. However, synthetic grafts, such as ePTFE, Dacron (polyester), and polyurethane induce initial stenosis and thrombosis due to low blood flow, compliance mismatch, and high shear stress, and therefore, are not suitable clinically small-diameter (<6 mm) vascular grafts (Chong et al. 2014; Ravi and Chaikof 2010).

One of the significant issues of synthetic grafts is their rigidity, particularly in small-diameter grafts compared with native vessels (Tiwari et al. 2003) because viscoelasticity of the graft is required to minimize the impedance of flow and risk of stasis and thrombosis (Ahmed et al. 2014; Sarkar et al. 2006). For this reason, novel materials convenient for biologically active small-diameter vessels are required. The fibrous network formed by BC has been extensively studied for artificial blood vessel formation due to its morphology, water-holding capacity, high purity, higher tensile strength, ultrafine fibrous network, foldability, moldability, and biocompatibility (Leitão et al. 2013; Klemm et al. 2001; Svensson et al. 2005; Zang et al. 2015).

In this concept, the development of a medical product called Bacterial Synthesized Cellulose (BASYC®) presents a breakthrough in biomedical applications of BC. This product has been successfully demonstrated the replacement of the carotid arteries (small-diameter blood vessel) in several in vivo studies and showed one-year follow-up in the rat model (Klemm et al. 2001). Similarly, a 90-day follow-up of BASYC® pig prostheses is displayed, and the prosthesis has revealed epithelialization and shaped the structure with three district layered, like the nature of a blood vessel (Schumann et al. 2009). It should be noted that besides the artificial blood vessels, the fact that BASYC® is alone or loaded with a neuro-regenerative drug gives promising results regarding tissue innervation in in vitro models following neurosurgery (Klemm et al. 2001, 2005). In another study, BC-fibrin composites crosslinked with glutaraldehyde (GLA) offered better mechanical properties similar to the natural small-diameter blood vessels (Brown et al. 2011).

Andrade et al. (2010) have shown the possible of BC improved with chimeric proteins containing cellulose-binding module and connection peptide for the heart vessel graft. The improved BC was able to develop endothelial cell adhesion and stimulate angiogenesis. On the other hand, the migration and ingrowth of cells on BC were decreased (Andrade et al. 2010). Besides depending on the growth factors and their receptors, angiogenesis is also affected by ECM protein receptors. The results showed that the RGD-containing recombinant peptide (RGD-CBM) stimulated the early formation of cellular cord-like structures on BC in comparison to recombinant peptide without the adhesion sequence (CBM) or buffer (Andrade et al. 2010).

In a latest report, Leitão et al. (2016) showed better mechanical properties and suture retention of the BC graft when compared to native vessels. Another significant result is that the BC graft offers a good match in mechanical properties of native vessels and grafts because the moldability and flexibility of the graft allow it to absorb the pressure change. The authors suggested that the open pore areas of the graft provide cavities that would enable fiber movement within the graft wall in response to pressure changes (Leitão et al. 2016).

Furthermore, to evaluate the potential of small-diameter arterial grafts of bacterial cellulose, the BC tubes were applied replacement of carotid arteries in sheep at 12 weeks postoperatively, compared with native arteries. A confluent luminal endothelial cell layer was revealed along with the migration of vascular smooth muscle cells into the BC matrix and a homogeneous endothelialization of the inner graft surface with no signs of prothrombogenic or inflammatory potential (Scherner et al. 2014).

BC-fibrin composites had similar time-dependent viscoelastic behavior as the normal blood vessel showing mechanical properties similar to a small-diameter blood vessel. But, the strain at break was lower than the normal blood vessel. As Young's modulus of the porcine carotid artery was reported to be around 1.04 Mpa (Backdahl et al. 2006), indicating that BC tubes had good tensile strength when compared with natural blood vessels (Zang et al. 2015). Based on the tensile strength measurement results, BC tubes have shown better deformation behavior compared to the porcine carotid artery, as well as Putra's result of 0.59 Mpa (Putra et al. 2008).

Moreover, BC tubes have good biocompatibility which indicates high potential for the production of artificial blood vessel materials (Zang et al. 2015). Based on the cytotoxicity and cell compatibility evaluation, BC tubes show no toxic or side effects on vascular cells cultured on their surface. This result demonstrated that the surface of BC tubes offers a suitable template for cell attachment, proliferation, and ingrowth, as characterized for the first time for all the vessel cell types, including endothelial cells, smooth muscle cells, and fibroblast (Backdahl et al. 2006; Zang et al. 2015).

Cai and Kim (2010) produced bacterial cellulose (BC)/poly(ethylene glycol) (PEG) composite by immersing wet BC pellicle in PEG aqueous solution followed by the freeze-drying process. The result revealed that PEG particles were not only covered on the BC fibrils surface but also encapsulated into BC fiber networks. According to tensile results, Young's modulus and tensile strength declined. Fibroblast cells (3T3) were enhanced on PEG-modified BC composite, and the result demonstrated much better biocompatibility compared to the pure BC (Cai and Kim 2010).

Recently, Jang et al. fabricated PVA blend BC nanocrystal composite tubes and evaluated in comparison with the BNC tubes for the potential use in the artificial blood vessel. The results demonstrated low cytotoxicity, enhanced tensile strength, and the suture retention that is an essential feature of the composite to maintain their long-term integrity (Tang et al. 2015).

5.3 Cornea and Contact Lenses

The corneal disease is the primary cause of blindness, and 10 million people are estimated worldwide to have lost their eyesight due to corneal disease or the pathological conditions where the corneal transparency is lost (Foster 2003; Whitcher et al. 2001). Therefore, the need for corneal transplants has promoted the search of a wide variety of biomaterials to prepare bioengineered corneas. To regain the vision, keratoplasty is a conventional technique to replace damaged cornea, although associated with adverse effects, including xerophthalmia, corneal neovascularization, donor shortage, and graft rejection (Wang et al. 2013). Recently, both tissue and bioengineering techniques have been used to design and engineer corneal replacement including the synthesis of artificial cornea and tissue engineering for cornea production. Excellent mechanical properties of BC such as its nanoporous structure that aid to preserve the intraocular pressure and definite pellucidness render it a favorable material to be used as an artificial cornea. In addition, due to its light transmittance, and biocompatibility, BC has been considered as promising biomaterial for artificial cornea and corneal tissue engineering (Hui et al. 2009). Kharaghani et al. showed that corneal stromal cells were attached and proliferated on the BC scaffolds. The study also showed that BC material promoted the internal growth of corneal cells and therefore demonstrated their potential for corneal replacement and corneal tissue engineering. Likewise, PVA-nano hydroxyapatite (nHA) (BC-PVA-nHA)-modified BC composite revealed the water content capacity (82–84%) similar to the natural human cornea (78%) (Kharaghani et al. 2015).

In another study, the production of BC-based polyvinyl alcohol (PVA) hybrid membranes with biocompatible, flexible, and one-piece artificial corneal tissue was reported (Wang et al. 2010). In the study, BC-PVA hybrid membranes were produced by the freeze–thaw method as an artificial cornea. PVA was selected due to its stable character, while BC provides excellent biocompatibility and high mechanical features. A hydrogel composite was obtained by soaking different amounts of PVA into a BC scaffold and characterized its various properties. The study indicated that BC-PVA composites improved water-holding capacity, light transmission, mechanical characteristics, and thermal stability which were the requirements of artificial cornea materials. Therefore, the composites were an innovative candidate biomaterial for cornea tissue replacement (Wang et al. 2010). Furthermore, a large amount of hydroxyl groups that present on the two components maintained interface interactions via hydrogen bonding, resulting in the desired adhesion at the BC-PVA interface.

More than 2 billion people around the world affected from visionary defects induced by refractive error (RE) (Naidoo and Jaggernath 2012). Being the most common vision defect, RE results in poor vision, such as astigmatism, myopia, hyperopia, and presbyopia with severe economic and social implications, if left untreated (Bourne et al. 2004). It is easy to diagnose RE, measure, and adjust it using the refractive surgical procedures or devices, such as spectacles and contact lenses (Kempen et al. 2004).

Contact lenses which are available in hard and soft varieties are broadly used for the modification of visual defects in human. The hard variety eye implant lenses are produced from glass or a hard polymer, such as poly(hydroxyethyl methacrylate) (PHEMA), poly(methyl methacrylate) (PMMA), and polycarbonates, while the soft lenses are generally produced from a soft material (Levinson and Glonek 2010). Another interesting applicability of bacterial cellulose (BC) is eye implant lenses that are also few documented in the literature. BC can be used in fabrication of eye implant lenses owing to its transparency and a high degree of light transmittance. The BC-based eye implant lens was produced from a viscous BC solution in 1-butyl-3-methylimidazolium chloride. The BC solution was slipped casting, the solution was first exposed with isopropanol and then water, and a clear material precipitated. The synthetic eye implant lenses are naturally separated from the mold surface. The remaining solvent, 1-butyl-3-methylimidazolium chloride, was spread over contact lens to water. The hydrated BC eye implant lens remained stable for more than eight weeks, maintaining its transparency and shape (Levinson and Glonek 2010). Hydrogel composite containing transparent polymer was produced by combining water-insoluble nanofibers of BC with water-soluble PHEMA polymer. The produced sample had a water content of around 40% (w/w) and high mechanical properties (Li et al. 2010). BC-PHEMA nanocomposites structured can be allowed in hydrogel applications, especially in eye implant lenses.

Messaddeq et al. developed and patented a system that used BC to produce contact lenses with correct angle and shape (Messaddeq et al. 2008). In this mechanism, pure and wet BC is cut into a round form and then compressed with a stick forming a hemispherical tip at a fixed temperature of 150 °C. Thus, an eye implant lens internally formed by the compression stick and externally can be obtained from the mold.

BC-based contact lenses might be loaded with drugs to maintain adequate concentrations during the treatment of eye infections and allergies, as recently addressed by Cavicchioli et al. (2015) describing the incorporation of ciprofloxacin into the BC structure with or without 2-hydroxypropyl-γ- cyclodextrin before contact lenses were pressed. The study was indicated that the BC-ciprofloxacin membrane induced cytotoxicity, while the investigated materials, except for BC-CPX, present a high potential for biomaterial applications, mainly as a contact lens used for regeneration or protection against bacteria, wound dressing after eye surgery, replace antibiotics eye drops, or enhance the recovery of ocular burns (Cavicchioli et al. 2015).

5.4 Bone and Cartilage Regeneration

Bone tissues severely damaged by trauma, tumors, or bone-related diseases are frequently challenging to regenerate and stimulate by the different healing process. Over two million bone grafts reconstruction are performed worldwide every year, costing more than $15 billion. Bone tissue consists of an organic such as collagen and non-collagenous proteins and inorganic minerals such as hydroxyapatite (HA) and tricalcium phosphate (TCP) (Ao et al. 2017).

The scaffold in bone tissue engineering should exhibit high biocompatibility, biodegradability, high porosity, surface area and pore interconnectivity, chemical functionality. Besides, it should be supported cells proliferation, promoted cell differentiation to functional cells, and offered excellent mechanical strength to the dynamic remodeling of the engineered bone graft in vivo (Ao et al. 2017). The nanofibrous form of BC is natural, which imitates the extracellular matrix and gives the opportunity to chemical modifications. The three-dimensional structure of BC has also been explored in guided bone regeneration. Hydroxyapatite nanoparticles (HA - $Ca_5(PO_4)_3(OH)$) and additive BC membrane were produced during the BC fermentation process. The results indicated that bone defects in rat tibiae showed faster bone regeneration that was filled with new bone tissue at four weeks after implantation and no inflammatory responses around the implant (Saska et al. 2011).

Actually, a number of studies have investigated that tissue engineered biomaterial scaffolds should ideally possess a network of interconnected pores, which aids to provide rapid nutrient diffusion, keep intercellular communication, promote and guide cell migration and new tissue regeneration (Fang et al. 2009; Zaborowska et al. 2010; Loh and Choong 2013; Huang et al. 2017). Therefore, researchers demonstrated ex situ modifications of BC with hydroxyapatite (HA) for bone regeneration applications and investigated the relation between porosity and new tissue formation (Fang et al. 2009; Ramani and Sastry 2014; Wang et al. 2011; Luo et al. 2014; Huang et al. 2015).

Wan et al. and Fang et al. fabricated the BC-HA membranes that were used a phosphorylated BC pellet to induce calcium phosphate formation (Wan et al. 2007; Fang et al. 2009). The study indicated that the mean pore size of the BC-HA composite was 1 μm, while the unmodified BC had an average pore size of about 100 nm (Fang et al. 2009). Moreover, the BC-HA membrane demonstrated better cell adhesion than the pure BC, and it was showed for the first time that the BC-HA membrane promoted osteoblast proliferation and enhanced human bone marrow stem cells (hBMSC) differentiation even without osteogenic reagents (Fang et al. 2009).

Besides, Huang et al. reported the synthesis of the BC/PA/Gel/HAp scaffold for application in bone tissue engineering (Huang et al. 2017). The authors showed that the BC/PA/Gel/HAp scaffold displayed the highest mechanical property in all of Young's modulus, maximum load, and compressive strength when compared to pure BC, and Gel and PA/Gel-modified scaffolds. Compared with the scaffolds, the ratio of nanopores and macropores decreased slightly in modified BC scaffolds due to the reduced the porosity in the presence of Gel and HAp particles. In addition, the BC/PA/Gel/HAp scaffolds improved the cytocompatibility, increased cell adhesion, and showed good intercellular communications (Huang et al. 2017).

The interaction between BC matrix and calcium carbonate ($CaCO_3$) crystals could be an interesting research subject. Some research groups have used calcium chloride ($CaCl_2$) and sodium carbonate (Na_2CO_3) to stimulate calcium carbonate deposition on BC pellicles (Stoica-Guzun et al. 2012; Tazi et al. 2012). The BC-Hap composite membrane showed a potential candidate biomaterial for bone tissue engineering owing to inducing enhanced osteoblast cells, increased the amount of alkaline phosphatase activity, and better bone tissue regeneration (Tazi et al. 2012).

Some researchers have introduced that BC membranes synthesized in the presence of carboxymethylcellulose (CMC) showed better adsorption capacity of metal ions than membranes of pure BC (Grande et al. 2009; Chen et al. 2009; Sakairi et al. 1998). Zimmermann et al. reported the fabrication of BC pellicles and tubes with the adsorption of carboxymethyl cellulose (CMC) to initiate nucleation of calcium-deficient hydroxyapatite (cdHAp) might be potential implants for tissue engineering (Zimmermann et al. 2011). The authors showed that the presence of cdHAp crystals on the BC surface increased osteoprogenitor cells attachment. Besides, the BC-cdHAp scaffolds increased alkaline phosphatase (ALP) activity of bone cells. Thus, the results indicate that BC-cdHAp scaffold is an available bone scaffold material and needs further investigation (Shi et al. 2009; Zimmermann et al. 2011). Similarly, Grande et al. (2009) developed BC-Hap membranes for tissue engineering applications and achieved excellent findings concerning reconstruction of bone and connective tissues (Grande et al. 2009).

In the complex process of new bone formation, various growth factors (GFs) such as transforming growth factor-b (TGF-b), primary fibroblast growth factor (bFGF), and bone morphogenic proteins (BMPs) play an important role (Bostrom and Asnis 1998; Chen et al. 2007a, b; Shi et al. 2012). BMP-2 is delivered to skeletal defects due to a high osteoinductive activity. To eliminate rapid diffusion and increase the local concentration of BMP-2 in the wound site, Shi et al. loaded BMP-2 into BC scaffolds and studied the therapeutic potential of BC loaded with BMP-2 both in vitro (C2C12 cells) and vivo. It is shown that the differentiation of C2Cl2 into the osteoblast-like cell occurred only on BC in the presence of BMP-2. It expressed the concentration of BMP-2 loaded into BC membrane still retained the osteoconductive function. Furthermore, in vivo data indicated that BC scaffolds with BMP-2 showed new bone formation at two weeks after implantation. High concentrations of BMP-2-loaded BC supported fully of bone tissue, while low concentrations of the BMP-2 group were only covered with the bone tissue (Shi et al. 2012). The study offered BC scaffolds as a localized drug delivery template in TE applications.

Many researchers also prepared and characterized BC membrane consisting of osteogenic growth peptide (OGP), which is a native molecule with a main material identical to the C-terminal sequence of histone H4 to promote a better cell attachment, proliferation, and osteoinduction (Saska et al. 2012; Pigossi et al. 2015). It has been known that OGP and OGP derivatives, the C-terminal pentapeptide OGP [10–14] control the expression of growth factors, for example, transforming growth factors (TGF): TGF-b1, TGF-b2, TGF-b3, fibroblastic growth factor (FGF-2), and insulin-like growth factor (IGF-1) (Brager et al. 2000). Thus, these peptides (OGP and OGP [10–14]) adjust the marrow mesenchymal cell differentiation into osteoblasts, promoting new bone formation (Chen et al. 2007a, b; Vanella et al. 2010). The study investigated the BC membranes functionalized with OGP and its C-terminal pentapeptide OGP [10–14]. The results indicated that the membrane showed good flexibility for using surgical implantation and shown no in vitro cytotoxicity, genotoxicity, or mutagenicity properties (Saska et al. 2012).

Similarly, Pigossi et al. (2015) investigated the potential of BC-Hap scaffolds with the incorporation of osteogenic growth peptide (OGP) or pentapeptide OGP (10–14) on bone regeneration in critical-size calvarial defect model (Pigossi et al. 2015). In this study, the fabricated scaffolds (the BC-Hap, BC-Hap-OGP, and BC-Hap-OGP (10–14)) were analyzed in predetermined interval times to observe results at early periods (3, 7, 15 days) and late periods (30, 60, 90 days). In the late periods, the BC-HA scaffold indicated a high percentage of bone regeneration due to up-regulation of the Alpl, Spp1 genes. On the other hand, the combination of OGP and OGP (10–14) peptides in the BC-HA scaffolds had shown no advantage in bone formation (Pigossi et al. 2015).

Kumbhar et al. presented that the modification of BC with glycosaminoglycans (GAG) and hydroxyapatite (HA) demonstrated proliferation and growth of human adipose tissue-derived mesenchymal stem cells (hATMSCs). The observations indicate that the 3D scaffold offers a suitable template over a 2D surface (Kumbhar et al. 2017). Moreover, the differentiation of hATMSC into osteoblasts and chondrocytes was observed onto BC-HA and BC-GAG scaffolds, respectively. The reason could be that HA and GAGs are practically non-immunogenic and help in adhesion, migration, proliferation, and differentiation of cells. Therefore, BC-HA and BC-GAG scaffolds provide sufficient 3D support for cell proliferation and improve the functional activity of hATMSC (Kumbhar et al. 2017). Angiogenesis and neovascularization of the newly formed bone tissue are associated with implanted biomaterials features such as bioactive factors, porous nature of the membrane, and the metabolic activity of the infiltrating host tissue (Burugapalli and Pandit 2007). The in vivo study of BC-based scaffolds after subcutaneous implantation demonstrated that BC, BC-HA, and BC-GAG demonstrated host fibroblast infiltration, collagen deposition, and blood vessel formation at the implantation site (Kumbhar et al. 2017).

Similarly, cartilage is an avascular tissue composed of a single type of cells, chondrocytes, embedded in a matrix consisted of collagen and proteoglycans. Owing to the limited self-repair capacity of the adult cartilage tissue, the repair of cartilage defects is a significant clinical problem. Biomaterials for artificial cartilage are required to be not only tough and resistant but also biodegradable and suitable for new tissue formation. BC presents an appropriate biomaterial in the application of auricular cartilage tissue engineering due to its high strength in the wet state as well as easy to shape in situ, biocompatibility and relatively simple, cost-effective production. Svensson et al. (2005) demonstrated that unmodified BC supports chondrocyte proliferation. The study also showed that BC membrane supported chondrocytes differentiated form. Based on these results, BC membranes appear to have potential as a bio-mimicking scaffold for cartilage replacement (Svensson et al. 2005).

Newly, BC with 15% cellulose content has been suggested as an irresistible graft material for auricular cartilage replacement because it has mechanical properties and most vitally matches the host tissue response of human auricular cartilage (Ávila et al. 2014). Another study evaluated the friction and wore behaviors of bacterial cellulose (BC) against bovine articular cartilage (BAC). The produced BC biomaterials have low friction coefficient values (~0.05) combined with the preservation of the mating surfaces which indicate the potential usage of BC as artificial cartilage for articular joints (Lopes et al. 2011).

5.5 *Meniscus Implant*

Meniscal lesions caused by degenerative diseases, trauma, or a combination of both are the universal problems (Tienen et al. 2004). Extended or total meniscectomy may expose changing of degenerative in the trauma area. Such lesions often progress to osteoarthritis. The allogenic meniscus implantation was performed for symptomatic treatment after partial or total meniscectomy of middle-aged patients (Tienen et al. 2004). Bodin et al. reported a comparison study on BC gel properties and both different traditional collagen meniscal implant materials and real pig meniscus (Bodin et al. 2007). The work also indicated that Young's modulus of BC gel was so similar to the normal pig meniscus and 5-times higher than the collagen material. BC presented favorable cell proliferation and a controlled meniscus shape, indicating that BC could be useful material for using meniscus graft. On the other hand, the tensile strength and mechanical performance of BC can be further enhanced to form more suitable nanocomposites.

5.6 Dental Root Canal Treatment

The first usage attempt was associated with the regeneration of periodontal disease. Nanocellulose products (Gengiflex®-Gore-Tex®) were used in the dental industry to make to help periodontal tissue recovery. Novaes and Novaes (1992) reported a complete restoration related to osseointegrated implants using the commercial BC membrane, Gengiflex®. The advantages include the reestablishment of aesthetics and mouth function, reducing the number of surgeries. Gengiflex® contains two main layers: The inner layer is developed by making use of BC, which provides rigid to the membrane, while the external alkali–cellulose layer is chemically modified (Novaes and Novaes 1992). Salata and his colleagues (Salata et al. 1998) compared the biological effects of Gengiflex® and Gore-Tex® membranes using the in vivo non-healing bone-defect model suggested long before (Dahlin et al. 1988) to compare the biological effects of Gengiflex® and Gore-Tex® membranes. The study designated that Gore-Tex® membranes (a composite containing PTFE, nylon, and urethane) caused considerably decreasing inflammation; both membranes accelerated bone reconstruction equally in the same effect. A significant amount of bone growth was detected by the Gore-Tex® or BC membrane in comparison to the control groups. In addition, when compared to Gengiflex®, Gore-Tex® shows improved properties, and the most importantly, the structure is tolerated the wide range of different tissue structures.

Another study conducted by Matuda et al. (2004) investigated BC and poly-tetrafluoroethylene (PTFE) used to improve bone deficiencies. In the study, two osseous defects (8 mm in diameter) were performed in both posterior feet of four adult rabbits, using surgical blades with continuous irrigation using a sterile saline solution. PTFE membranes were applied to the defects located on right posterior feet, while Gengiflex was utilized over the injuries created in the left posterior feet. After three-month follow-up, the histological evaluation showed that PTFE application was utterly reconstructed bone defect, and on the other hand, Gengiflex® membranes was not as effective as PTFE in repairing bone defects.

In conventional treatment, paper point and cotton pellet made of plant cellulose have been used for dental root canal drying and sterilization (Taneja et al. 2010; Ullah et al. 2016). When compared to normal paper point materials, BC showed greater biological features for dental root canal treatment in respect of preservation superior tensile strength under wet condition, higher expansion, tenfold greater absorption, biocompatibility, and a significantly higher drug release. Also, it exhibited maintenance of physical integrity, and the foreign body reaction was found to be very low (Yoshino et al. 2013; Ullah et al. 2016). Considering the results of BC study, BC composite has excellent potential for applications in dental root canal treatment.

5.7 Urinary Conduits

American Cancer Society (ACS) mentions that bladder cancer is the fourth most common cancer in men; however, it is very less common in women. Various congenital and acquired pathologies such as hypospadias, epispadias, strictures, fistulas, and malignancies disrupt the normal functioning of the urethra and require urethral reconstruction. Short, non-complex urethral defects are repaired by end-to-end anastomosis approach. However, repair of long-segment urethral limits remains one of the main problems associated with possible complications including increased risk of infection, swelling, pain, and limitations of the oral opening which have been reported (Das et al. 2009).

Given the limitations associated with conventional surgical approaches, a substantial need for the development of alternative biomaterials for urethral reconstruction exists. It is shown that with the development in the biomaterials, naturally occurring regenerative stem cells can be employed as scaffolds to latch onto to regrow the bladder smooth muscle and epithelium. Such engineered tissues have important urologic tissue regeneration, but still there are very big challenges to overcome. The study discussed that a three-dimensional (3D) porous bacterial cellulose material offers suitable conditions to differentiate urine-derived stem cell into urothelial and smooth muscle cells that were related in the improvement of a scaffold urinary conduit (Bodin et al. 2010). Huang et al. used gelatin sponge scaffold to interfere with cellulose assembly during BC fermentation process to produce a 3D porous BC scaffold and seeded with lingual keratinocytes to repair urethral defects in a rabbit model. The results showed that 3D porous BC exhibited good biocompatibility with lingual keratinocytes and improved urethral tissue regeneration compared with BC. These results confirmed that 3D porous BC scaffold might be a promising alternative scaffold for urethra reconstruction (Huang et al. 2015).

5.8 Neural Implants and Dura Mater

Reconstruction of the nerve tissue is a certainly difficult challenge. Self-renewal may be difficult, depending on the size of the damage. Bacterial cellulose was evaluated as a replacement for dura mater in mongrel dogs. The macroscopic examination of the grafts showed good acceptance and attachment to bone fragments (Mello et al. 1997). Pértile et al. (2012) studied the BC membrane as an artificial application of nerve tissue engineering. The study indicated that mesenchymal stem cells were seeded to bacterial cellulose scaffold exhibit proliferation and expression of nerve growth factor neurotrophin, thus creating a microenvironment that supports neuronal regeneration (Pértile et al. 2012). In one of the studies, BC has also been developed as a biomaterial for regeneration of peripheral nerve injury through cellulosic guidance channels. In vivo experiments were conducted on the femoral nerve of Wistar rats for three-month follow-up. Evaluation of the results obtained from histological analysis and postoperative motor recovery observation was observed that the operated tubes did not touch the original dimension and enabled the accumulation of neurotrophic factors inside and allowed implanted of nerve regeneration (Kowalska-Ludwicka et al. 2013).

Freshly, BC was investigated to fabricate the nerve conduits for healing of peripheral nerve damages. The work presented that BC showed good biocompatibility in vitro study using with Schwann cells and demonstrated no contrary hematological and histopathological effects on in vivo implantation study in Sprague-Dawley rats (Zhu et al. 2015). Bacterial cellulose was also investigated for the repair of dural defects (Xu et al. 2014). BC showed reduced inflammatory response compared to traditional dura mater substituents (NormalGEN®), and on the other hand, it needs to be validated for long-term effects in larger animals (Xu et al. 2014).

6 Conclusions and Future Perspectives

In vitro- and in vivo-based trials have been dominated tissue engineering researches during recent decades. Since then, investigations have focused more on closer development and representation of end applications of biomaterials such as wound healing, artificial skin, and skin tissue repair. Use of some laboratory tests and animal trials are conducted under controlled and field conditions to developed biomaterials and structures. Production, structure, and purpose of the BC are reviewed in this chapter. The mechanisms of BC involved in healing process have not yet been well established; however, as discussed clearly, it has a great potential as biomaterial-based product research and development. Some of the applications of BC are highlighted and discussed in current chapter which are cardiovascular system, artificial cornea and contact lenses, bone and cartilage regeneration, meniscus implant, dental root canal treatment, urinary conduits, and neural implants/dura mater. We encourage researchers to consider issues raised in this chapter, particularly the interaction between material and the application area needs to be examined in detail for further steps which are human trials. With biomaterials being so much a part of international trade due to need in healthcare sector, we expect increased interest in compliance with international standards and regional agreements on test methods to investigate and analyze the developed biomaterials.

References

Andrade FK, Costa R, Domingues L, Soares R, Gama M (2010) Improving bacterial cellulose for blood vessel replacement: functionalization with a chimeric protein containing a cellulose-binding module and an adhesion peptide. Acta Biomater 6(10):4034–4041

Ahmed M, Hamilton G, Seifalian AM (2014) The performance of a small-calibre graft for vascular reconstructions in a senescent sheep model. Biomaterials 35(33):9033–9040

Ashton JH, Mertz JA, Harper JL, Slepian MJ, Mills JL, McGrath DV, Geest JV (2011) Polymeric endoaortic paving: mechanical, thermoforming, and degradation properties of polycaprolactone/polyurethane blends for cardiovascular applications. Acta Biomater 7(1):287–294

Ao C, Niu Y, Zhang X, He X, Zhang W, Lu C (2017) Fabrication and characterization of electrospun cellulose/nano-hydroxyapatite nanofibers for bone tissue engineering. Int J Biol Macromol 97:568–573

Ávila MH, Schwarz S, Feldmann EM, Mantas A, Von Bomhard A, Gatenholm P et al (2014) Biocompatibility evaluation of densified bacterial nanocellulose hydrogel as an implant material for auricular cartilage regeneration. Appl Microbiol Biotechnol 98:7423–7435

Backdahl H, Helenius G, Bodin A, Nannmark U, Johansson BR, Risberg B, Gatenholm P (2006) Mechanical properties of bacterial cellulose and interactions with smooth muscle cells. Biomaterials 27(9):2141–2149

Balasubramani M, Kumar TR, Babu M (2001) Skin substitutes: a review. Burns 27:534–544

Barozzi L, Brizard CP, Galati JC, Konstantinov IE, Bohuta L, d'Udekem Y (2011) Side-to-side aorto-GoreTex central shunt warrants central shunt patency and pulmonary arteries growth. Ann Thoracic Surg 92(4):1476–1482

Bielecki SE, Krystynowicz AE, Turkiewicz M, Kalinowska HE (2005) Bacterial cellulose. Biopolymers Online. 5. Polysaccharides from prokaryotes. Wiley Online Library. https://doi.org/10.01002/3527600035.bpol5003

Bodin A, Concaro S, Brittberg M, Gatenholm P (2007) Bacterial cellulose as a potential meniscus implant. J Tissue Eng Regenerative Med 1(5):406–408

Bodin A, Bharadwaj S, Wu S, Gatenholm P, Atala A, Zhang Y (2010) Tissue-engineered conduit using urine-derived stem cells seeded bacterial cellulose polymer in urinary reconstruction and diversion. Biomaterials 31:8889–8901

Bostrom MP, Asnis P (1998) Transforming growth factor beta in fracture repair. Clin Orthop Relat Res® 355:S124–S131

Bourne RR, Dineen BP, Huq DMN, Ali SM, Johnson GJ (2004) Correction of refractive error in the adult population of Bangladesh: meeting the unmet need. Invest Ophthalmol Vis Sci 45(2):410–417

Brager MA, Patterson MJ, Connolly JF, Nevo Z (2000) Osteogenic growth peptide normally stimulated by blood loss and marrow ablation has local and systemic effects on fracture healing in rats. J Orthop Res 18(1):133–139

Brown AJ (1886) J Chem Soc 49, 51:172, 432, 643

Brown EE, Laborie M-PG, Zhang J (2011) Glutaraldehyde treatment of bacterial cellulose/fibrin composites: impact on morphology, tensile and viscoelastic properties. Cellulose 19(1):127–137

Burugapalli K, Pandit A (2007) Characterization of tissue response and in vivo degradation of cholecyst-derived extracellular matrix. Biomacromol 8(11):3439–3451

Cai Z, Kim J (2010) Bacterial cellulose/poly (ethylene glycol) composite: characterization and first evaluation of biocompatibility. Cellulose 17(1):83–91

Cai ZJ, Yang G (2011) Bacterial cellulose/collagen composite: characterization and first evaluation of cytocompatibility. J Appl Polym Sci 1205:2938–2944

Cavicchioli M, Corso CT, Coelho F, Mendes L, Saska S, Soares CP et al (2015) Characterization and cytotoxic, genotoxic and mutagenic evaluations of bacterial cellulose membranes incorporated with ciprofloxacin: a potential material for use as therapeutic contact lens. World J Pharm Pharm Sci 4:1626–1647

Chawla PR, Bajaj IB, Survase SA, Singhal RS (2009) Microbial cellulose: fermentative production and applications. Food Technol Biotechnol 47(2)

Chen B, Lin H, Wang J, Zhao Y, Wang B, Zhao W, Dai J et al (2007a) Homogeneous osteogenesis and bone regeneration by demineralized bone matrix loading with collagen-targeting bone morphogenetic protein-2. Biomaterials 28(6):1027–1035

Chen S, Zou Y, Yan Z, Shen W, Shi S, Zhang X, Wang H (2009) Carboxymethylated-bacterial cellulose for copper and lead ion removal. J Hazard Mater 161(2–3):1355–1359

Chen ZX, Chang M, Peng YL, Zhao L, Zhan YR, Wang LJ, Wang R (2007b) Osteogenic growth peptide C-terminal pentapeptide [OGP (10–14)] acts on rat bone marrow mesenchymal stem cells to promote differentiation to osteoblasts and to inhibit differentiation to adipocytes. Regul Pept 142(1–2):16–23

Chong DST, Lindsey B, Dalby MJ, Gadegaard N, Seifalian AM, Hamilton G (2014) Luminal surface engineering, 'Micro and nanopatterning': potential for self endothelialising vascular grafts? Eur J Vasc Endovasc Surg 47(5):566–576

Czaja W, Romanovicz D, Malcolm Brown R (2004) Structural investigations of microbial cellulose produced in stationary and agitated culture. Cellulose 11(3–4):403–411

Czaja W, Krystynowicz A, Bielecki S, Brown RM Jr (2006) Microbial cellulose—the natural power to heal wounds. Biomaterials 27(2):145–151

Dahlin C, Linde A, Gottlow J, Nyman S (1988) Healing of bone defects by guided tissue regeneration. Plastic Reconstr Surg 81(5):672–676

Das SK, Kumar A, Sharma GK, Pandey AK, Bansal H, Trivedi S, Singh PB et al (2009) Lingual mucosal graft urethroplasty for anterior urethral strictures. Urology 73(1):105–108

Eming SA, Smola H, Krieg T (2002) The treatment of chronic wounds: current concepts and future aspects. Cells Tissues Organs 172:105–117

Fang B, Wan YZ, Tang TT, Gao C, Dai KR (2009) Proliferation and osteoblastic differentiation of human bone marrow stromal cells on hydroxyapatite/bacterial cellulose nanocomposite scaffolds. Tissue Eng Part A 15(5):1091–1098

Fernandes SCM, Oliveira L, Freire CSR, Silvestre AJD, Neto CP, Gandini A, Desbriéres J (2009) Novel transparent nanocomposite films based on chitosan and bacterial cellulose. Green Chem 11:2023–2029

Foster A (2003) Vision 2020—the right to sight. Trop Doct 33(4):193–194

Grande CJ, Torres FG, Gomez CM, Bañó MC (2009) Nanocomposites of bacterial cellulose/hydroxyapatite for biomedical applications. Acta Biomater 5(5):1605–1615

Grasl C, Bergmeister H, Stoiber M, Schima H, Weigel G (2010) Electrospun polyurethane vascular grafts: in vitro mechanical behavior and endothelial adhesion molecule expression. J Biomed Mater Res Part A: Official J Soc Biomater Jpn Soc Biomater Aust Soc Biomater Korean Soc Biomater 93(2):716–723

Guhados G, Wan W, Hutter JL (2005) Measurement of the elastic modulus of single bacterial cellulose fibers using atomic force microscopy. Langmuir 21(14):6642–6646

Hestrin S, Schramm M (1954) Synthesis of cellulose by *Acetobacter xylinum*. 2. Preparation of freeze-dried cells capable of polymerizing glucose to cellulose. Biochem J 58(2):345

Hu W, Chen S, Li X, Shi S, Shen W, Zhang X, Wang H (2009) In situ synthesis of silver chloride nanoparticles into bacterial cellulose membranes. Mater Sci Eng, C 29(4):1216–1219

Hui J, Yuanyuan J, Jiao W, Yuan H, Yuan Z, Shiru J (2009) Potentiality of bacterial cellulose as the scaffold of tissue engineering of cornea. In: 2nd international conference on biomedical engineering and informatics, BMEI'09, pp 1–5

Huang JW, Lv XG, Li Z, Song LJ, Feng C, Xie MK et al (2015) Urethral reconstruction with a 3D porous bacterial cellulose scaffold seeded with lingual keratinocytes in a rabbit model. Biomed Mater 10:055005

Huang Y, Wang J, Yang F, Shao Y, Zhang X, Dai K (2017) Modification and evaluation of micro-nano structured porous bacterial cellulose scaffold for bone tissue engineering. Mater Biol Appl 75:1034 (Materials Science & Engineering C)

Isenberg BC, Williams C, Tranquillo RT (2006) Small-diameter artificial arteries engineered in vitro. Circ Res 98:25–35

Jonas R, Farah LF (1998) Production and application of microbial cellulose. Polym Degrad Stabil 59:101–106

Kempen JH, Mitchell P, Lee KE, Tielsch JM, Broman AT, Taylor HR et al (2004) The prevalence of refractive errors among adults in the United States, Western Europe and Australia. Arch Ophthalmol 122:495–505

Khajavi R, Esfahani JE, Sattari M (2011) Crystalline structure of microbial cellulose compared with native and regenerated cellulose. Int J Polym Mater 60(14):1178–1192. https://doi.org/10.01080/00914037.2010551372

Kim J, Cai Z, Lee HS, Choi GS, Lee DH, Jo C (2011) Preparation and characterization of a bacterial cellulose/chitosan composite for potential biomedical application. J Polym Res 18(4):739–744

Kharaghani D, Meskinfam M, Rezaeikanavi M, Balagholi S, Fazili N (2015) Synthesis and characterization of hybrid nanocomposite via biomimetic method as an artificial cornea. Invest Ophthalmol Vis Sci 56(7):5024

Khan S, Ul-Islam M, Ikram M, Ullah MW, Israr M, Subhan F, Park JK et al (2016) Three-dimensionally microporous and highly biocompatible bacterial cellulose–gelatin composite scaffolds for tissue engineering applications. RSC Adv 6(112):110840–110849

Khan S, Ul-Islam M, Ikram M, Islam SU, Ullah MW, Israr M, Park JK et al (2018) Preparation and structural characterization of surface modified microporous bacterial cellulose scaffolds: a potential material for skin regeneration applications in vitro and in vivo. Int J Biol Macromol 117:1200–1210

Klemm D, Schumann D, Udhardt U, Marsch S (2001) Bacterial synthesized cellulose—artificial blood vessels for microsurgery. Prog Polym Sci 26(9):1561–1603

Klemm D, Heublein B, Fink H-P, Bohn A (2005) Cellulose: fascinating biopolymer and sustainable raw material. Angew Chem Int Edn 44:3358–3393

Kolpak FJ, Blackwell J (1976) Determination of the structure of cellulose II. Macromolecules 9(2):273–278

Kowalska-Ludwicka K, Cala J, Grobelski B, Sygut D, Jesionek-Kupnicka D, Kolodziejczyk M et al (2013) Modified bacterial cellulose tubes for regeneration of damaged peripheral nerves. Arch Med Sci 9:527–534

Kudo FA, Nishibe T, Miyazaki K, Flores J, Yasuda K (2002) Albumin-coated knitted Dacron aortic prostheses: study of postoperative inflammatory reactions. Int Angiol 21(3):214

Kuga S, Jr Malcolm Brown R (1988) Silver labeling of the reducing ends of bacterial cellulose. Carbohydr Res 180(2):345–350

Kumbhar JV, Jadhav SH, Bodas DS, Barhanpurkar-Naik A, Wani MR, Paknikar KM, Rajwade JM (2017) In vitro and in vivo studies of a novel bacterial cellulose-based acellular bilayer nanocomposite scaffold for the repair of osteochondral defects. Int J Nanomed 12:6437

Lessim S, Oughlis S, Lataillade JJ, Migonney V, Changotade S, Lutomski D, Poirier F (2015) Protein selective adsorption properties of a polyethylene terephtalate artificial ligament grafted with poly (sodium styrene sulfonate)(polyNaSS): correlation with physicochemical parameters of proteins. Biomed Mater 10(6):065021

Lee SE, Park YS (2017) The role of bacterial cellulose in artificial blood vessels. Mol Cell Toxicol 13(3):257–261

Leitão AF, Silva JP, Dourado F, Gama M (2013) Production and characterization of a new bacterial cellulose/poly (vinyl alcohol) nanocomposite. Materials 6(5):1956–1966

Leitão AF, Faria MA, Faustino AM, Moreira R, Mela P, Loureiro L, Gama M et al (2016) A novel small-caliber bacterial cellulose vascular prosthesis: production, characterization, and preliminary in vivo testing. Macromol Biosci 16(1):139–150

Levinson DJ, Glonek T (2010) Microbial cellulose contact lens. US Patent, US7832857 B2

Li X, Wan W, Panchal CJ (2010) Transparent bacterial cellulose nanocomposite hydrogels. US Patent, 8940337 B2

Lin WC, Lien CC, Yeh HJ, Yu CM, Hsu SH (2013a) Bacterial cellulose and bacterial cellulose–chitosan membranes for wound dressing applications. Carbohydr Polym 94(1):603–611

Lin SP, Calvar IL, Catchmark JM, Liu JR, Demirci A, Cheng KC (2013b) Biosynthesis, production and applications of bacterial cellulose. Cellulose 20(5):2191–2219

Loh QL, Choong C (2013) Three-dimensional scaffolds for tissue engineering applications: role of porosity and pore size. Tissue Eng Part B Rev 19(6):485–502

Luo H, Xiong G, Huang Y, He F, Wang Y, Wan Y (2008) Preparation and characterization of a novel COL/BC composite for potential tissue engineering scaffolds. Mater Chem Phys 110:193–196

Luo H, Zhang J, Xiong G, Wan Y (2014) Evolution of morphology of bacterial cellulose scaffolds during early culture. Carbohydr Polym 111:722–728

Lopes JL, Machado JM, Castanheira L, Granja PL, Gama FM, Dourado F et al (2011) Friction and wear behaviour of bacterial cellulose against articular cartilage. Wear 271:2328–2333

Maneerung T, Tokura S, Rujiravanit R (2008) Impregnation of silver nanoparticles into bacterial cellulose for antimicrobial wound dressing. Carbohydr Polym 72(1):43–51

Matuda FS, Macedo lGS, Valera MC, Carvalho R, Monteiro ASF et al (2004) Evaluation of two membranes in guided bone tissue regeneration: histological study in rabbits. Braz J Oral Sci 3(8):395–400

Mello LR, Feltrin LT, Neto PTF, Ferraz FAP (1997) Duraplasty with biosynthetic cellulose: an experimental study. J Neurosurg 86:143–150

Messaddeq Y, Ribeiro SJL, Thomazini W (2008) Trigger, Pesquisa & Desenvolvimentos Biotecnologicos Ltda. (TRIG-Non-standard), assignee. Contact lens for therapy, method and apparatus for their production and use. Brazil patent BR, PI0603704-6

Naidoo KS, Jaggernath J (2012) Uncorrected refractive errors. Indian J Ophthalmol 60(5):432–437

Nakayama A, Kakugo A, Gong JP, Osada Y, Takai M, Erata T, Kawano S (2004) High mechanical strength double network hydrogel with bacterial cellulose. Adv Funct Mater 14:1124–1128

Novaes Jr AB, Novaes AB (1992) IMZ implants placed into extraction sockets in association with membrane therapy (Gengiflex) and porous hydroxyapatite: a case report. Int J Oral Maxillofac Implant 7(4)

Park S, Baker JO, Himmel ME, Parilla PA, Johnson DK (2010) Cellulose crystallinity index: measurement techniques and their impact on interpreting cellulase performance. Biotechnol Biofuels 3(1):10

Pértile R, Moreira S, Andrade F, Domingues L, Gama M (2012) Bacterial cellulose modified using recombinant proteins to improve neuronal and mesenchymal cell adhesion. Biotechnol Prog 28:526–532

Petersen N, Gatenholm P (2011) Bacterial cellulose-based materials and medical devices: current state and perspectives. Appl Microbiol Biotechnol 91(5):1277

Pigossi SC, de Oliveira GJPL, Finoti LS, Nepomuceno R, Spolidorio LC, Rossa C Jr et al (2015) Bacterial cellulose-hydroxyapatite composites with osteogenic growth peptide (OGP) or pentapeptide OGP on bone regeneration in critical-size calvarial defect model. J Biomed Mater Res, Part A 103:3397–3406

Ramani D, Sastry TP (2014) Bacterial cellulose-reinforced hydroxyapatite functionalized graphene oxide: a potential osteoinductive composite. Cellulose 21(5):3585–3595

Ravi S, Chaikof EL (2010) Biomaterials for vascular tissue engineering. Regenerative Med 5(1):107–120

Putra A, Kakugo A, Furukawa H, Gong JP, Osada Y (2008) Tubular bacterial cellulose gel with oriented fibrils on the curved surface. Polymer 49(7):1885–1891

Ross P, Mayer R, Benziman M (1991) Cellulose biosynthesis and function in bacteria. Microbiol Rev 55:35–58

Saha SP, Muluk S, Schenk W III, Burks SG, Grigorian A, Ploder B, Hantak E et al (2011) Use of fibrin sealant as a hemostatic agent in expanded polytetrafluoroethylene graft placement surgery. Ann Vasc Surg 25(6):813–822

Sakairi N, Suzuki S, Ueno K, Han SM, Nishi N, Tokura S (1998) Biosynthesis of hetero-polysaccharides by *Acetobacter xylinum*-Synthesis and characterization of metal-ion adsorptive properties of partially carboxymethylated cellulose. Carbohydr Polym 37(4):409–414

Salata LA, Craig GT, Brook IM (1998) Bone healing following the use of hydroxyapatite or ionomeric bone substitutes alone or combined with a guided bone regeneration technique: an animal study. Int J Oral Maxillofac Implant 13(1)

Sarkar S, Salacinski HJ, Hamilton G, Seifalian AM (2006) The mechanical properties of infrainguinal vascular bypass grafts: their role in influencing patency. Eur J Vasc Endovasc Surg 31(6):627–636

Saska S, Barud HS, Gaspar AMM, Marchetto R, Ribeiro SJL, Messaddeq Y (2011) Bacterial cellulose-hydroxyapatite nanocomposites for bone regeneration. Int J Biomater

Saska S, Scarel-Caminaga RM, Teixeira LN, Franchi LP, Dos Santos RA, Gaspar AMM, Ribeiro SJL et al (2012) Characterization and in vitro evaluation of bacterial cellulose membranes functionalized with osteogenic growth peptide for bone tissue engineering. J Mater Sci Mater Med 23(9):2253–2266

Scherner M, Reutter S, Klemm D, Sterner-Kock A, Guschlbauer M, Richter T, Wippermann J et al (2014) In vivo application of tissue-engineered blood vessels of bacterial cellulose as small arterial substitutes: proof of concept? J Surg Res 189(2):340–347

Schluesener JK, Schluesener HJ (2013) Nanosilver: application and novel aspects of toxicology. Arch Toxicol 87(4):569–576

Schultz GS, Mast BA (1998) Molecular analysis of the environment of healing and chronic wounds: cytokines, proteases and growth factors. Wounds 10(SupplF):1F–9F

Schumann DA, Wippermann J, Klemm DO, Kramer F, Koth D, Kosmehl H et al (2009) Artificial vascular implants from bacterial cellulose: preliminary results of small arterial substitutes. Cellulose 16:877–885

Shao W, Liu H, Liu X, Wang S, Wu J, Zhang R, Huang M et al (2015) Development of silver sulfadiazine loaded bacterial cellulose/sodium alginate composite films with enhanced antibacterial property. Carbohydr Polym 132:351–358

Sheykhnazari S, Tabarsa T, Ashori A, Shakeri A, Golalipour M (2011) Bacterial synthesized cellulose nanofibers; Effects of growth times and culture mediums on the structural characteristics. Carbohydr Polym 86(3):1187–1191. https://doi.org/10.01016/j.carbpol.2011.06.011

Shi S, Chen S, Zhang X, Shen W, Li X, Hu W, Wang H (2009) Biomimetic mineralization synthesis of calcium-deficient carbonate-containing hydroxyapatite in a three-dimensional network of bacterial cellulose. J Chem Technol Biotechnol Int Res Process Environ Clean Technol 84(2):285–290

Shi Q, Li Y, Sun J, Zhang H, Chen L, Chen B, Wang Z et al (2012) The osteogenesis of bacterial cellulose scaffold loaded with bone morphogenetic protein-2. Biomaterials 33(28):6644–6649

Shi Z, Zhang Y, Phillips GO, Yang G (2014) Utilization of bacterial cellulose in food. Food Hydrocoll 35:539–545

Stoica-Guzun A, Stroescu M, Jinga S, Jipa I, Dobre T, Dobre L (2012) Ultrasound influence upon calcium carbonate precipitation on bacterial cellulose membranes. Ultrason Sonochem 19(4):909–915

Svensson A, Nicklasson E, Harrah T, Panilaitis B, Kaplan DL, Brittberg M, Gatenholm P (2005) Bacterial cellulose as a potential scaffold for tissue engineering of cartilage. Biomaterials 26(4):419–431

Tammelin T, Saarinen T, Österberg M, Laine J (2006) Preparation of Langmuir/Blodgett-cellulose surfaces by using horizontal dipping procedure. Application for polyelectrolyte adsorption studies performed with QCM-D. Cellulose 13(5):519

Taneja S, Kumari M, Parkash H (2010) Nonsurgical healing of large periradicular lesions using a triple antibiotic paste: a case series. Contemp Clin Dent 1(1):31–35

Tang J, Bao L, Li X, Chen L, Hong FF (2015) Potential of PVA-doped bacterial nano-cellulose tubular composites for artificial blood vessels. J Mater Chem B 3(43):8537–8547

Tang J, Li X, Bao L, Chen L, Hong FF (2017) Comparison of two types of bioreactors for synthesis of bacterial nanocellulose tubes as potential medical prostheses including artificial blood vessels. J Chem Technol Biotechnol 92(6):1218–1228

Tarr HLA, Hibbery H (1931) Can J Res 4:372

Tazi N, Zhang Z, Messaddeq Y, Almeida-Lopes L, Zanardi LM, Levinson D, Rouabhia M (2012) Hydroxyapatite bioactivated bacterial cellulose promotes osteoblast growth and the formation of bone nodules. Amb Express 2(1):61

Theron JP, Knoetze JH, Sanderson RD, Hunter R, Mequanint K, Franz T, Bezuidenhout D et al (2010) Modification, crosslinking and reactive electrospinning of a thermoplastic medical polyurethane for vascular graft applications. Acta Biomater 6(7):2434–2447

Tienen TG, Verdonschot N, Heijkants RGJC, Buma P, Scholten JGF, Van Kampen A et al (2004) Prosthetic replacement of the medial meniscus in cadaveric knees does the prosthesis mimic the functional behavior of the native meniscus? Am J Sports Med 32(5):1182–1188

Tiwari A, Cheng KS, Salacinski H, Hamilton G, Seifalian AM (2003) Improving the patency of vascular bypass grafts: the role of suture materials and surgical techniques on reducing anastomotic compliance mismatch. Eur J Vasc Endovasc Surg 25(4):287–295

Tonouchi N, Tsuchida T, Yoshinaga F, Beppu T, Horinouchi S (1996) Characterization of the biosynthetic pathway of cellulose from glucose and fructose in *Acetobacter xylinum*. Biosci Biotechnol Biochem 60:1377–1379

Trengove NJ, Stacey MC, Macauley S, Bennett N, Gibson J, Burslem F, Murphy G, Schultz G (1999) Analysis of the acute and chronic wound environments: the role of proteases and their inhibitors. Wound Rep Reg 7:442–452

Ueno H, Yamada H, Tanaka I, Kaba N, Matsuura M, Okumura M et al (1999) Accelerating effects of chitosan for healing at early phase of experimental open wound in dogs. Biomaterials 20(15):1407–1414

Ul-Islam M, Shah N, Ha JH, Park JK (2011) Effect of chitosan penetration on physico-chemical and mechanical properties of bacterial cellulose. Korean J Chem Eng 28(8):1736

Ul-Islam M, Khan S, Ullah MW, Park JK (2015) Bacterial cellulose composites: synthetic strategies and multiple applications in bio-medical and electro-conductive fields. Biotechnol J 10(12):1847–1861

Ullah H, Wahid F, Santos HA, Khan T (2016) Advances in biomedical and pharmaceutical applications of functional bacterial cellulose-based nanocomposites. Carbohydr Polym 150:330–352

Uzun M, Anand S, Shah T (2013) Study of the pH and physical performance characteristics of silver treated absorbent wound dressings. J Ind Text 42(3):231–243

Vanella L, Kim DH, Asprinio D, Peterson SJ, Barbagallo I, Vanella A, Abraham NG et al (2010) HO-1 expression increases mesenchymal stem cell-derived osteoblasts but decreases adipocyte lineage. Bone 46(1):236–243

Wan YZ, Huang Y, Yuan CD, Raman S, Zhu Y, Jiang HJ, Gao C et al (2007) Biomimetic synthesis of hydroxyapatite/bacterial cellulose nanocomposites for biomedical applications. Mater Sci Eng, C 27(4):855–864

Wang S, Gupta AS, Sagnella S, Barendt PM, Kottke-Marchant K, Marchant RE (2009) Biomimetic fluorocarbon surfactant polymers reduce platelet adhesion on PTFE/ePTFE surfaces. J Biomater Sci Polym Edn 20(5–6):619–635

Wang W, Li HY, Zhang DW, Jiang J, Cui YR, Qiu S, Zhang XX et al (2010) Fabrication of bienzymatic glucose biosensor based on novel gold nanoparticles-bacteria cellulose nanofibers nanocomposite. Electroanalysis 22(21):2543–2550

Wang J, Wan Y, Han J, Lei X, Yan T, Gao C (2011) Nanocomposite prepared by immobilising gelatin and hydroxyapatite on bacterial cellulose nanofibres. Micro Nano Lett 6(3):133–136

Wang H-Y, Wei R-H, Zhao SZ (2013) Evaluation of corneal cell growth on tissue engineering materials as artificial cornea scaffolds. Int J Ophthalmol 6(6):873–878

WHO. Cardiovascular diseases (CVDs). http://www.who.int/cardiovascular_diseases/en/

Whitcher JP, Srinivasan M, Upadhyay MP (2001) Corneal blindness: a global perspective. Bull World Health Organ 79(3):214–221

Wiegand C, Elsner P, Hipler U-C, Klemm D (2006) Protease and ROS activities influenced by a composite of bacterial cellulose and collagen type Iin vitro. Cellulose 13(6):689–696

Xu C, Ma X, Chen S, Tao M, Yuan L, Jing Y (2014) Bacterial cellulose membranes used as artificial substitutes for dural defection in rabbits. Int J Mol Sci 15(6):10855–10867

Yang G, Xie J, Hong F, Cao Z, Yang X (2012) Antimicrobial activity of silver nanoparticle impregnated bacterial cellulose membrane: effect of fermentation carbon sources of bacterial cellulose. Carbohydr Polym 87(1):839–845

Yashiro B, Shoda M, Tomizawa Y, Manaka T, Hagiwara N (2012) Long-term results of a cardiovascular implantable electronic device wrapped with an expanded polytetrafluoroethylene sheet. J Artif Organs 15(3):244–249

Yoshino A, Tabuchi M, Uo M, Tatsumi H, Hideshima K, Kondo S et al (2013) Applicability of bacterial cellulose as an alternative to paper points in endodontic treatment. Acta Biomater 9:6116–6122

Yu X, Atalla RH (1996) Production of cellulose II by *Acetobacter xylinum* in the presence of 2,6-dichlorobenzonitrile. Int J Biol Macromol 19:145–146

Yu HC, Chen LJ, Cheng KC, Li YX, Yeh CH, Cheng JT (2012) Silymarin inhibits cervical cancer cell through an increase of phosphatase and tensin homolog. Phytother Res 26(5):709–715

Zaborowska M, Bodin A, Bäckdahl H, Popp J, Goldstein A, Gatenholm P (2010) Microporous bacterial cellulose as a potential scaffold for bone regeneration. Acta Biomater 6(7):2540–2547

Zang S, Zhang R, Chen H, Lu Y, Zhou J, Chang X et al (2015) Investigation on artificial blood vessels prepared from bacterial cellulose. Mater Sci Eng, C 46:111–117

Zhijiang C, Guang Y (2011) Bacterial cellulose/collagen composite: characterization and first evaluation of cytocompatibility. J Appl Polym Sci 120(5):2938–2944

Zimmermann KA, LeBlanc JM, Sheets KT, Fox RW, Gatenholm P (2011) Biomimetic design of a bacterial cellulose/hydroxyapatite nanocomposite for bone healing applications. Mater Sci Eng, C 31(1):43–49

Zhu W, Li W, He Y, Duan T (2015) In-situ biopreparation of biocompatible bacterial cellulose/graphene oxide composites pellets. Appl Surf Sci 338:22–26

Printed by Printforce, the Netherlands